AF356161

Application of Plant Biotechnology in Forestry

Application of Plant Biotechnology in Forestry

Editors

Ricardo Javier Ordás
José Manuel Álvarez Díaz

Basel • Beijing • Wuhan • Barcelona • Belgrade • Novi Sad • Cluj • Manchester

Editors
Ricardo Javier Ordás
Department of Organisms
and Systems Biology, Institute
of Biotechnology of Asturias,
University of Oviedo
Oviedo
Spain

José Manuel Álvarez Díaz
Department of Organisms
and Systems Biology, Institute
of Biotechnology of Asturias,
University of Oviedo
Oviedo
Spain

Editorial Office
MDPI
St. Alban-Anlage 66
4052 Basel, Switzerland

This is a reprint of articles from the Special Issue published online in the open access journal *Forests* (ISSN 1999-4907) (available at: https://www.mdpi.com/journal/forests/special_issues/Application_Plant_Biotechnology).

For citation purposes, cite each article independently as indicated on the article page online and as indicated below:

Lastname, A.A.; Lastname, B.B. Article Title. *Journal Name* **Year**, *Volume Number*, Page Range.

ISBN 978-3-7258-0121-3 (Hbk)
ISBN 978-3-7258-0122-0 (PDF)
doi.org/10.3390/books978-3-7258-0122-0

Contents

Editorial

Application of Plant Biotechnology in Forestry

José Manuel Alvarez * and Ricardo Javier Ordás

Área de Fisiología Vegetal, Departamento de Biología de Organismos y Sistemas, Instituto Universitario de Biotecnología de Asturias (IUBA), Universidad de Oviedo, 33071 Oviedo, Spain; rordas@uniovi.es

* Correspondence: alvarezmanuel@uniovi.es

Forests, often referred to as the lungs of our planet, stand as a testament to the incredible diversity and significance of our terrestrial ecosystems. These majestic giants, the forest trees, constitute the primary component of continental biomass and are guardians of terrestrial biodiversity. They provide us with a multitude of ecological services that sustain life on Earth and serve as a vital source of raw materials essential for various purposes.

In recent years, the global demand for wood and its derivatives has grown unabated and shows no sign of subsiding. As we navigate the complex landscape of environmental sustainability, it becomes imperative to address this surging demand with a concerted effort to minimize its ecological footprint. It is here that the fascinating realm of biotechnology comes into play, offering us a path forward that can harmonize our need for forest resources with the preservation of our fragile ecosystems.

Biotechnological tools have emerged as our allies in this endeavor. Genomic selection, micropropagation, and genetic engineering are poised to revolutionize how we manage and harness the potential of our forests. These tools promise not only to meet the increasing demand for wood but also to do so while minimizing its impact on our environment.

Forest regeneration, a process once left predominantly to natural forces, is now evolving through the strategic implementation of artificial regeneration with selected genotypes. This approach emerges as the most effective means to enhance forest yield and, by extension, our sustainability efforts. However, it is crucial to acknowledge the unique challenges posed by forest species in this context. Their sheer size, extended generation times, and prolonged juvenile stages make traditional plant breeding a demanding and time-consuming task. This is where genomic selection enters the stage, armed with the potential to accelerate breeding cycles, intensify selection, and enhance the accuracy of breeding values.

With this backdrop, the objective of this Special Issue is to delve deep into the latest biotechnological approaches in forestry. From rejuvenation through clonal propagation to somatic embryogenesis, cryopreservation of germplasm, and the use of molecular techniques and genetic engineering, we aim to unravel the potential and promise of these cutting-edge technologies.

In the pages that follow, leading experts in the field share their insights, research findings, and visions for the future. Together, we embark on a journey to unlock the full potential of biotechnology in forestry, seeking sustainable solutions that balance our insatiable need for forest resources with our responsibility to preserve the delicate balance of our planet's ecosystems.

The Special Issue comprises 10 papers. They represent a wide range of aspects related to the application of plant biotechnology in forestry and give timely examples of research activities that can be observed around the globe. An overview on conifer biotechnology is firstly presented [1]. Among the arsenal of biotechnological approaches, micropropagation techniques coupled with the rooting of cuttings [2–8] are currently recognized as the most potent tools for the large-scale propagation of elite forest varieties. These techniques harness the remarkable developmental plasticity of plants, enabling them to adapt to diverse environmental conditions while retaining their high regeneration capacity.

Citation: Alvarez, J.M.; Ordás, R.J. Application of Plant Biotechnology in Forestry. *Forests* **2023**, *14*, 2148. https://doi.org/10.3390/f14112148

Received: 16 October 2023
Accepted: 25 October 2023
Published: 28 October 2023

Molecular studies and genetic engineering emerge as a game-changer, offering the power to analyze gene functions and transfer specific traits into selected genotypes without compromising their desirable genetic background [9,10]. This revolutionary tool holds the key to swiftly increase forest yield and wood quality, drastically shortening the traditional breeding process.

Join us in this exploration, as we pave the way towards a greener, more sustainable future where forests thrive, and humanity prospers in harmony with nature.

Conflicts of Interest: The authors declare no conflict of interest.

References

1. Rodríguez, S.M.; Ordás, R.J.; Alvarez, J.M. Conifer Biotechnology: An Overview. *Forests* **2022**, *13*, 1061. [CrossRef]
2. Hazubska-Przybył, T.; Wawrzyniak, M.K.; Kijowska-Oberc, J.; Staszak, A.M.; Ratajczak, E. Somatic Embryogenesis of Norway Spruce and Scots Pine: Possibility of Application in Modern Forestry. *Forests* **2022**, *13*, 155. [CrossRef]
3. Varis, S.; Tikkinen, M.; Edesi, J.; Aronen, T. How to Capture Thousands of Genotypes; Initiation of Somatic Embryogenesis in Norway Spruce. *Forests* **2023**, *14*, 810. [CrossRef]
4. Martínez, M.T.; Corredoira, E. Efficient Procedure for Induction Somatic Embryogenesis in Holm Oak: Roles of Explant Type, Auxin Type, and Exposure Duration to Auxin. *Forests* **2023**, *14*, 430. [CrossRef]
5. Liu, X.; Liu, Y.; Yu, X.; Tretyakova, I.N.; Nosov, A.M.; Shen, H.; Yang, L. Improved Method for Cryopreservation of Embryogenic Callus of *Fraxinus mandshurica* Pupr. by Vitrification. *Forests* **2023**, *14*, 28. [CrossRef]
6. Rojas-Vargas, A.; Montalbán, I.A.; Moncaleán, P. Adult Trees *Cryptomeria japonica* (Thunb. ex L.f.) D. Don Micropropagation: Factors Involved in the Success of the Process. *Forests* **2023**, *14*, 743. [CrossRef]
7. Ioannidis, K.; Tomprou, I.; Panayiotopoulou, D.; Boutsios, S.; Daskalakou, E.N. Potential and Constraints on In Vitro Micropropagation of *Juniperus drupacea* Labill. *Forests* **2023**, *14*, 142. [CrossRef]
8. Abshahi, M.; García-Morote, F.A.; Zarei, H.; Zahedi, B.; Nejad, A.R. Improvement of Rooting Performance in Stem Cuttings of Savin Juniper (*Juniperus sabina* L.) as a Function of IBA Pretreatment, Substrate, and Season. *Forests* **2022**, *13*, 1705. [CrossRef]
9. Colavolpe, M.B.; Vaz Dias, F.; Serrazina, S.; Malhó, R.; Lourenço Costa, R. *Castanea crenata* Ginkbilobin-2-like Recombinant Protein Reveals Potential as an Antimicrobial against Phytophthora cinnamomi, the Causal Agent of Ink Disease in European Chestnut. *Forests* **2023**, *14*, 785. [CrossRef]
10. Alvarez, J.M.; Rodríguez, S.M.; Fuente-Maqueda, F.; Feito, I.; Ordás, R.J.; Cuesta, C. Identification of Candidate Genes Involved in Bud Growth in *Pinus pinaster* through Knowledge Transfer from *Arabidopsis thaliana* Models. *Forests* **2023**, *14*, 1765. [CrossRef]

Review

Conifer Biotechnology: An Overview

Sonia María Rodríguez, Ricardo Javier Ordás and José Manuel Alvarez *

Área de Fisiología Vegetal, Departamento de Biología de Organismos y Sistemas, Instituto Universitario de Biotecnología de Asturias (IUBA), Universidad de Oviedo, 33071 Oviedo, Spain; uo270227@uniovi.es (S.M.R.); rordas@uniovi.es (R.J.O.)
* Correspondence: alvarezmanuel@uniovi.es

Abstract: The peculiar characteristics of conifers determine the difficulty of their study and their great importance from various points of view. However, their study faces numerous important scientific, methodological, cultural, economic, social, and legal challenges. This paper presents an approach to several of those challenges and proposes a multidisciplinary scientific perspective that leads to a holistic understanding of conifers from the perspective of the latest technical, computer, and scientific advances. This review highlights the deep connection that all scientific contributions to conifers can have in each other as fully interrelated communicating vessels.

Keywords: biotechnological research in conifers; genomics; cell biology and biochemistry research in conifers; micropropagation techniques; transgenesis; CRISPR/Cas9

1. General Traits, Distribution, and Diversity

Conifers are a group of plants that encompasses the oldest living trees and shrubs on our planet. They have existed for more than 300 million years [1,2], starting from a common ancestor of gymnosperms and angiosperms. Conifers comprise two-thirds of gymnosperms [3] and include species of high forest interest, such as pines, spruces, cypresses, or sequoias [4]. Conifers constitute the largest and most diverse group of gymnosperms (for a complete review, see [5]). Conifers and other gymnosperms were the dominant trees during the Mesozoic Era, which is also known as the Age of the Conifers, although they posteriorly declined and were replaced by angiosperms as the dominant group.

It is very complex to gather all the characteristics of conifers into one definition; they typically have simple needle-shaped or scale-shaped evergreen leaves, even though some deciduous species have been described. In general, conifers are large woody plants with strong apical dominance, although there are shrubby species too. Its main characteristic is to develop cones or strobil es, which are primitive reproductive structures. The highly variable fruiting structures reflect strong selective pressures associated with modes of seed dispersal [5]. Regarding the mating system, conifers are predominantly allogamous. This fact, together with the long-distance pollen dispersal by wind, is responsible for the high gene flow among distant populations leading to the low levels of genetic differentiation between them and the great genetic diversity observed in multiple species [6].

Like many other green plants, they have a diplohaplonic life cycle with the particularity that the dominant diploid sporophyte phase and the annual gametophyte phase occur on the same plant (for a complete revision, see [7]). Three phases can be distinguished in the sporophyte: the juvenile stage, during which they are not reproductively competent; the reproductive onset stage, where cones are only produced in response to certain external stimuli; and the reproductive competence stage, during which cones develop annually under almost any conditions. Most conifers produce woody cones, and seeds are dispersed mainly by wind and gravity, although some species have developed edible fleshy structures for favoring animal seed dispersal.

Citation: Rodríguez, S.M.; Ordás, R.J.; Alvarez, J.M. Conifer Biotechnology: An Overview. *Forests* **2022**, *13*, 1061. https://doi.org/10.3390/f13071061

Academic Editor: Carol Loopstra

Received: 30 May 2022
Accepted: 2 July 2022
Published: 5 July 2022

Conifers are distributed worldwide in a great variety of ecosystems, especially in the boreal and temperate forests of North America and Eurasia. This fact reveals great adaptability to variable environmental conditions, although they are practically absent in deserts, steppes, the Arctic tundra, and some tropical rainforests [5].

Although there are some discrepancies about their taxonomic classification, it is currently accepted based on morphological and molecular studies that conifers include the Class Coniferae or Pinosida, the subclass Pinidae, and three different taxa at the level of order: the Pinales, which only includes the family Pinaceae; the Araucariales, which includes the Araucariaceae and Podocarpaceae families; and the Cupressales, which comprises the Sciadopityaceae, Cupressaceae, and Taxaceae families, although some authors include in this order the two additional families Cephalotaxaceae and Phyllocladaceae [2,8].

The family Pinaceae is the largest in terms of species, as it includes about 232 species distributed in 11 genera (*Abies, Cathaya, Cedrus, Keteleeria, Larix, Nothotsuga, Picea, Pinus, Pseudolarix, Pseudotsuga*, and *Tsuga*). *Pinus* is the largest genus in the family, with about 119 recognized species (The Gymnosperm Database: https://www.conifers.org/) (accessed on 28 May 2022).

Conifers comprise a probably monophyletic group of highly branched trees or shrubs with simple leaves, this being a possible apomorphy of the group. Phylogenetically they are a paraphyletic group with respect to Gnetales (taxon Pinidae, Coniferophyta or others). Different genera such as *Picea* are believed to have originated in North America and then dispersed across the Bering land bridge, showing how the place of origin does not determine the center of diversity [3]. Some results even suggest that, before the division between angiosperms and gymnosperms occurred, a functional specialization had already taken place. The current conifer species map is not yet fully known. It has been reported that the discovery of new genuine conifer species is unsettled, especially in varied and inaccessible ecosystems in the southern hemisphere [2].

Scientific analysis of phylogenetic relationships at various taxonomic levels, mechanisms of evolution at the molecular level of lineages and genomes, and biogeographic dispersal in the development of intercontinental disjunctions or patterns of species diversification [3] are challenges at this point. Understanding the genetic basis of biological processes, evolution, or variation in certain traits is also key to the molecular improvement of specimens [9]. Comparative transcriptomics and genomics of conifer developmental studies are currently being used for these purposes [10], and a database of research related to the study of conifer genera and species [1,2] will be useful for further studies on a more comprehensive and solid basis. These aspects are not unique to conifer studies, and they also have great importance in other fields [11]. However, in conifers, which have such a varied global distribution in disparate environments, and which maintain enormous adaptability and diversification, the challenge is even greater. In conclusion, the use of new technologies is key to scientific innovation in the field of conifer research.

2. Ecological and Economic Importance of Conifers

From an ecological and economic point of view, conifers are the most important group of gymnosperms [3]. Coniferous forests account for 31% of the world's total forest plantation area (FAOSTAT), covering vast areas in the Northern hemisphere, and constitute one of the largest terrestrial carbon sinks and play an important role in climate change mitigation. Conifers, in addition to being widely used for ornamental purposes, have an enormous economic importance, as they are a renewable source of timber, both for the elaboration of manufactured product, and to produce energy (50% of the global timber obtained is supplied by conifers, mainly by *Pinus*, as they generate higher and faster economic yield than angiosperms [2]), paper pulp, and other non-wood products such as resins, natural oils, edible seeds, and products with medical use (for example, the anti-cancer drug Taxol).

It has been reiterated that coniferous forests have a relevant role as carbon sinks [12,13] and are expected to increase their prevalence in the current century [13]. However, soil respiration in coniferous forest systems also releases greenhouse gases. Understanding gas release will allow scientific and efficient management of coniferous forests and maintaining forest reserves [14] without neglecting care in urban environments [15]. In addition, being able to accurately monitor the intensity of conifer burning on a large scale would be key to a good analysis of climatic and biological changes in ecosystems [16]. The challenge of understanding all the mechanisms governing the plant biomass and organic carbon stocks behavior will subsequently allow for improved soil organic carbon projections [17] and for combating air pollution and its consequences on health and the economy. The results of this research will, in turn, lead to the sustainability and efficiency of biotechnological conifer forestry.

Habitat deterioration, particularly fires, is depleting conifer ecosystems, a situation aggravated by ineffective conservation methods [18]. Certain conifer groups are already seriously threatened with extinction due to this combination of factors [2]. For example, *Araucaria angustifolia* is listed by the International Union for Conservation of Nature as a critically endangered species [18]. Not only this, but the preservation of conifers has a direct impact on other species so that a decrease in their numbers can have repercussions on the global functionality of the ecosystem, causing changes in trophic cascades and even the loss of biodiversity [2]. For all these reasons, the discipline of restoration ecology has arisen, which, due to the growing importance of the above, is requiring great efforts in current research [2]. Knowing the mechanisms that allow the fastest and most efficient reconfiguration of each deteriorated ecosystem, in our case, coniferous forests, will require a very complex diagnostic analysis that will involve different fields: genetic, molecular, tissue, organic, etc. It is a great challenge for the international scientific community to coordinate its efforts in ecological restoration processes, in which bioinformatics and technological innovations will play a crucial role.

Globally, conifers are currently suffering increased mortality due to recent droughts [19,20]. Understanding their ability to adjust their physiology to adapt to drought is another research challenge. Studies in *Pinus sylvestis*, *Pinus halepensis*, or *Pinus pinaster* have shown that in drier and warmer conditions, there is reduced growth and a higher mortality rate, leading to a loss of tree productivity. This can also be used in order to identify forest regions, such as those under Mediterranean conditions, with increased vulnerability so that the responses of different forest ecosystems to ongoing aridification can be monitored [19]. One of the first studies addressing the physiological and biochemical dynamics under extreme drought stress, and subsequent recovery of a gymnosperm species, *Pinus massoniana*, has now been carried out [20]. It is highly likely that, as the climate continues to warm, forest ecosystems will increase their susceptibility to severe drought, leading to community deterioration, death, or reduced growth of individuals. It is, therefore, essential to conduct different studies that assess the response peculiarities of different conifer species [21].

To face the new challenges brought about by climate change and increased pressure due to land use for agriculture, livestock, or urban development, actions must be taken to improve the use and conservation of genetic resources [22]. The seeds have great importance at this point. In this regard, experiments in *Pinus sylvestrys* have been carried out to evaluate the effect of humidity reduction on the quality of seeds obtained from cones in order to observe a faster and more intensive scale opening of cone scales [23]. Research has also been initiated on the performance of seedlings from color-sorted seeds of Scots pine [24]. In addition, the impacts of seed source in pine, and the possibility of selecting provenances to improve growth rates and physical and anatomical wood quality attributes related to tracheid dimensions, have also been analyzed in *Pinus banksiana* [25].

On the other hand, the sustainability of forest ecosystems is seriously altered by pine plantation forestry [13], this being one of the most interesting challenges in the immediate future. Furthermore, understanding ecological relationships in forestry could also prevent the proliferation of diseases in conifers in the face of increasing pests [14] and facilitate research on the suitability of seed and plant production systems, leading to better use of their breeding programs. Due to the increasing wood demand, the establishment of high-yield plantations with enhanced biomass production is necessary. For that purpose, breeding programs for the identification and selection of superior genotypes with improved production traits, such as growth rate, wood quality, and tolerance to biotic and abiotic stresses, have been developed. However, the domestication of conifer species through traditional genetic improvement techniques is much more difficult than in other crops due to their long generation times and the fact that some traits that are important for production cannot be evaluated during the juvenile stage.

Reforestation and conifer plantations sometimes have to deal with poor, eroded, or degraded soils; it is, therefore, necessary to understand the factors that facilitate the reabsorption of nutrients despite the existence of a shortage or limitation of nutrients [26], which could open interesting perspectives for the articulation of the productive system, and the obtaining of resilient conifers. Soil fertilization and the type of containers used in nurseries have been shown to improve yield and crop quality [27,28]. Future studies should address the different macronutrients present in soils at different depths and their variations in order to understand their relationship with conifer growth and development, which will contribute to sustainable *Cunninghamia lanceolata* plantations [26]. One of the factors that has allowed conifers to survive in suboptimal conditions has been the establishment of symbiotic relationships with fungi [2]. In this way, they manage to increase up to ten thousand times the area that allows them to absorb water and nutrients. The effect of inoculated native ectomycorrhizal strains and compatible fungus–conifer combinations for inoculation in seedling nurseries should be increased, even under real field conditions. This would ensure root colonization before transplanting to the field, thus, reducing seedling mortality due to water stress in *Pinus hartwegii* and *Abies religiosa* [29]. Current studies also address the usefulness of mycorrhizae to biologically control different diseases, such as pine wilt [30]. Alternatively, to improve sustainable pest management in the field of conifer bioprotection, and obtaining safe and highly effective insecticides, numerous benefits of soil fumigation for forest conifer seedling production have been described [31,32]. It is foreseeable that new and promising lines of research in this field will open up in the near future. This holistic study will require multidisciplinary and integrative research in order to encompass all interacting microbial communities [33].

At the same time, it is essential to preserve the genetic diversity of native conifer forests, which is essential to conserve the capacity for stress resilience and adaptation to variable environmental conditions of an ecosystem [12,34]. Thus, sustainable forest management requires the development of efficient breeding programs and alternative strategies for the conservation of conifer's genetic diversity.

A summary of the main ecological problems faced by conifers is listed in Table 1.

Table 1. Summary of the main ecological problems faced by conifers.

Problem	References
Diseases and pests	[14]
Habitat deterioration	[18]
Drought	[19–21]
Climate change and human pressure	[22]

Biotechnology would have a strong impact on conifer breeding, propagation programs, and their adaptation to the environmental settings that support their development, such as soils, light, or temperature. New biotechnological tools, such as genomic, micropropagation, and genetic engineering, would also offer the possibility to overcome these problems. Nevertheless, the application of these techniques requires a better knowledge of conifer biology. For that purpose, it is necessary to better understand the molecular basis of traits and processes that are important for production and adaptation and the development of reliable experimental systems for their study.

3. Genomic Research in Conifers

The availability of full genomes is key to the identification and characterization of gene networks controlling multiple processes, as well as to elucidating the relationship between genotypic and phenotypic diversity in populations. The flowering plant *Arabidopsis thaliana* (*Arabidopsis*), the model plant species par excellence, was the first plant genome sequenced (The Arabidopsis Genome Initiative, 2000). Since then, several angiosperm genomes with high economic importance, as well as model species from other plant groups, have been sequenced. The first tree genome sequenced was the *Populus trichocarpa* (black cottonwood) in 2006 [35].

Conifers are characterized by extraordinarily large genomes [36] with high heterozygosity levels and high repetitive DNA content [37,38]; that is why full genome sequencing of conifers was not technically or economically viable before 2013 (Table 2). The development of next-generation DNA sequencing (NGS) technologies and powerful bioinformatics methods for the assembly and annotation of the genome sequence allowing the obtaining of the full genome and/or transcriptome from several conifer species (for a complete review, see [39,40]).

Table 2. A comparison between the genome size of different species.

Species	Genome Size	Released in
Arabidopsis thaliana	119.1 Mb	2000
Populus thrichocarpa	434.1 Mb	2006
Picea glauca	26.6 Gb	2013
Picea abies	12.0 Gb	2013
Pinus taeda	22.1 Gb	2014
Pinus lambertiana	27.6 Gb	2016
Pseudotsuga menziesii	14.7 Gb	2017

The genome drafts from several conifers such as white spruce (*Picea glauca*) [41], Norway spruce (*Picea abies*) [42], loblolly pine (*Pinus taeda*) [43,44], sugar pine (*Pinus lambertiana* Dougl.) [45], and Douglas fir (*Pseudotsuga menziesii*) [46] are already available (for a complete review, see [39]). *Pinus pinaster* and *P. sylvestris* genomes have also been sequenced. A reference transcriptome was also obtained by RNA sequencing (RNA-seq) in maritime pine (*P. pinaster*) [47] and sugar pine (*P. lambertiana*) [48]. There is also a lot of transcriptome data in public databases, such as CONGENIE (https://congenie.org accessed on 28 May 2022) or Gymnoplaza (https://bioinformatics.psb.ugent.be/plaza/versions/gymno-plaza/ accessed on 28 May 2022).

Annotated transcripts of *Pinus elliotti* using third-generation technologies are key information for phylogenetic research and breeding of other non-referenced species [9]. Diverse algorithms have also been developed to process data related to different processes of plant development, such as caulogenesis and rhizogenesis, which can be extrapolated to conifers in a consistent and compatible way; this has a high impact on the research of naturogenic processes and anthropogenic influences on tree growth and development [49], e.g., to refine selection criteria for germination-competent mature embryos in conifers [4,50].

To promote the reproduction, biodiversity, and conservation of conifers, different comparative genome analyses are carried out with an emphasis on the evolution of key traits. This might help to understand why the genomes of these organisms are so large [42]. This is another pending challenge deeply linked to the development of bioinformatics and new technologies. First, comparative analyses show that the estimated number of coding genes in conifers is similar to or slightly higher than the one from model angiosperms [39]. Furthermore, it has been observed that there is considerable conservation of gene families among seed plants, although there are genes that are unique in conifers, and there are notable differences in relative abundances.

Many important milestones have already been achieved, but there is still a long way to go in the area of genetic information. Their big data pose challenges such as storage, management, integration, security, and confidentiality [51]. Efficient improvements and developments will be needed in the computational methods and technologies used in research. The projection of bioinformatics in conifer science seems undeniable. An important and encouraging factor is that most of these biotechnological innovations are trying to be made accessible to the scientific community in appropriate databases, which, in turn, are key for subsequent studies [52]. One factor that must be taken into account in the immediate future is that these technological resources, in addition to being accessible to researchers globally, must be affordable [38]. Universal and real accessibility (simple and inexpensive) to the latest technological, scientific innovations will give a formidable boost to conifer research in all its facets and, above all, will open up a feedback loop that will allow a constant progression of knowledge. Consolidating and progressing the use of technological innovations is one of the cornerstones of the forthcoming development of conifer research.

4. Breeding Programs and Biotechnological Alternatives

Pines show, in general, a great variation among individuals for productivity of forests traits. This aspect provides an interesting opportunity for the establishment of breeding programs to obtain an increase in yield for these traits. The development of breeding plans should combine three general objectives: (1) Conserve genetic resources in forestry and manage naturally regenerated forests; (2) improve the production in the most productive areas through specific treatments; and (3) define a breeding line oriented to obtain and propagate highly productive genotypes for use in new plantations and forest crops.

Nowadays, pine genetic improvement programs are aimed at increasing productivity. These programs are based on field identification of outstanding specimens, their establishment in clonal banks to evaluate their behavior under different environmental conditions and to select the most productive specimens [53]. Once the best individuals have been determined, their vegetative propagation is planned in order to establish high-production plantations or seed orchards [53]. Vegetative propagation is an optimal method to capture all the genetic gain within a given generation since it allows exploiting all the components of genetic variance (dominant, additive, and epistatic) without the need to carry out crossing and selection procedures [54].

Unfortunately, the conventional breeding of trees is not as straightforward as for herbaceous plants [55]. Trees have long life cycles, are self-incompatible, highly heterozygous, and many relevant traits of interest in conifers cannot be adequately assessed until a mature stage is reached [56]. This makes the fixation of an allele of commercial interest very difficult and time-consuming [57]. Maturation induces changes in the meristem behavior, reducing the propagation potential of the tree [56,58]. For this reason, it would be desirable to develop a more efficient and economic technology to facilitate the propagation of selected adult trees for the establishment of clonal banks and plantations. In vitro cultures can restore this regenerative competence, either through a transient increase in vigor (reinvigoration) or through a rejuvenation that allows the recovery of characteristics of juvenile individuals, such as rapid growth and rooting capacity [58]. Therefore, breeders and biotechnologists should work together and focus on traits that improve productivity, sustainability, and wood quality [59]. Nowadays, forest regeneration after harvest is often left to natural processes, although prompt artificial regeneration with selected genotypes provides the most effective means to increase forest yield [59–61].

Biotechnological tools, such as in vitro asexual propagation, are suitable procedures for mass and commercial clonal production of trees in both coniferous and hardwood species. Micropropagation is the in vitro multiplication and/or regeneration of plant material under aseptic and controlled environmental conditions. Micropropagation techniques are often used successfully in most species. However, it is complex in conifers if tissues from adult individuals are used due to their recalcitrance [62]. If it were possible to use material from mature trees, the time needed to obtain improved varieties would be reduced [12,63]. This is why their use for forest breeding is currently limited [22], and new avenues are opening up.

However, the success of in vitro culture applied to adult conifers is variable. In most cases, regeneration of new plants from adult material is either impossible or too difficult to be practically applied. In any case, there are species where promising results have been obtained, such as *Taxus mairei* [64], *Larix* [65], and *Pinus pinaster* [66]. Moreover, in the case of *Pinus radiata*, research has led to a practical application, with the establishment of a company that offers micropropagation of adult individuals among its services (The Tree Lab, Rotorua, New Zealand).

There are three micropropagation procedures for regenerating a plant; the production of plantlets via axillary shoots growth with the least somaclonal variation, adventitious shoot induction via direct or indirect organogenesis, and somatic embryogenesis [67]. Organogenesis is the initiation of a unipolar structure, shoots, or roots in response to a treatment or to an appropriate culture conditions. Somatic embryogenesis (SE) can be described as an asexual process where somatic cells develop bipolar structures similar to zygotic embryos to form a whole plant without a vascular connection with the parental tissue.

Micropropagation is carried out by taking small sections of tree tissue called primary explants and growing them under artificial conditions. Then, the different types of explants begin a process of growth stimulation of axillary preformed buds, or morphogenesis that will produce adventitious buds—through de novo organogenesis (DNO)—or somatic embryos—through SE [12,68].

For the clonal propagation of trees considered recalcitrant, factors such as the culture medium, the time of year, and the position of the primary explant on the mother tree (topophysis and cyclophysis) must be considered because they have a great influence on the regeneration response [69,70]. Apart from successful cases, most clonal propagation protocols present difficulties, often in the culture media. Problems may include changes in morphology accompanied by hyperhydricity (previously known as vitrification), lack of elongation, or occurrence of necrosis, poor rooting efficiency, poor regeneration, and excessive phenolic exudation.

The axillary bud multiplication involves the development of the axillary buds. The primary explants are usually isolated from the tip of young shoots that develop axillary bud under the effect of cytokinins. The role of cytokinin is to suppress apical dominance and promote the development of axillary buds. The axillary bud method is often combined with the single node method. A concrete example is the micropropagation protocol for *Juniperus thurifera* L., using microcuts with axillary buds from young shoots. This is relevant as this plant is endangered precisely because of the lack of viable seeds [71]. There have been several reports dealing with in vitro culture of mature conifers in the last twenty years [64–66,72–76].

However, in most cases, plant regeneration from adult trees showed severe problems that limit its practical application. The authors of [77] presented a plant regeneration method for producing clonal plants from mature trees of *P. pinea* via shoot development from winter-dormant buds. The low rooting percentage and the lack of axillary bud proliferation despite the juvenile appearance of the shoots indicate that in vitro culture induced reinvigoration (transient appearance of juvenile characteristics) in the brachyblast meristems rather than the desirable rejuvenation [58]. Nevertheless, the results showed that it is possible to obtain rooted shoots from a mature origin, encouraging further investigation into the elongation and rooting phases of the protocol.

Micropropagation of conifers is usually limited to juvenile materials [69], being the in vitro amplification of progeny common by DNO or SE from seeds selected in seed orchards [78]. Mass vegetative propagation of selected families is a useful adjunct to improve programs based on recurrent and non-recurrent selection [79]. Experimentally, statistical efficiency is increased when treatments are applied to clones instead of families due to the absence of genetic variance within the clones. In breeding, clonal banks are mainly used in populations with large economic (or ecological) value. The purpose of using clonal propagation is to capitalize on both the time saving of large-scale implementation and the relatively larger genetic gains available through clonal tree improvement programs [80].

De novo organogenesis generally begins using juvenile or embryonic primary explants, embryonic axes, and cotyledons [81–83]. The DNO process consists of four or five different stages: initiation, proliferation, elongation/rooting, and acclimation. During initiation, usually in the presence of cytokinins (CKs), explants acquire morphogenetic competence, cell identity is determined, and meristemoids are induced and differentiated, which later, in the absence of the stimulus, develop shoots [81]. The multiplication phase, carried out with shoots separated from the cotyledonary explants and elongated by sequential subculturing on hormone-free medium and preferably with activated charcoal, led to the production of axillary shoots, which were excised and subcultured. Axillary buds and brachyblasts formed during the elongation phase can be used to produce new shoots [83]. One of the main bottlenecks of this method is the rooting of micro-shoots already developed and preferably elongated; the efficiency of the process is usually not high and depends on the auxin treatment, degree of juvenility of the primary explants used, the species, and even the genotype of the seed [12]. More roots imply a better performance upon outplanting, but the need still exists to understand what effect root system quality plays in long-term growth and development [83]. The goal is that this procedure can be used in full-sib family forestry with high predicted performance for deployment to forestry. The performance of a full-sib family can be predicted either based on the performance of the family itself and/or on the performance of its parents in crosses with other genotypes.

As commented above, superior genotypes can be propagated by vegetative multiplication using in vitro techniques, such as organogenesis [83–86] and somatic embryogenesis [63,87–90], both considered to have greater potential than traditional rooting of cuttings [91]. Somatic embryogenesis enables clonal propagation for forestry and forest research and is a key tool for genetic transformation [92]. It allows the production of plants with known, uniform, and desirable characteristics [62]. In addition, the resulting plants closely resemble those from seed due to the development of zygotic embryos with a strong root–shoot connection [4]. All this has a great impact on species of great economic interest, such as Norway spruce (*Picea abies* L. Karst.) and Scots pine (*Pinus sylvestris* L.), as this would ensure their stable production even in situations of climate stress or biodiversity crisis [93]. Another advantage of the method is that it facilitates automation and amplification, thereby reducing the personnel costs required and increasing the reliability of the entire process as a whole [4]. Somatic embryogenesis offers, in short, significant advantages and reduced difficulties and costs.

Somatic embryogenesis involves the formation of proembryogenic masses (PEM) at the early stage, which later gives rise to plants [68,94]. Such PEMs are usually initiated from immature zygotic embryos but have also been studied from shoot explants and mature embryos in *Picea abies* and *Picea glauca*. The SE technique in conifers is multi-stage: embryogenic cultures are initiated from explants, then somatic embryo induction occurs; later, embryogenic masses proliferate and multiply; then, somatic embryos mature and develop from the previous embryogenic masses. Somatic embryo maturation is a critical process that affects the subsequent germination ability of embryos [12,95]. Maturation and conversion of somatic embryos in plants are two crucial steps that hamper the development of efficient somatic embryogenesis systems.

The successful induction of embryonic tissues and SE depends on the genotypes, explant types, date of seed collection, and the media compositions at each step of production. Phytohormones play a key role in embryo formation, particularly auxins; these enhance regenerative responses in vitro because they facilitate the activation of specific developmental programs, which could also be induced by stress factors (temperature, osmotic stress, starvation, heavy metal ions) or wounding [96].

Somatic embryo maturation in pine species is stimulated by the transfer of proliferating tissue to a medium devoid of auxins and cytokinins and supplemented by abscisic acid (ABA). The optimum concentration of ABA varies depending on species [97,98] and may differ between two cell lines of the same species [99]. Despite the potential offered by SE and despite the progress made in the last few years, the main bottleneck of the technique continues to be the progression from immature embryogenic cultures into mature cotyledonary embryos able to develop properly into well-growing plants [89].

In conifers, maturation into cotyledonary embryos is stimulated by the exogenous application of abscisic acid (ABA) and osmotic stress through the use of PEG [100] or by reduced water availability through the use of a gelling agent [101]. Changes in the composition of the maturation medium have been reported as a significant improvement in mature embryos in *P. radiata* [98]. In conifers, there is an inverse correlation between maturation capacity and subculture number; in the case of *P. pinaster*, this loss of capacity occurs in less than 10 months [102]. Another problem is the passage from immature embryogenic cultures to mature cotyledonary embryos, their acclimatization, and their conversion into plants [90,103].

An important objective for improvement is to overcome the conversion of PEMs into plants due to low maturation rate, low germination frequencies, and poor quality of somatic embryos [12]. Solutions have been investigated to overcome disadvantages related to the low production rate of somatic embryos, not to mention the reduction in yields that occur at each step during further development up to conversion [68]. Research on the incidence of the initiation environment and the effect on subsequent conversion to somatic seedlings should be further investigated [104].

Another unknown is the knowledge of the mechanisms responsible for periderm establishment and formation, which, despite being so relevant, remain largely unknown [105], or the implications of telomere shortening in explants on the frequency of SE induction [106]. The effect of different types of auxins on the physiological reaction of plant materials during SE has been studied. This is a starting point for further studies on the mechanisms of SE induction [107].

A key solution and challenge have been the automation of the process in order to reduce labor, which is still required in these processes today, and which is costly in areas where conifer forestry is of high relevance [4]. The development of universal protocols for coniferous species is very difficult, as the efficiency of the SE procedure varies greatly, and it remains a challenge [93].

Factors that until recently seemed of little relevance are now showing great promise for advancing research. One example is the case of temperatures during the different stages of the SE of conifers. It has recently been reported that high temperature in SE alters the subsequent stages of the process and the ex vitro behavior of the resulting somatic plants [108].

Another example is the importance of light, its intensity, and spectra on the particular stages of SE. It has recently been reported that somatic embryos germinate differentially under exposure to different light spectra. The differences lie in the shoot, root growth, and their survival [93,109]. Here, too, the origin of responses to different temperatures in SE in conifers will have to be analyzed [110]. The effects of the timing of sample collection, its family components, and the means of induction of embryogenic lines have now been successfully demonstrated [111]. It is a challenge for researchers to permanently question all solutions that might seem definitively adopted only a short time ago.

Somatic embryogenesis technology is usually associated with cryopreservation, which offers an appropriate tool to overcome these problems since all the metabolic and physical processes are arrested and require minimal equipment and maintenance [112]. Additionally, cryopreservation can establish dormancy and help enable massive clonal propagation [18,95]. During the juvenile stage, embryogenic tissues can be stored for long periods of time (up to hundreds of years) in liquid nitrogen containers [93]. This method allows the greatest genetic gains to be made, as cryopreservation enables long-term field trials to be carried out [113], and maintenance costs can be reduced [104,114]. This enables the establishment of field tests to evaluate the different lines during their adult stage. Only the best performing clones are massively propagated from the cryopreserved stock of embryogenic tissue [115–117]. Thus, more studies developing a cryopreservation protocol that ensures a continuous supply of juvenile mass are desirable.

In addition, when the cultures are maintained for longer durations, the frequency of mutation will be higher in in vitro regenerated plants [118,119]. That is a significant advantage of SE in rapport to DNO, which lacks cryopreservation methods to maintain the long-term juvenility of the material [12,113], contrasting with the possibility of cryopreservation that does occur when using SE [110]. The development of effective cryopreservation protocols and appropriate genetic markers would make DNO as promising as SE. This is another challenge that remains open to the scientific community. An example of this is the vegetative propagation of Norway spruce, formerly using rooted cuttings and more currently by SE [114]. This possibility is key, given the scarcity of high-quality forest regeneration materials, as the flowering of the species is irregular, and there have also been pest problems that hinder seed generation [62].

Somatic embriogenesis has been shown to be very successful in other genera of the Pinaceae family: *Abies*, *Larix*, *Picea*, *Pinus*, and *Pseudotsuga*, and others of the families Cupressaceae, Taxaceae, Cephalotaxaceae, and Araucariaceae [68,120]. Currently, SE has been achieved for almost 30 pine species [111].

Although the SE system has been developed for a large number of conifers, such as spruce or pine, some desirable adult trees with known characteristics cannot be propagated through SE and can only be initiated from juvenile plants [62,69]. Its extensive use in practice is limited as it can only be applied to certain genotypes. Furthermore, a potential link between biotic defense and SE induction recalcitrance has been observed. In addition, a relevant problem is that production management costs are very high in this case, which also hinders its use [93].

In order to provide an efficient and abundant supply of somatic embryos for industrial applications, the molecular mechanisms of SE will need to be further studied [121], and protocols for the efficient induction of embryogenic cell lines will need to be improved [122]. The most pressing challenge is to understand the molecular regulation of embryogenesis in conifers, the knowledge of which remains very limited [123], largely due to the lack of identified embryonic defective mutants [124]. In conclusion, it would be desirable to further develop clear, detailed, and reliable protocols, which are essential for the mechanization and homogenization of experimental systems and, to this end, all mechanisms and resources for vegetative propagation and plant regeneration should be investigated [125].

Somatic embryogenesis is suitable for studying the molecular basis of embryogenesis processes in conifers. Primarily, the model species under study are angiosperms; therefore, the identification of key proteins in the control and regulation of SE processes in conifers is limited for now, and the study of their structural domains will be very relevant for the understanding and monitoring of the process. Much more research is needed on the role of different genes [126,127], particularly homeoboxes, in conifer development [124,128–131].

The molecular basis and signaling events in conifers SE need to be understood [12,132]. In addition, detailed genetic research is needed for conifers because genes involved in SE suppression (such as PICKLE) are unknown [68,133]. In the near future, genetics and transcriptomics research will be boosted by single-cell 'omics' [134].

Every step forward in the SE process will contribute to the final success rate [93]. Improved robotization and automation will even lead to the identification of weak regulatory interactions, the identification of rare intermediate cell states, the understanding of histone modifications or methylation, and many other advances [134]. It is very likely that we will witness, in a few years, a new scenario in research and knowledge of the conifer genome and its underlying mechanisms and an unprecedented advance of knowledge in these fields. We will foreseeably witness a shortening of phases and an improvement in the quality and yield of wood by bringing advances to conventional breeding [135]. There will also be a strengthening in the combined use of methods, as has already occurred in conifers with the combination of SE with reverse genetics as a model for studying the regulation of embryonic development. If the system is well-coordinated, it can lead to abundant somatic embryos at different stages of development [136].

Genetic engineering (discussed below) has been reported recently as a valuable tool to overcome some tissue culture limitations. The ectopic expression of growth regulator genes may bypass tissue culture-based regeneration and allow direct regeneration in a wide range of species [137–139].

Thanks to advances in molecular research and technology, genomics, cell biology, and biochemistry will also converge [134,140] in the quantitative analysis of the whole genome of individual conifer cells. This convergence will be part of a new era in biotechnological research.

5. Genetic Transformation

Genetic transformation, genetic modification, genetic engineering, or transgenesis is defined as 'the use of recombinant DNA and asexual gene transfer methods to alter the structure or expression of specific genes and traits' (FAO, 2004). The genes from an organism that are inserted into another are called transgenes and have the ability to confer to the latter a particular trait. Often, but not always, the transgene is obtained from a different species than that of the recipient. Successful genetic transformation depends on the stable incorporation of the novel gene in the genome of the recipient, leading to the transmission of the input gene (transgene) to successive generations.

Genetic improvement of conifers by traditional breeding is time-consuming due to the long juvenile phase and genome complexity. The ability to rapidly transfer new traits from one species to another has the potential to enhance traditional tree breeding and improvement since generation times for forest tree species are rather prolonged. Plant transformation technology has become a versatile platform for tree improvement, as well as for studying gene function in plants.

In general, the introduction of new genes via genetic transformation is fully justified if there is a difficulty in transferring a trait from one species or variety to another without the risk of altering the rest of the phenotypic characteristics or an excessive complexity and duration of crossing, backcrossing, and selection programs. It is evident that a large part of these conditions are present in the improvement of species with long reproductive cycles, as well as in species with vegetative propagation, in which there is generally a high degree of heterozygosis, which, associated with the need to maintain the characteristics of the variety unaltered, limit the breeding programs.

Therefore, from the above, we can deduce the possible potential of this methodology in agroforestry and woody species in general, which would perfectly complement existing breeding programs. Genetic engineering methods increase the diversity of genes and germplasm available for incorporation into a given species and reduce the time required to produce new varieties and hybrids [141,142]. Progress in the development of genetic engineering protocols for conifers has been rapid at the end of the last century, and there are numerous reports of conifers that have been transformed using biolistic and *Agrobacterium*-mediated techniques.

Agrobacterium tumefaciens, now named *Rhizobium radiobacter*, is a gram-negative soil bacterium that belongs to the family *Rhizobiaceae* and is the causal agent of crown gall disease (the formation of tumors) in dicots. Tumorigenesis is caused by the insertion of a small segment of DNA (known as the T-DNA, for 'transfer DNA') from a plasmid called Ti (for 'Tmour-inducing'), which is incorporated at a semi-random location into the plant genome (resulting in genetic manipulation of the host). The genes within the T-DNA region responsible for tumorous growth can be removed and replaced by DNA segments of interest. Strains are considered 'disarmed' if they do not contain oncogenes that could be transferred to a plant. This capacity to transfer genes into plants has been used to develop *A. tumefaciens* as a vector for genetic manipulation. *Agrobacterium* transformation has been demonstrated in several conifers [143–147]. *Larix decidua* was the first conifer from which transgenic plants regenerated and transformed with *A. tumefaciens* were obtained [148].

In early transformation studies, conifers appeared to be recalcitrant and less susceptible to *Agrobacterium* infection, and, therefore, direct gene transfer protocols such as polyethylene glycol-based methods, electroporation, and particle bombardment were also developed [149–152].

The biolistic method consists of bombarding competent cells and tissues with metal microparticles, preferably gold, which is less toxic, coated with DNA. Biolistics, also named the gun gen method, can be used for the routine transformation of many conifers. Regenerated transgenic *P. radiata* plants were obtained from bombarded embryogenic calli explants [153]. It has certain disadvantages, such as the integration of many gene copies, which can lead to undesirable effects, such as gene fragmentation or silencing of gene expression [154,155]. *Agrobacterium tumefaciens*-mediated transformation results in transgenic lines with lower transgene copy number compared to particle bombardment but show more stable transgene expression in subsequent generations and fewer cases of transcriptional gene silencing (TGS) and post-transcriptional gene silencing (PTGS). In general, as mentioned above, the biolistic techniques are estimated to be less efficient than *Agrobacterium*-mediated in the case of conifers.

Modifications introduced and demonstrated in conifers include insect resistance [156,157], herbicide tolerance [158,159], wood pulp efficiency [160,161], stress tolerance [162], and sterility [163]. These technologies that allow the design of modified conifers to produce biochemicals and biomass for specific purposes [164], however, have not yet been commercialized.

6. Genome Editing with CRISPR/Cas9

Functional gene research in gymnosperms lags behind that in angiosperms. As already mentioned, the absence of an efficient transformation system and a genome-wide mutant library, together with the complexity of conifer life cycles, has hindered progress in its plant biology knowledge. Furthermore, the extrapolation of research and data based on angiosperm model systems, such as *Arabidopsis*, to conifers is often less informative and confusing, as gymnosperms and angiosperms started to diverge 300 million years ago [165]. Therefore, other approaches and molecular methodologies suitable for conifer species are needed.

Genome editing is an effective technology for functional gene research and trait improvement and has opened up a promising alternative. To date, various tools have been applied successfully in genome editing, such as zinc finger nucleases (ZFNs), transcription activator-like effector nucleases (TALENs), and clustered regularly interspaced short palindromic repeats/CRISPR-associated proteins (CRISPR/Cas); of which, CRISPR/Cas is the most widely used tool owing to its high mutagenesis efficiency and easy application [166].

The CRISPR/Cas system can be categorized into different classes depending on the effector proteins. Among the different CRISPR/Cas systems, CRISPR/Cas9 was the first to be used for plant genome editing [167]. Thanks to this technology, it is possible to generate site-specific mutations and obtain a theoretically very agile route to study processes or improve and introduce traits in conifers [168]. Since it was released in 2012 that CRISPR/Cas9 could be used for targeted genome editing and that it allowed the introduction of targeted modifications at a locus in the genome of any living entity [169], these technologies have boosted the genomics of forest trees and made the scientific determination of gene function possible [170]. In addition, the availability of RNA seq and NGS methods, together with other directions in genome editing, such as CRISPR/Cas9 double cutting, will contribute substantially to shortening the breeding period of conifers [168].

On the other hand, in the last decade, the CRISPR/Cas9 system has been used for both fundamental research and precision breeding [171–173]. Novel traits that are difficult to achieve through breeding, such as resistance to biotic and abiotic stresses [174–176] and sterility [177], can be generated through knockout-mediated trait improvement. Desirable traits can be fine-tuned by generating a range of alleles through genome editing or base editing [178–181].

To date, this technology has already been used in more than 45 genera from 24 plant families, such as *Arabidopsis*, economically relevant crop plants or plants with medical uses [170,182]. Genome editing has also been employed in trees such as poplar and eucalyptus [183–186]. However, scarce applications in gymnosperms have been published, so the use of this technique in conifers remains a challenge [168]. For example, using larch (*Larix gmelinii*) protoplasts, a Cas9 variant without PAM SpRY, has been found to possess genome editing capacity, but no plants were regenerated [187]. An efficient CRISPR/Cas9 system based on SE suitable for conifers has also been published [188].

The possibility of applying the CRISPR/Cas9 technique to conifers to rapidly modify key traits of interest has recently been demonstrated in *Pinus radiata* [168]. Very recently, the pioneering case of targeted mutagenesis using CRISPR/Cas in a conifer species, *Cryptomeria japonica*, has been described, and genome editing studies using the improved vector and producing edited male sterile lines have been announced [189]. However, these publications, while interesting, still showed a high occurrence of chimeras. Based on prior reports, the application of transgenic conifers still requires a lot of development, but the technology is progressing.

A summary of biotechnological alternatives to traditional breeding programs in conifers, their effectiveness, and their emerging problems can be found in Table 3.

Table 3. Biotechnological alternatives to traditional breeding programs in conifers.

Alternative	Effectiveness	Emerging Problems
Organogenesis	High in juvenile explants Difficult in adult material in most cases	Some recalcitrant species Somaclonal variation Rooting
Somatic embryogenesis	High in juvenile explants Difficult in adult material in most cases	Some recalcitrant species Somaclonal variation Maturation and germination are a bottleneck
Genetic transformation (including gene editing)	Insect resistance Herbicide tolerance Wood pulp efficiency Stress tolerance Sterility	Some recalcitrant species Genotype-dependent Chimeras in some cases Gene silencing

The challenges facing genetic modification and genome editing technologies are not only scientific. As mentioned above, genetic modification (GM) offers the opportunity to make transformational changes in shorter time frames but is challenged by current genetically modified organism (GMO) regulations. Legislation and social consideration can be and are, in many cases, barriers as strong as the scientific and technical ones. The important discrepancy is whether the process or the product is focused [190]. Thus far, there is no complete assent on the regulation of gene editing that develop after the current regulatory frameworks were established.

The time and cost of developing and obtaining regulatory approval to commercialize GMOs are usually prohibitive. The global social and regulatory landscape around GM crops is complex, with many different regulatory systems in different countries [191,192]. Several nations, including the United States, Canada, and Argentina, have resolved that gene-editing technologies where the cultivated or commercialized plant does not contain introduced DNA will not be regulated [190,193]. In contrast, the European Union has recently decided that all gene editing methodologies will be regulated in the same way as conventional transgenic organisms [194,195]. Others, such as China and Australia, have not yet decided on their regulatory approach [172].

7. Conclusions

Conifers have always fascinated researchers, but the complexity of their genomes and their peculiar characteristics determine the difficulty of their study. This review tries to make an enunciative approach to some of the innumerable challenges that conifers suggest in current research. It seeks to offer a synthesis of some of the most urgent challenges and refers to some of the latest biotechnological advances from a multidisciplinary perspective.

Author Contributions: All authors contributed equally to this work. All authors have read and agreed to the published version of the manuscript.

Funding: This research received no external funding.

Data Availability Statement: Not applicable.

Acknowledgments: We thank all collaborators and former members for their contribution in the knowledge here presented. We apologize to colleagues whose work is not cited in this review owing to space limitations.

Conflicts of Interest: The authors declare no conflict of interest.

References

1. Smith, S.A.; Beaulieu, J.M.; Donoghue, M.J. An uncorrelated relaxed-clock analysis suggests an earlier origin for flowering plants. *Proc. Natl. Acad. Sci. USA* **2010**, *107*, 5897–5902. [CrossRef] [PubMed]
2. Farjon, A. The Kew Review: Conifers of the World. *Kew Bull.* **2018**, *73*, 8. [CrossRef]
3. Wang, X.Q.; Ran, J.H. Evolution and biogeography of gymnosperms. *Mol. Phylogenet. Evol.* **2014**, *75*, 24–40. [CrossRef]
4. Egertsdotter, U.; Ahmad, I.; Clapham, D. Automation and scale up of somatic embryogenesis for commercial plant production, with emphasis on conifers. *Front. Plant Sci.* **2019**, *10*, 109. [CrossRef] [PubMed]
5. Neale, D.B.; Wheeler, N.C. The Conifers. In *The Conifers: Genomes, Variation and Evolution*; Springer: Cham, Switzerland, 2019; pp. 1–21. [CrossRef]
6. O'Connell, L.M.; Mosseler, A.; Rajora, O.P. Extensive long-distance pollen dispersal in a fragmented landscape maintains genetic diversity in white spruce. *J. Hered.* **2007**, *98*, 640–645. [CrossRef]
7. Williams, C.G. Conifer reproductive biology. In *Conifer Reproductive Biology*; Springer: Dordrecht, The Netherlands, 2009. [CrossRef]
8. Christenhusz, M.J.M.; Reveal, J.L.; Farjon, A.; Gardner, M.F.; Mill, R.R.; Chase, M.W. A new classification and linear sequence of extant gymnosperms. *Phytotaxa* **2011**, *19*, 55. [CrossRef]
9. Diao, S.; Ding, X.; Luan, Q.; Jiang, J. A complete transcriptional landscape analysis of *Pinus elliottii* Engelm. Using third-generation sequencing and comparative analysis in the *Pinus* phylogeny. *Forests* **2019**, *10*, 942. [CrossRef]
10. Onyenedum, J.G.; Pace, M.R. The role of ontogeny in wood diversity and evolution. *Am. J. Bot.* **2021**, *108*, 2331–2355. [CrossRef]
11. Li, Z.; Jiao, Y.; Zhang, C.; Dou, M.; Weng, K.; Wang, Y.; Xu, Y. VvHDZ28 positively regulate salicylic acid biosynthesis during seed abortion in Thompson Seedless. *Plant Biotechnol. J.* **2021**, *19*, 1824–1838. [CrossRef]
12. Bueno, N.; Cuesta, C.; Centeno, M.L.; Ordás, R.J.; Alvarez, J.M. In vitro plant pegeneration in conifers: The role of *WOX* and *KNOX* gene families. *Genes* **2021**, *12*, 438. [CrossRef]
13. Vogel, J.G.; Bracho, R.; Akers, M.; Amateis, R.; Bacon, A.; Burkhart, H.E.; Gonzalez-Benecke, C.A.; Grunwald, S.; Jokela, E.J.; Kane, M.B.; et al. Regional assessment of crbon pool response to intensive silvicultural practices in loblolly pine plantations. *Forests* **2021**, *13*, 36. [CrossRef]
14. Zhang, X.; Zhao, Z.; Chen, T.; Zhao, T.; Song, L.; Mei, L. Fertilization and clear-cutting effects on greenhouse gas emissions of pinewood nematode damaged Masson pine plantation. *Ecosyst. Health Sustain.* **2021**, *7*, 1868271. [CrossRef]
15. Hoppa, A.; Sikorska, D.; Przybysz, A.; Melon, M.; Sikorski, P. The role of trees in winter air purification on children's routes to school. *Forests* **2022**, *13*, 40. [CrossRef]
16. Singh, S.; Singh, H.; Sharma, V.; Shrivastava, V.; Kumar, P.; Kanga, S.; Sahu, N.; Meraj, G.; Farooq, M.; Singh, S.K. Impact of forest fires on air quality in Wolgan valley, New South Wales, Australia; A mapping and monitoring study using Google Earth engine. *Forests* **2022**, *13*, 4. [CrossRef]
17. Terrer, C.; Phillips, R.P.; Hungate, B.A.; Rosende, J.; Pett-Ridge, J.; Craig, M.E.; van Groenigen, K.J.; Keenan, T.F.; Sulman, B.N.; Stocker, B.D.; et al. A trade-off between plant and soil carbon storage under elevated CO_2. *Nature* **2021**, *591*, 599–603. [CrossRef]
18. Dorigan de Matos Furlanetto, A.L.; Kaziuk, F.D.; Martinez, G.R.; Donatti, L.; Merlin Rocha, M.E.; Dos Santos, A.L.W.; Floh, E.I.S.; Cadena, S. Mitochondrial bioenergetics and enzymatic antioxidant defense differ in Parana pine cell lines with contrasting embryogenic potential. *Free Radic. Res.* **2021**, *55*, 255–266. [CrossRef]
19. Valeriano, C.; Gazol, A.; Colangelo, M.; Camarero, J.J. Drought drives growth and mortality rates in three pine species under mediterranean conditions. *Forests* **2021**, *12*, 1700. [CrossRef]

20. Shao, C.; Duan, H.; Ding, G.; Luo, X.; Fu, Y.; Lou, Q. Physiological and biochemical dynamics of *Pinus massoniana* Lamb. Seedlings under extreme drought stress and during recovery. *Forests* **2022**, *13*, 65. [CrossRef]

21. Qi, C.; Jiao, L.; Xue, R.; Wu, X.; Du, D. Timescale effects of radial growth responses of two dominant coniferous trees on climate change in the Eastern Qilian mountains. *Forests* **2022**, *13*, 72. [CrossRef]

22. Flores, A.; López-Upton, J.; Rullán-Silva, C.D.; Olthoff, A.E.; Alía, R.; Sáenz-Romero, C.; Garcia del Barrio, J.M. Priorities for Conservation and Sustainable Use of Forest Genetic Resources in Four Mexican Pines. *Forests* **2019**, *10*, 675. [CrossRef]

23. Aniszewska, M.; Gendek, A.; Tulska, E.; Pęska, P.; Moskalik, T. Influence of the duration of microwave irradiation of Scots pine (*Pinus sylvestris* L.) cones on the quality of harvested seeds. *Forests* **2019**, *10*, 1108. [CrossRef]

24. Novikov, A.; Sokolov, S.; Drapalyuk, M.; Zelikov, V.; Ivetić, V. Performance of Scots pine seedlings from seeds graded by colour. *Forests* **2019**, *10*, 1064. [CrossRef]

25. Mvolo, C.S.; Koubaa, A.; Beaulieu, J.; Cloutier, A. Effect of seed transfer on selected wood quality attributes of jack pine (*Pinus banksiana* Lamb.). *Forests* **2019**, *10*, 985. [CrossRef]

26. Wu, H.; Xiang, W.; Ouyang, S.; Xiao, W.; Li, S.; Chen, L.; Lei, P.; Deng, X.; Zeng, Y.; Zeng, L.; et al. Tree growth rate and soil nutrient status determine the shift in nutrient-use strategy of Chinese fir plantations along a chronosequence. *For. Ecol. Manag.* **2020**, *460*, 117896. [CrossRef]

27. Sung, S.-J.S.; Dumroese, R.K.; Pinto, J.R.; Sayer, M.A.S. The persistence of container nursery treatments on the field performance and root system morphology of longleaf pine seedlings. *Forests* **2019**, *10*, 807. [CrossRef]

28. Liu, Q.; Xu, H.; Yi, H. Impact of fertilizer on crop yield and C:N:P stoichiometry in arid and semi-arid soil. *Int. J. Environ. Res. Public Health* **2021**, *18*, 4341. [CrossRef]

29. Flores-Rentería, D.; Barradas, V.L.; Álvarez-Sánchez, J. Ectomycorrhizal pre-inoculation of *Pinus hartwegii* and *Abies religiosa* is replaced by native fungi in a temperate forest of central Mexico. *Symbiosis* **2018**, *74*, 131–144. [CrossRef]

30. Vicente, C.S.L.; Soares, M.; Faria, J.M.S.; Ramos, A.P.; Inacio, M.L. Insights into the role of fungi in pine wilt disease. *J. Fungi.* **2021**, *7*, 780. [CrossRef]

31. Stejskal, V.; Vendl, T.; Aulicky, R.; Athanassiou, C. Synthetic and natural insecticides: Gas, liquid, gel and solid formulations for stored-product and food-industry pest control. *Insects* **2021**, *12*, 590. [CrossRef]

32. Balla, A.; Silini, A.; Cherif-Silini, H.; Chenari Bouket, A.; Moser, W.K.; Nowakowska, J.A.; Oszako, T.; Benia, F.; Belbahri, L. The threat of pests and pathogens and the potential for biological control in forest ecosystems. *Forests* **2021**, *12*, 1579. [CrossRef]

33. Rama, T.; Quandt, C.A. Improving fungal cultivability for natural products discovery. *Front. Microbiol.* **2021**, *12*, 706044. [CrossRef] [PubMed]

34. Potter, K.M.; Riitters, K. A national multi-scale assessment of regeneration deficit as an indicator of potential risk of forest genetic variation loss. *Forests* **2021**, *13*, 19. [CrossRef]

35. Tuskan, G.A.; Difazio, S.; Jansson, S.; Bohlmann, J.; Grigoriev, I.; Hellsten, U.; Putnam, N.; Ralph, S.; Rombauts, S.; Salamov, A.; et al. The genome of black cottonwood, *Populus trichocarpa* (Torr. & Gray). *Science* **2006**, *313*, 1596–1604. [CrossRef] [PubMed]

36. Zonneveld, B.J.M. Conifer genome sizes of 172 species, covering 64 of 67 genera, range from 8 to 72 picogram. *Nord. J. Bot.* **2012**, *30*, 490–502. [CrossRef]

37. De La Torre, A.R.; Lin, Y.C.; Van de Peer, Y.; Ingvarsson, P.K. Genome-wide analysis reveals diverged patterns of codon bias, gene expression, and rates of sequence evolution in *Picea* gene families. *Genome Biol. Evol.* **2015**, *7*, 1002–1015. [CrossRef] [PubMed]

38. Cañas, R.A.; Pascual, M.B.; de la Torre, F.N.; Ávila, C.; Cánovas, F.M. Resources for conifer functional genomics at the omics era. *Adv. Bot. Res.* **2019**, *89*, 39–76. [CrossRef]

39. Neale, D.B.; Wheeler, N.C. Gene and genome sequencing in conifers: Modern era. In *The Conifers: Genomes, Variation and Evolution*; Springer: Cham, Switzerland, 2019; pp. 43–60. [CrossRef]

40. Neale, D.B.; Martinez-Garcia, P.J.; De La Torre, A.R.; Montanari, S.; Wei, X.X. Novel Insights into Tree Biology and Genome Evolution as Revealed Through Genomics. *Annu. Rev. Plant Biol.* **2017**, *68*, 457–483. [CrossRef]

41. Birol, I.; Raymond, A.; Jackman, S.D.; Pleasance, S.; Coope, R.; Taylor, G.A.; Yuen, M.M.; Keeling, C.I.; Brand, D.; Vandervalk, B.P.; et al. Assembling the 20 Gb white spruce (*Picea glauca*) genome from whole-genome shotgun sequencing data. *Bioinformatics* **2013**, *29*, 1492–1497. [CrossRef]

42. Nystedt, B.; Street, N.R.; Wetterbom, A.; Zuccolo, A.; Lin, Y.C.; Scofield, D.G.; Vezzi, F.; Delhomme, N.; Giacomello, S.; Alexeyenko, A.; et al. The Norway spruce genome sequence and conifer genome evolution. *Nature* **2013**, *497*, 579–584. [CrossRef]

43. Zimin, A.; Stevens, K.A.; Crepeau, M.W.; Holtz-Morris, A.; Koriabine, M.; Marcais, G.; Puiu, D.; Roberts, M.; Wegrzyn, J.L.; de Jong, P.J.; et al. Sequencing and assembly of the 22-gb loblolly pine genome. *Genetics* **2014**, *196*, 875–890. [CrossRef]

44. Zimin, A.V.; Stevens, K.A.; Crepeau, M.W.; Puiu, D.; Wegrzyn, J.L.; Yorke, J.A.; Langley, C.H.; Neale, D.B.; Salzberg, S.L. An improved assembly of the loblolly pine mega-genome using long-read single-molecule sequencing. *Gigascience* **2017**, *6*, 1–4. [CrossRef] [PubMed]

45. Stevens, K.A.; Wegrzyn, J.L.; Zimin, A.; Puiu, D.; Crepeau, M.; Cardeno, C.; Paul, R.; Gonzalez-Ibeas, D.; Koriabine, M.; Holtz-Morris, A.E.; et al. Sequence of the sugar pine megagenome. *Genetics* **2016**, *204*, 1613–1626. [CrossRef] [PubMed]

46. Neale, D.B.; McGuire, P.E.; Wheeler, N.C.; Stevens, K.A.; Crepeau, M.W.; Cardeno, C.; Zimin, A.V.; Puiu, D.; Pertea, G.M.; Sezen, U.U.; et al. The Douglas-fir genome sequence reveals specialization of the photosynthetic apparatus in Pinaceae. *G3* **2017**, *7*, 3157–3167. [CrossRef] [PubMed]

47. Canales, J.; Bautista, R.; Label, P.; Gomez-Maldonado, J.; Lesur, I.; Fernandez-Pozo, N.; Rueda-Lopez, M.; Guerrero-Fernandez, D.; Castro-Rodriguez, V.; Benzekri, H.; et al. *De novo* assembly of maritime pine transcriptome: Implications for forest breeding and biotechnology. *Plant Biotechnol. J.* **2014**, *12*, 286–299. [CrossRef] [PubMed]
48. Gonzalez-Ibeas, D.; Martinez-Garcia, P.J.; Famula, R.A.; Delfino-Mix, A.; Stevens, K.A.; Loopstra, C.A.; Langley, C.H.; Neale, D.B.; Wegrzyn, J.L. Assessing the gene content of the megagenome: Sugar pine (*Pinus lambertiana*). *G3* **2016**, *6*, 3787–3802. [CrossRef] [PubMed]
49. Newton, P.F. Examining naturogenic processes and anthropogenic influences on tree growth and development via stem analysis: Data processing and computational analytics. *Forests* **2019**, *10*, 1058. [CrossRef]
50. Le, K.-C.; Weerasekara, A.B.; Ranade, S.S.; Egertsdotter, E.M.U. Evaluation of parameters to characterise germination-competent mature somatic embryos of Norway spruce (*Picea abies*). *Biosyst. Eng.* **2021**, *203*, 55–59. [CrossRef]
51. Lassoued, R.; Macall, D.M.; Smyth, S.J.; Phillips, P.W.B.; Hesseln, H. Data challenges for future plant gene editing: Expert opinion. *Transgenic Res.* **2021**, *30*, 765–780. [CrossRef]
52. Uddenberg, D.; Akhter, S.; Ramachandran, P.; Sundstrom, J.F.; Carlsbecker, A. Sequenced genomes and rapidly emerging technologies pave the way for conifer evolutionary developmental biology. *Front. Plant Sci.* **2015**, *6*, 970. [CrossRef]
53. Catalán, G. Current situation and prospects of the stonepine as nut producer. *FAO-Nucis-Newsletter* **1998**, *7*, 28–32.
54. Ahuja, M.R.; Libby, W.J. Genetics, Biotechnology and Clonal Forestry. In *Clonal Forestry I*; Springer: Berlin/Heidelberg, Germany, 1993; pp. 1–4. [CrossRef]
55. Merkle, S.A.; Dean, J.F.D. Forest tree biotechnology. *Curr. Opin. Biotechnol.* **2000**, *11*, 298–302. [CrossRef]
56. Greenwood, M.S. Juvenility and maturation in conifers: Current concepts. *Tree Physiol.* **1995**, *15*, 433–438. [CrossRef] [PubMed]
57. Campbell, M.M.; Brunner, A.M.; Jones, H.M.; Strauss, S.H. Forestry's fertile crescent: The application of biotechnology to forest trees. *Plant Biotechnol. J.* **2003**, *1*, 141–154. [CrossRef] [PubMed]
58. von Aderkas, P.; Bonga, J.M. Influencing micropropagation and somatic embryogenesis in mature trees by manipulation of phase change, stress and culture environment. *Tree Physiol.* **2000**, *20*, 921–928. [CrossRef] [PubMed]
59. Harfouche, A.; Meilan, R.; Kirst, M.; Morgante, M.; Boerjan, W.; Sabatti, M.; Scarascia Mugnozza, G. Accelerating the domestication of forest trees in a changing world. *Trends Plant Sci.* **2012**, *17*, 64–72. [CrossRef]
60. Trontin, J.-F.; Harvengt, L.; Garin, E.; Lopez-Vernaza, M.; Arancio, L.; Hoebeke, J.; Canlet, F.; Pâques, M. Towards genetic engineering of maritime pine (*Pinus pinaster* Ait.). *Ann. For. Sci.* **2002**, *59*, 687–697. [CrossRef]
61. Niskanen, A.M.; Lu, J.; Seitz, S.; Keinonen, K.; Von Weissenberg, K.; Pappinen, A. Effect of parent genotype on somatic embryogenesis in Scots pine (*Pinus sylvestris*). *Tree Physiol.* **2004**, *24*, 1259–1265. [CrossRef]
62. Varis, S.; Klimaszewska, K.; Aronen, T. Somatic embryogenesis and plant regeneration from primordial shoot explants of *Picea abies* (L.) H. Karst. somatic trees. *Front. Plant Sci.* **2018**, *9*, 1551. [CrossRef]
63. Lelu-Walter, M.-A.; Bernier-Cardou, M.; Klimaszewska, K. Simplified and improved somatic embryogenesis for clonal propagation of *Pinus pinaster* (Ait.). *Plant Cell Rep.* **2006**, *25*, 767–776. [CrossRef]
64. Chang, S.H.; Ho, C.K.; Chen, Z.Z.; Tsay, J.Y. Micropropagation of *Taxus mairei* from mature trees. *Plant Cell Rep.* **2001**, *20*, 496–502. [CrossRef]
65. Ewald, D. Advances in tissue culture of adult larch. *In Vitro Cell. Dev. Biol. Plant* **1998**, *34*, 325–330. [CrossRef]
66. Dumas, E.; Monteuuis, O. In vitro rooting of micropopagated shoots from juvenile and mature *Pinus pinaster* explants: Influence of activated charcoal. *Plant Cell Tissue Organ Cult.* **1995**, *40*, 231–235. [CrossRef]
67. Loberant, B.; Altman, A. Micropropagation of Plants. In *Encyclopedia of Industrial Biotechnology: Bioprocess, Bioseparation, and Cell Technology*; Wiley: New York, NY, USA, 2010; pp. 1–17.
68. Ranade, S.S.; Egertsdotter, U. In silico characterization of putative gene homologues involved in somatic embryogenesis suggests that some conifer species may lack LEC2, one of the key regulators of initiation of the process. *BMC Genom.* **2021**, *22*, 392. [CrossRef] [PubMed]
69. Bonga, J.M.; Klimaszewska, K.K.; von Aderkas, P. Recalcitrance in clonal propagation, in particular of conifers. *Plant Cell Tissue Organ Cult.* **2010**, *100*, 241–254. [CrossRef]
70. Sarmast, M.K.; Salehi, H.; Khosh-Khui, M. Nano silver treatment is effective in reducing bacterial contaminations of *Araucaria excelsa* R. Br. var. glauca explants. *Acta Biol. Hung.* **2011**, *62*, 477–484. [CrossRef]
71. Khater, N.; Benbouza, H. Preservation of *Juniperus thurifera* L.: A rare endangered species in Algeria through in vitro regeneration. *J. For. Res.* **2019**, *30*, 77–86. [CrossRef]
72. Abdullah, A.A.; Yeoman, M.M.; Grace, J. Micropropagation of mature Calabrian pine (*Pinus brutia* Ten.) from fascicular buds. *Tree Physiol.* **1987**, *3*, 123–136. [CrossRef]
73. Horgan, K.J. *Pinus radiata*. In *Tissue Culture in Forestry*; Bonga, J.M., Durzan, D., Eds.; Martinus Nijhalf: Dordrecht, The Netherlands, 1987; Volume 3, pp. 128–145.
74. Parasharami, V.A.; Poonawala, I.S.; Nadgauda, R.S. Bud break and plantlet regeneration in vitro from mature trees of *Pinus roxburghii* Sarg. *Curr. Sci.* **2003**, *84*, 203–208.
75. Prehn, D.; Serrano, C.; Mercado, A.; Stange, C.; Barrales, L.; Arce-Johnson, P. Regeneration of whole plants from apical meristems of *Pinus radiata*. *Plant Cell Tissue Organ Cult.* **2003**, *73*, 91–94. [CrossRef]
76. Renau-Morata, B.; Ollero, J.; Arrillaga, I.; Segura, J. Factors influencing axillary shoot proliferation and adventitious budding in cedar. *Tree Physiol.* **2005**, *25*, 477–486. [CrossRef]

77. Cortizo, M.; de Diego, N.; Moncaleán, P.; Ordás, R.J. Micropropagation of adult Stone Pine (*Pinus pinea* L.). *Trees* **2009**, *23*, 835–842. [CrossRef]

78. Cuesta, C.; Ordás, R.; Fernández, B.; Rodríguez, A. Clonal micropropagation of six selected half-sibling families of *Pinus pinea* and somaclonal variation analysis. *Plant Cell Tissue Organ Cult.* **2008**, *95*, 125–130. [CrossRef]

79. Danusevicius, D.; Lindgren, D. Efficiency of selection based on phenotype, clone and progeny testing in long-term breeding. *Silvae Genet.* **2002**, *51*, 19–25.

80. Foster, G.S.; Shaw, D.V. A tree improvement program to develop clones of loblolly pine for reforestation. In Proceedings of the Southern Forest Tree Improvement Conference (USA), College Station, TX, USA, 16–18 June 1987. [CrossRef]

81. López, M.; Pacheco, J.; Rodríguez, R.; Ordás, R.J. Regeneration of plants from insolated cotyledons of Salgareño Pine (*Pinus nigra* Arn. ssp. salzmannii (Dunal) Franco). *In Vitro Cell. Dev. Biol. Plant* **1996**, *32*, 109–114. [CrossRef]

82. Alonso, P.; Moncaleán, P.; Fernández, B.; Rodríguez, A.; Centeno, M.L.; Ordás, R.J. An improved micropropagation protocol for stone pine (*Pinus pinea* L.). *Ann. For. Sci.* **2006**, *63*, 879–885. [CrossRef]

83. Alvarez, J.M.; Majada, J.; Ordás, R.J. An improved micropropagation protocol for maritime pine (*Pinus pinaster* Ait.) isolated cotyledons. *Forestry* **2009**, *82*, 175–184. [CrossRef]

84. Calixto, F.; Pais, M.S. Adventitious shoot formation and plant regeneration from *Pinus pinaster* Sol. ex Aiton. *In Vitro Cell. Dev. Biol. Plant* **1997**, *33*, 119–124. [CrossRef]

85. Tereso, S.; Gonçalves, S.; Marum, L.; Oliveira, M.; Maroco, J.; Miguel, C. Improved axillary and adventitious bud regeneration from Portuguese genotypes of *Pinus pinaster* Ait. *Propag. Ornam. Plant.* **2006**, *6*, 24–33.

86. De Diego, N.; Montalbán, I.A.; Fernandez de Larrinoa, E.; Moncaleán, P. *In vitro* regeneration of *Pinus pinaster* adult trees. *Can. J. For. Res.* **2008**, *38*, 2607–2615. [CrossRef]

87. Bercetche, J.; Pâques, M. Somatic embryogenesis in maritime pine (*Pinus pinaster*). In *Somatic Embryogenesis in Woody Plants*; Jain, S.M., Gupta, P., Newton, R.J., Eds.; Kluwer Academic Publishers: Dordrecht, The Netherlands, 1995; Volume Gymnosperms, pp. 269–285.

88. Miguel, C.; Gonçalves, S.; Tereso, S.; Marum, L.; Maroco, J.; Oliveira, M. Somatic embryogenesis from 20 open-pollinated families of portuguese plus trees of maritime pine. *Plant Cell Tissue Organ Cult.* **2004**, *76*, 121–130. [CrossRef]

89. Harvengt, L. Somatic embryogenesis in maritime pine (*Pinus pinaster* Ait.). In *Protocol of Somatic Embryogenesis in Woody Plants*; Jain, S.M., Gupta, P.K., Eds.; Springer: Berlin, Germany, 2005; Volume 77, pp. 107–120.

90. Alvarez, J.M.; Bueno, N.; Cortizo, M.; Ordás, R.J. Improving plantlet yield in *Pinus pinaster* somatic embryogenesis. *Scand. J. For. Res.* **2013**, *28*, 613–620. [CrossRef]

91. Wise, F.C.; Caldwell, T.D. Macropropagation of conifers by stem cuttings. In *Applications of Vegetative Propagation in Forestry*; Foster, G.S., Diner, A.M., Eds.; General Technical Report SO-108; USDA Forest Service, Southern Forest Experiment Station: New Orleans, LA, USA, 1994; pp. 51–73.

92. Zhang, S.; Yan, S.; An, P.; Cao, Q.; Wang, C.; Wang, J.; Zhang, H.; Zhang, L. Embryogenic callus induction from immature zygotic embryos and genetic transformation of *Larix kaempferi* 3x *Larix gmelinii* 9. *PLoS ONE* **2021**, *16*, e0258654. [CrossRef] [PubMed]

93. Hazubska-Przybył, T.; Wawrzyniak, M.K.; Kijowska-Oberc, J.; Staszak, A.M.; Ratajczak, E. Somatic embryogenesis of Norway spruce and Scots pine: Possibility of application in modern forestry. *Forests* **2022**, *13*, 155. [CrossRef]

94. von Arnold, S.; Clapham, D.; Abrahamsson, M. Chapter Five—Embryology in conifers. *Adv. Bot. Res.* **2019**, *89*, 157–184.

95. Corredoira, E.; Merkle, S.A.; Martínez, M.T.; Toribio, M.; Canhoto, J.M.; Correia, S.I.; Ballester, A.; Vieitez, A.M. Non-zygotic embryogenesis in hardwood species. *Crit. Rev. Plant Sci.* **2019**, *38*, 29–97. [CrossRef]

96. Ikeuchi, M.; Favero, D.S.; Sakamoto, Y.; Iwase, A.; Coleman, D.; Rymen, B.; Sugimoto, K. Molecular mechanisms of plant regeneration. *Annu. Rev. Plant Biol.* **2019**, *70*, 377–406. [CrossRef] [PubMed]

97. Aronen, T.; Pehkonen, T.; Ryynänen, L. Enhancement of somatic embryogenesis from immature zygotic embryos of *Pinus sylvestris*. *Scand. J. For. Res.* **2009**, *24*, 372–383. [CrossRef]

98. Montalbán, I.A.; De Diego, N.; Moncaleán, P. Bottlenecks in *Pinus radiata* somatic embryogenesis: Improving maturation and germination. *Trees* **2010**, *24*, 1061–1071. [CrossRef]

99. Carneros, E.; Toribio, M.; Celestino, C. Effect of ABA, the auxin antagonist PCIB and partial desiccation on stone pine somatic embryo maturation. *Plant Cell Tissue Organ Cult.* **2017**, *131*, 445–458. [CrossRef]

100. Attree, S.; Moore, D.; Sawhney, V.; Fowke, L. Enhanced maturation and desiccation tolerance of white spruce [*Picea glauca* (Moench) Voss] somatic embryos: Effects of a non-plasmolysing water stress and abscisic acid. *Ann. Bot.* **1991**, *68*, 519–525. [CrossRef]

101. Klimaszewska, K.; Smith, D.R. Maturation of somatic embryos of *Pinus strobus* is promoted by a high concentration of gellan gum. *Physiol. Plant* **1997**, *100*, 949–957. [CrossRef]

102. Breton, D.; Harvengt, L.; Trontin, J.-F.; Bouvet, A.; Favre, J.-M. Long-term subculture randomly affects morphology and subsequent maturation of early somatic embryos in maritime pine. *Plant Cell Tissue Organ Cult.* **2006**, *87*, 95–108. [CrossRef]

103. Peng, C.; Gao, F.; Wang, H.; Shen, H.; Yang, L. Optimization of maturation process for somatic embryo production and cryopreservation of embryogenic tissue in *Pinus koraiensis*. *Plant Cell Tissue Organ Cult.* **2020**, *144*, 185–194. [CrossRef]

104. Montalbán, I.A.; Castander-Olarieta, A.; Hargreaves, C.L.; Gough, K.; Reeves, C.B.; van Ballekom, S.; Goicoa, T.; Ugarte, M.D.; Moncaleán, P. Hybrid pine (*Pinus attenuata* × *Pinus radiata*) somatic embryogenesis: What do you prefer, mother or nurse? *Forests* **2020**, *12*, 45. [CrossRef]

105. Wunderling, A.; Ripper, D.; Barra-Jimenez, A.; Mahn, S.; Sajak, K.; Targem, M.B.; Ragni, L. A molecular framework to study periderm formation in Arabidopsis. *New Phytol.* **2018**, *219*, 216–229. [CrossRef] [PubMed]
106. Aronen, T.; Virta, S.; Varis, S. Telomere length in Norway spruce during somatic embryogenesis and cryopreservation. *Plants* **2021**, *10*, 416. [CrossRef] [PubMed]
107. Hazubska-Przybyl, T.; Ratajczak, E.; Obarska, A.; Pers-Kamczyc, E. Different roles of auxins in somatic embryogenesis efficiency in two *Picea* species. *Int. J. Mol. Sci.* **2020**, *21*, 3394. [CrossRef]
108. Pereira, C.; Castander-Olarieta, A.; Sales, E.; Montalban, I.A.; Canhoto, J.; Moncalean, P. Heat stress in *Pinus halepensis* somatic embryogenesis induction: Effect in DNA methylation and differential expression of stress-related genes. *Plants* **2021**, *10*, 2333. [CrossRef]
109. Varis, S.; Tikkinen, M.; Välimäki, S.; Aronen, T. Light spectra during somatic embryogenesis of Norway spruce—Impact on growth, embryo productivity and embling survival. *Forests* **2021**, *12*, 301. [CrossRef]
110. Cui, Y.; Gao, Y.; Zhao, R.; Zhao, J.; Li, Y.; Qi, S.; Zhang, J.; Kong, L. Transcriptomic, metabolomic, and physiological analyses reveal that the culture temperatures modulate the cryotolerance and embryogenicity of developing somatic embryos in *Picea glauca*. *Front. Plant Sci.* **2021**, *12*, 694229. [CrossRef]
111. Gao, F.; Peng, C.; Wang, H.; Tretyakova, I.N.; Nosov, A.M.; Shen, H.; Yang, L. Key techniques for somatic embryogenesis and plant regeneration of *Pinus koraiensis*. *Forests* **2020**, *11*, 912. [CrossRef]
112. Salaj, T.; Panis, B.; Swennen, R.; Salaj, J. Cryopreservation of embryogenic tissues of *Pinus nigra* Arn. by a slow freezing method. *Cryo. Lett.* **2007**, *28*, 69–76.
113. Bonga, J. Conifer clonal propagation in tree improvement programs. In *Vegetative Propagation of Forest Trees*; National Institute of Forest Science (NIFoS): Seoul, Korea, 2016; pp. 3–31.
114. Rosvall, O. Using Norway spruce clones in Swedish forestry: General overview and concepts. *Scand. J. For. Res.* **2019**, *34*, 336–341. [CrossRef]
115. Park, Y.-S. Implementation of conifer somatic embryogenesis in clonal forestry: Technical requirements and deployment considerations. *Ann. For. Sci.* **2002**, *59*, 651–656. [CrossRef]
116. Klimaszewska, K.; Trontin, J.F.; Becwar, M.; Devillard, C.; Park, Y.S.; Lelu-Walter, M.A. Recent progress on somatic embryogenesis of four *Pinus* spp. *Tree For. Sci. Biotechnol.* **2007**, *1*, 11–25.
117. Alvarez, J.M.; Cortizo, M.; Ordás, R. Cryopreservation of somatic embryogenic cultures of *Pinus pinaster*: Effects on regrowth and embryo maturation. *Cryo. Lett.* **2012**, *33*, 476–484.
118. Phillips, R.L.; Kaeppler, S.M.; Olhoft, P. Genetic instability of plant tissue cultures: Breakdown of normal controls. *Proc. Natl. Acad. Sci. USA* **1994**, *91*, 5222–5226. [CrossRef]
119. Pischke, M.S.; Huttlin, E.L.; Hegeman, A.D.; Sussman, M.R. A transcriptome-based characterization of habituation in plant tissue culture. *Plant Physiol.* **2006**, *140*, 1255–1278. [CrossRef]
120. von Arnold, S.; Egertsdotter, U.; Ekberg, I.; Gupta, P.; Mo, H.; Nörgaard, J. Somatic embryogenesis in Norway spruce (*Picea abies*). In *Somatic Embryogenesis in Woody Plants*; Jain, S.M., Gupta, P.K., Newton, R.J., Eds.; Springer: Dordrecht, The Netherlands, 1995; Volume 3—Gymnosperms, pp. 17–36.
121. Izuno, A.; Maruyama, T.E.; Ueno, S.; Ujino-Ihara, T.; Moriguchi, Y. Genotype and transcriptome effects on somatic embryogenesis in *Cryptomeria japonica*. *PLoS ONE* **2020**, *15*, e0244634. [CrossRef]
122. Maruyama, T.E.; Ueno, S.; Hosoi, Y.; Miyazawa, S.I.; Mori, H.; Kaneeda, T.; Bamba, Y.; Itoh, Y.; Hirayama, S.; Kawakami, K.; et al. Somatic embryogenesis initiation in sugi (Japanese cedar, *Cryptomeria japonica* D. Don): Responses from male-fertile, male-sterile, and polycross-pollinated-derived seed explants. *Plants* **2021**, *10*, 398. [CrossRef]
123. Rensing, S.A.; Weijers, D. Flowering plant embryos: How did we end up here? *Plant Reprod* **2021**, *34*, 365–371. [CrossRef]
124. Zhu, T.; Moschou, P.N.; Alvarez, J.M.; Sohlberg, J.J.; von Arnold, S. *WUSCHEL-RELATED HOMEOBOX 2* is important for protoderm and suspensor development in the gymnosperm Norway spruce. *BMC Plant Biol.* **2016**, *16*, 19. [CrossRef] [PubMed]
125. Yan, J.; Peng, P.; Duan, G.; Lin, T.; Bai, Y.E. Multiple analyses of various factors affecting the plantlet regeneration of *Picea mongolica* (H. Q. Wu) W.D. Xu from somatic embryos. *Sci. Rep.* **2021**, *11*, 6694. [CrossRef] [PubMed]
126. Pathak, P.K.; Zhang, F.; Peng, S.; Niu, L.; Chaturvedi, J.; Elliott, J.; Xiang, Y.; Tadege, M.; Deng, J. Structure of the unique tetrameric STENOFOLIA homeodomain bound with target promoter DNA. *Acta Crystallogr. D Struct. Biol.* **2021**, *77*, 1050–1063. [CrossRef] [PubMed]
127. Chano, V.; Sobrino-Plata, J.; Collada, C.; Soto, A. Wood development regulators involved in apical growth in *Pinus canariensis*. *Plant Biol.* **2021**, *23*, 438–444. [CrossRef]
128. Bueno, N.; Alvarez, J.M.; Ordás, R.J. Characterization of the *KNOTTED1-LIKE HOMEOBOX* (*KNOX*) gene family in *Pinus pinaster* Ait. *Plant Sci.* **2020**, *301*, 110691. [CrossRef]
129. Zhu, T.; Moschou, P.N.; Alvarez, J.M.; Sohlberg, J.J.; von Arnold, S. *WUSCHEL-RELATED HOMEOBOX 8/9* is important for proper embryo patterning in the gymnosperm Norway spruce. *J. Exp. Bot.* **2014**, *65*, 6543–6552. [CrossRef]
130. Alvarez, J.M.; Sohlberg, J.; Engström, P.; Zhu, T.; Englund, M.; Moschou, P.N.; von Arnold, S. The *WUSCHEL-RELATED HOMEOBOX 3* gene *PaWOX3* regulates lateral organ formation in Norway spruce. *New Phytol.* **2015**, *208*, 1078–1088. [CrossRef]
131. Tvorogova, V.E.; Krasnoperova, E.Y.; Potsenkovskaia, E.A.; Kudriashov, A.A.; Dodueva, I.E.; Lutova, L.A. What does the WOX say? Review of regulators, targets, partners. *Mol. Biol.* **2021**, *55*, 311–337. [CrossRef]

132. Subban, P.; Kutsher, Y.; Evenor, D.; Belausov, E.; Zemach, H.; Faigenboim, A.; Bocobza, S.; Timko, M.P.; Reuveni, M. Shoot regeneration is not a single cell event. *Plants* **2020**, *10*, 58. [CrossRef]
133. Jha, P.; Ochatt, S.J.; Kumar, V. WUSCHEL: A master regulator in plant growth signaling. *Plant Cell Rep.* **2020**, *39*, 431–444. [CrossRef] [PubMed]
134. Junker, J.P.; van Oudenaarden, A. Every cell is special: Genome-wide studies add a new dimension to single-cell biology. *Cell* **2014**, *157*, 8–11. [CrossRef]
135. Alvarez, J.M.; Ordas, R.J. Stable *Agrobacterium*-mediated transformation of maritime pine based on kanamycin selection. *Sci. World J.* **2013**, *2013*, 681792. [CrossRef] [PubMed]
136. von Arnold, S.; Zhu, T.; Larsson, E.; Uddenberg, D.; Clapham, D. Regulation of somatic embryo development in Norway spruce. *Methods Mol. Biol.* **2020**, *2122*, 241–255. [CrossRef] [PubMed]
137. Luo, G.; Palmgren, M. GRF-GIF chimeras boost plant regeneration. *Trends Plant Sci.* **2021**, *26*, 201–204. [CrossRef]
138. Debernardi, J.M.; Tricoli, D.M.; Ercoli, M.F.; Hayta, S.; Ronald, P.; Palatnik, J.F.; Dubcovsky, J. A GRF–GIF chimeric protein improves the regeneration efficiency of transgenic plants. *Nat. Biotechnol.* **2020**, *38*, 1274–1279. [CrossRef]
139. Deng, W.; Luo, K.; Li, Z.; Yang, Y. A novel method for induction of plant regeneration via somatic embryogenesis. *Plant Sci.* **2009**, *177*, 43–48. [CrossRef]
140. Monson, R.K.; Trowbridge, A.M.; Lindroth, R.L.; Lerdau, M.T. Coordinated resource allocation to plant growth-defense tradeoffs. *New Phytol.* **2022**, *233*, 1051–1066. [CrossRef]
141. Gasser, C.S.; Fraley, R.T. Genetically engineering plants for crop improvement. *Science* **1989**, *244*, 1293–1299. [CrossRef]
142. Manders, G.; Dos Santos, A.V.P.; Vaz, U.; Davey, M.R.; Power, J.B. Transient gene expression in electroporated protoplasts of *Eucalyptus citriodora* Hook. *Plant Cell Tissue Organ Cult.* **1992**, *30*, 69–75. [CrossRef]
143. Sederoff, R.; Stomp, A.M.; Chilton, W.S.; Moore, L.W. Gene transfer into loblolly pine by *Agrobacterium tumefaciens*. *Nat. Biotechnol.* **1986**, *4*, 647–649. [CrossRef]
144. Loopstra, C.A.; Stomp, A.M.; Sederoff, R.R. *Agrobacterium*-mediated DNA transfer in sugar pine. *Plant Mol. Biol.* **1990**, *15*, 1–9. [CrossRef] [PubMed]
145. Stomp, A.M.; Loopstra, C.; Chilton, W.S.; Sederoff, R.R.; Moore, L.W. Extended host range of *Agrobacterium tumefaciens* in the genus *Pinus*. *Plant Physiol.* **1990**, *92*, 1226–1232. [CrossRef] [PubMed]
146. Charest, P.J.; Michel, M.-F. *Basics of Plant Genetic Engineering and Its Potential Applications to Tree Species*; Canadian Forest Service Publications, Petawawa National Forestry Institute: Chalk River, ON, Canada, 1991; Volume 104.
147. Tzfira, T.; Yarnitzky, O.; Vainstein, A.; Altman, A. *Agrobacterium rhizogenes*-mediated DNA transfer in *Pinus halepensis* Mill. *Plant Cell Rep.* **1996**, *16*, 26–31. [CrossRef] [PubMed]
148. Huang, Y.; Diner, A.M.; Karnosky, D.F. *Agrobacterium rhizogenes*-mediated genetic transformation and regeneration of a conifer: *Larix decidua*. *In Vitro Cell. Dev. Biol. Plant* **1991**, *27*, 201–207. [CrossRef]
149. Bekkaoui, F.; Pilon, M.; Laine, E.; Raju, D.S.; Crosby, W.L.; Dunstan, D.I. Transient gene expression in electroporated *Picea glauca* protoplasts. *Plant Cell Rep.* **1988**, *7*, 481–484. [CrossRef]
150. Tautorus, T.E.; Bekkaoui, F.; Pilon, M.; Datla, R.S.; Crosby, W.L.; Fowke, L.C.; Dunstan, D.I. Factors affecting transient gene expression in electroporated black spruce (*Picea mariana*) and jack pine (*Pinus banksiana*) protoplasts. *Theor. Appl. Genet.* **1989**, *78*, 531–536. [CrossRef]
151. Wilson, S.M.; Thorpe, T.A.; Moloney, M.M. PEG-mediated expression of *GUS* and *CAT* genes in protoplasts from embryogenic suspension cultures of *Picea glauca*. *Plant Cell Rep.* **1989**, *7*, 704–707. [CrossRef]
152. Goldfarb, B.; Strauss, S.H.; Howe, G.T.; Zaerr, J.B. Transient gene expression of microprojectile-introduced DNA in Douglas-fir cotyledons. *Plant Cell Rep.* **1991**, *10*, 517–521. [CrossRef]
153. Walter, C.; Grace, L.J.; Wagner, A.; White, D.W.R.; Walden, A.R.; Donaldson, S.S.; Hinton, H.; Gardner, R.C.; Smith, D.R. Stable transformation and regeneration of transgenic plants of *Pinus radiata* D. Don. *Plant Cell Rep.* **1998**, *17*, 460–468. [CrossRef]
154. Block, M.D.; Botterman, J.; Vandewiele, M.; Dockx, J.; Thoen, C.; Gossele, V.; Movva, N.R.; Thompson, C.; Montagu, M.V.; Leemans, J. Engineering herbicide resistance in plants by expression of a detoxifying enzyme. *EMBO J.* **1987**, *6*, 2513–2518. [CrossRef]
155. Hadi, M.Z.; McMullen, M.D.; Finer, J.J. Transformation of 12 different plasmids into soybean via particle bombardment. *Plant Cell Rep.* **1996**, *15*, 500–505. [CrossRef] [PubMed]
156. Grace, L.J.; Charity, J.A.; Gresham, B.; Kay, N.; Walter, C. Insect-resistant transgenic *Pinus radiata*. *Plant Cell Rep.* **2005**, *24*, 103–111. [CrossRef] [PubMed]
157. Lachance, D.; Hamel, L.-P.; Pelletier, F.; Valéro, J.; Bernier-Cardou, M.; Chapman, K.; Van Frankenhuyzen, K.; Séguin, A. Expression of a *Bacillus thuringiensis* cry1Ab gene in transgenic white spruce and its efficacy against the spruce budworm (*Choristoneura fumiferana*). *Tree Genet. Genomes* **2007**, *3*, 153–167. [CrossRef]
158. Bishop-Hurley, S.L.; Zabkiewicz, R.J.; Grace, L.; Gardner, R.C.; Wagner, A.; Walter, C. Conifer genetic engineering: Transgenic *Pinus radiata* (D. Don) and *Picea abies* (Kasrt) plants are resistant to the herbicide Buster. *Plant Cell Rep.* **2001**, *20*, 235–243. [CrossRef]
159. Charity, J.A.; Holland, L.; Donaldson, S.S.; Grace, L.; Walter, C. *Agrobacterium*-mediated transformation of *Pinus radiata* organogenic tissue using vacuum-infiltration. *Plant Cell Tissue Organ Cult.* **2002**, *70*, 51–60. [CrossRef]

160. Wadenback, J.; von Arnold, S.; Egertsdotter, U.; Walter, M.H.; Grima-Pettenati, J.; Goffner, D.; Gellerstedt, G.; Gullion, T.; Clapham, D. Lignin biosynthesis in transgenic Norway spruce plants harboring an antisense construct for cinnamoyl CoA reductase (CCR). *Transgenic Res.* **2008**, *17*, 379–392. [CrossRef]

161. Wagner, A.; Tobimatsu, Y.; Phillips, L.; Flint, H.; Torr, K.; Donaldson, L.; Pears, L.; Ralph, J. CCoAOMT suppression modifies lignin composition in *Pinus radiata*. *Plant J.* **2011**, *67*, 119–129. [CrossRef]

162. Tang, W.; Peng, X.; Newton, R. Enhanced tolerance to salt stress in transgenic loblolly pine simultaneously expressing two genes encoding mannitol-1-phosphate dehydrogenase and glucitol-6-phosphate dehydrogenase. *Plant Physiol. Biochem.* **2005**, *43*, 139–146. [CrossRef]

163. Zhang, C.; Norris-Caneda, K.H.; Rottmann, W.H.; Gulledge, J.E.; Chang, S.; Kwan, B.Y.; Thomas, A.M.; Mandel, L.C.; Kothera, R.T.; Victor, A.D.; et al. Control of pollen-mediated gene flow in transgenic trees. *Plant Physiol.* **2012**, *159*, 1319–1334. [CrossRef]

164. Myburg, A.A.; Hussey, S.G.; Wang, J.P.; Street, N.R.; Mizrachi, E. Systems and synthetic biology of forest trees: A bioengineering paradigm for woody biomass feedstocks. *Front. Plant Sci.* **2019**, *10*, 775. [CrossRef] [PubMed]

165. De La Torre, A.R.; Piot, A.; Liu, B.; Wilhite, B.; Weiss, M.; Porth, I. Functional and morphological evolution in gymnosperms: A portrait of implicated gene families. *Evol. Appl.* **2020**, *13*, 210–227. [CrossRef] [PubMed]

166. Mao, Y.; Botella, J.R.; Liu, Y.; Zhu, J.K. Gene editing in plants: Progress and challenges. *Natl. Sci. Rev.* **2019**, *6*, 421–437. [CrossRef] [PubMed]

167. Feng, Z.; Zhang, B.; Ding, W.; Liu, X.; Yang, D.L.; Wei, P.; Cao, F.; Zhu, S.; Zhang, F.; Mao, Y.; et al. Efficient genome editing in plants using a CRISPR/Cas system. *Cell Res.* **2013**, *23*, 1229–1232. [CrossRef] [PubMed]

168. Poovaiah, C.; Phillips, L.; Geddes, B.; Reeves, C.; Sorieul, M.; Thorlby, G. Genome editing with CRISPR/Cas9 in *Pinus radiata* (D. Don). *BMC Plant Biol.* **2021**, *21*, 363. [CrossRef]

169. Jinek, M.; Chylinski, K.; Fonfara, I.; Hauer, M.; Doudna, J.A.; Charpentier, E. A programmable dual-RNA–guided DNA endonuclease in adaptive bacterial immunity. *Science* **2012**, *337*, 816–821. [CrossRef]

170. Fernandez i Marti, A.; Dodd, R.S. Using CRISPR as a gene editing tool for validating adaptive gene function in tree landscape genomics. *Front. Ecol. Evol.* **2018**, *6*, 76. [CrossRef]

171. Urnov, F.D.; Rebar, E.J.; Holmes, M.C.; Zhang, H.S.; Gregory, P.D. Genome editing with engineered zinc finger nucleases. *Nat. Rev. Genet.* **2010**, *11*, 636–646. [CrossRef]

172. Fritsche, S.; Poovaiah, C.; MacRae, E.; Thorlby, G. A New Zealand perspective on the application and regulation of gene editing. *Front. Plant Sci.* **2018**, *9*, 1323. [CrossRef]

173. Arora, L.; Narula, A. Gene editing and crop improvement using CRISPR-Cas9 system. *Front. Plant Sci.* **2017**, *8*, 1932. [CrossRef]

174. Wang, M.-M.; Liu, M.-M.; Ran, F.; Guo, P.-C.; Ke, Y.-Z.; Wu, Y.-W.; Wen, J.; Li, P.-F.; Li, J.-N.; Du, H. Global Analysis of WOX Transcription Factor Gene Family in Brassica napus Reveals Their Stress- and Hormone-Responsive Patterns. *Int. J. Mol. Sci.* **2018**, *19*, 3470. [CrossRef] [PubMed]

175. Peng, A.; Chen, S.; Lei, T.; Xu, L.; He, Y.; Wu, L.; Yao, L.; Zou, X. Engineering canker-resistant plants through CRISPR/Cas9-targeted editing of the susceptibility gene *CsLOB1* promoter in citrus. *Plant Biotechnol. J.* **2017**, *15*, 1509–1519. [CrossRef] [PubMed]

176. Joshi, R.K.; Bharat, S.S.; Mishra, R. Engineering drought tolerance in plants through CRISPR/Cas genome editing. *3 Biotech* **2020**, *10*, 400. [CrossRef] [PubMed]

177. Zhou, H.; He, M.; Li, J.; Chen, L.; Huang, Z.; Zheng, S.; Zhu, L.; Ni, E.; Jiang, D.; Zhao, B.; et al. Development of commercial thermo-sensitive genic male sterile rice accelerates hybrid rice breeding using the CRISPR/Cas9-mediated TMS5 editing system. *Sci. Rep.* **2016**, *6*, 37395. [CrossRef]

178. Sun, Y.; Zhang, X.; Wu, C.; He, Y.; Ma, Y.; Hou, H.; Guo, X.; Du, W.; Zhao, Y.; Xia, L. Engineering herbicide-resistant rice plants through CRISPR/Cas9-mediated homologous recombination of acetolactate synthase. *Mol. Plant* **2016**, *9*, 628–631. [CrossRef]

179. Soyk, S.; Lemmon, Z.H.; Oved, M.; Fisher, J.; Liberatore, K.L.; Park, S.J.; Goren, A.; Jiang, K.; Ramos, A.; van der Knaap, E.; et al. Bypassing negative epistasis on yield in tomato imposed by a domestication gene. *Cell* **2017**, *169*, 1142–1155.e12. [CrossRef]

180. Tian, S.; Jiang, L.; Cui, X.; Zhang, J.; Guo, S.; Li, M.; Zhang, H.; Ren, Y.; Gong, G.; Zong, M.; et al. Engineering herbicide-resistant watermelon variety through CRISPR/Cas9-mediated base-editing. *Plant Cell Rep.* **2018**, *37*, 1353–1356. [CrossRef]

181. Zhang, H.; Si, X.; Ji, X.; Fan, R.; Liu, J.; Chen, K.; Wang, D.; Gao, C. Genome editing of upstream open reading frames enables translational control in plants. *Nat. Biotechnol.* **2018**, *36*, 894–898. [CrossRef]

182. Shan, S.; Soltis, P.S.; Soltis, D.E.; Yang, B. Considerations in adapting CRISPR/Cas9 in nongenetic model plant systems. *Appl. Plant Sci.* **2020**, *8*, e11314. [CrossRef]

183. Fan, D.; Liu, T.; Li, C.; Jiao, B.; Li, S.; Hou, Y.; Luo, K. Efficient CRISPR/Cas9-mediated targeted mutagenesis in *Populus* in the first generation. *Sci. Rep.* **2015**, *5*, 12217. [CrossRef]

184. Bruegmann, T.; Deecke, K.; Fladung, M. Evaluating the efficiency of gRNAs in CRISPR/Cas9 mediated genome editing in poplars. *Int. J. Mol. Sci.* **2019**, *20*, 3623. [CrossRef] [PubMed]

185. Dai, Y.; Hu, G.; Dupas, A.; Medina, L.; Blandels, N.; Clemente, H.S.; Ladouce, N.; Badawi, M.; Hernandez-Raquet, G.; Mounet, F.; et al. Implementing the CRISPR/Cas9 technology in eucalyptus hairy roots using wood-related genes. *Int. J. Mol. Sci.* **2020**, *21*, 3408. [CrossRef] [PubMed]

186. Muller, N.A.; Kersten, B.; Leite Montalvao, A.P.; Mahler, N.; Bernhardsson, C.; Brautigam, K.; Carracedo Lorenzo, Z.; Hoenicka, H.; Kumar, V.; Mader, M.; et al. A single gene underlies the dynamic evolution of poplar sex determination. *Nat. Plants* **2020**, *6*, 630–637. [CrossRef] [PubMed]

187. Ren, Q.; Sretenovic, S.; Liu, S.; Tang, X.; Huang, L.; He, Y.; Liu, L.; Guo, Y.; Zhong, Z.; Liu, G.; et al. PAM-less plant genome editing using a CRISPR-SpRY toolbox. *Nat. Plants* **2021**, *7*, 25–33. [CrossRef]

188. Cui, Y.; Zhao, J.; Gao, Y.; Zhao, R.; Zhang, J.; Kong, L. Efficient multi-sites genome editing and plant regeneration via somatic embryogenesis in *Picea glauca*. *Front. Plant Sci.* **2021**, *12*, 751891. [CrossRef]

189. Nanasato, Y.; Mikami, M.; Futamura, N.; Endo, M.; Nishiguchi, M.; Ohmiya, Y.; Konagaya, K.I.; Taniguchi, T. CRISPR/Cas9-mediated targeted mutagenesis in Japanese cedar (*Cryptomeria japonica* D. Don). *Sci. Rep.* **2021**, *11*, 16186. [CrossRef]

190. Ishii, T.; Araki, M. A future scenario of the global regulatory landscape regarding genome-edited crops. *GM Crops Food* **2017**, *8*, 44–56. [CrossRef]

191. Wolt, J.D.; Wang, K.; Yang, B. The regulatory status of genome-edited crops. *Plant Biotechnol. J.* **2016**, *14*, 510–518. [CrossRef]

192. Davison, J.; Ammann, K. New GMO regulations for old: Determining a new future for EU crop biotechnology. *GM Crops Food* **2017**, *8*, 13–34. [CrossRef]

193. Whelan, A.I.; Lema, M.A. Regulatory framework for gene editing and other new breeding techniques (NBTs) in Argentina. *GM Crops Food* **2015**, *6*, 253–265. [CrossRef]

194. Callaway, E. CRISPR plants now subject to tough GM laws in European Union. *Nature* **2018**, *560*, 16. [CrossRef] [PubMed]

195. Kupferschmidt, K. EU verdict on CRISPR crops dismays scientists. *Science* **2018**, *361*, 435–436. [CrossRef] [PubMed]

forests

Somatic Embryogenesis of Norway Spruce and Scots Pine: Possibility of Application in Modern Forestry

Teresa Hazubska-Przybył [1], Mikołaj Krzysztof Wawrzyniak [1,*], Joanna Kijowska-Oberc [1], Aleksandra Maria Staszak [2] and Ewelina Ratajczak [1]

[1] Institute of Dendrology, Polish Academy of Sciences, Parkowa 5, 62-035 Kornik, Poland; hazubska@man.poznan.pl (T.H.-P.); joberc@man.poznan.pl (J.K.-O.); eratajcz@man.poznan.pl (E.R.)
[2] Laboratory of Plant Physiology, Department of Plant Biology and Ecology, Faculty of Biology, University of Bialystok, Ciolkowskiego 1J, 15-245 Bialystok, Poland; a.staszak@uwb.edu.pl
*   Correspondence: mikwaw@man.poznan.pl

Abstract: Somatic embryogenesis (SE) is an important method for the vegetative propagation of trees. SE is the developmental in vitro process in which embryos are produced from somatic cells. This method can be integrated with other biotechnological techniques, genomic breeding and cryopreservation, which enables commercial-scale sapling production of selected high-yielding genotypes in wood production combined with fast breeding cycles. The SE is potential tool to improve plant stock in comparison with seed orchards. It can be useful for ecologically and economically important species, such as Norway spruce (*Picea abies* L. Karst.) and Scots pine (*Pinus sylvestris* L.), ensuring stable production in the era of climate change and biodiversity crisis. In this review, we summarize the current state of research on problems associated with somatic embryogenesis in *P. abies* and *P. sylvestris*.

Keywords: Norway spruce; Scots pine; somatic embryogenesis

Citation: Hazubska-Przybył, T.; Wawrzyniak, M.K.; Kijowska-Oberc, J.; Staszak, A.M.; Ratajczak, E. Somatic Embryogenesis of Norway Spruce and Scots Pine: Possibility of Application in Modern Forestry. *Forests* 2022, 13, 155. https://doi.org/10.3390/f13020155

Academic Editors: Ricardo Javier Ordás and José Manuel Álvarez Díaz

Received: 15 December 2021
Accepted: 17 January 2022
Published: 20 January 2022

1. Introduction

The demand for wood products is predicted to increase in response to a growing population and its needs [1]. This demand results from wood being a versatile, ecological and renewable resource used as an energy source, building compound or raw material in many industries [2,3]. Unfortunately, forest areas are still shrinking not only as a consequence of wood harvesting but also as a consequence of large-scale land operations, such as rapid urbanization and agriculture [4,5]. Deforestation leads to fragmentation of forest ecosystems, and the result is population division and modification of interspecies interactions [6]. This phenomenon generates both microclimate changes [7,8] and global-scale modifications of climate conditions [9–11], and consequently, the risks to biodiversity decrease [12,13]. Therefore, to obtain economically important timber without damaging natural forests, high-yield plantations have been established [2,14]. However, successful and efficient plantations need to be based on high-yielding genotypes of native timber species. This approach can be supported by genomic selection and somatic embryogenesis (SE), together with cryopreservation. Using vegetative propagation (e.g., SE) can possibly accelerate the breeding process around 20–30 years in comparison with seed orchards [15]. Integrating these methods can help in the near future with the climate crisis, providing clonal forests with new varieties of trees adapted to changed environmental conditions [16].

Two of the most important timber species in Europe are Norway spruce (*Picea abies* (L.) H. Karst) and Scots pine (*Pinus sylvestris* L.). Both belong to the Pinaceae family and are the primary species in forest management in many European countries (Sweden, Finland, France, Poland and the Baltic States). *P. abies* is distributed in the boreal forest zone covering Scandinavia to the Ural Mountains and in the mountainous areas of the temperate zone [17]. This species is also cultivated outside its natural range in warmer and drier regions of Europe for commercial purposes due to its valuable wood and is harvested mainly for production of cellulose and use by the construction industry. *P. abies* is wind-pollinated, *monoecious* and matures between 25–65 years of age. The species can be self-pollinated, although flower positioning on the trees and shift in flowering time minimize the process. In plantations, a strong relationship between the number single tree progenies and the probability of allele losses has been observed [18]. *P. sylvestris* is a very valuable forest tree species with a broad natural range that covers a large area of Eurasia [19]. Similar to spruce wood, pine wood is an important building material. It is also a source of cellulose and is used as fuel. *P. sylvestris* grows very well in poor habitats; thus, it is important as a foundational coniferous species commonly used in the reforestation of degraded lands and former arable lands [20,21]. Similarly to *P. abies*, *P. sylvestris* is wind-pollinated with male and female flowers present on the same tree. It mature between 25–30 years. Populations are genetically variable, due to effective geneflow and high mobility of pollen [22]. Unfortunately, over the past few decades, climate change has had a significant impact on the condition of Europe's boreal forests and forestry [23–25]. *P. abies* and *P. sylvestris* are among the species most threatened by the effects of climate change. Their distribution is modelled to decline over the next few years in Europe [26].

To ensure the desired breeding characteristics are retained in the offspring, propagation based on the use of vegetative methods is needed. A root cutting method is routinely used for the vegetative propagation of *P. abies*, e.g., in Finland and Sweden. However, due to high production costs, this method has not been introduced into mass plant production [27]. Such possibilities are potentially offered by the SE. SE is potentially the most efficient method of vegetative propagation of trees that could be used for breeding and selection of forest trees on a commercial scale, however its use is still usually limited to juvenile explants [28–30]. This technology is particularly useful both for tree species that are difficult to propagate using conventional methods and for vegetative propagation of valuable varieties [31]. It enables the production of many somatic seedlings in a relatively short time from the induction of embryogenic cultures [28]. Moreover, embryogenic tissues can be stored in liquid nitrogen (LN; −196 °C) in the juvenile stage for potentially hundreds of years, while the somatic seedlings obtained from them may be planted whenever necessary.

The SE has been intensively developed for spruce and pine species. Although the costs of somatic seedling production are still high, it is expected that the future automation of the process will allow a significant reduction in the costs generated and the implementation of SE in large-scale forest tree production. *P. sylvestris* is recalcitrant to vegetative propagation, and attempts to propagate it based on rooted cuttings have thus far not yielded satisfactory results [19,32]. SE may be the only way to obtain vegetative progeny for this coniferous species; hence, the development of propagation protocols based on this method seems justified. Unfortunately, although the SE system is advanced for many conifers (including spruces and pines), its extensive use for practical propagation is strictly limited due to its application to only a few selected genotypes and to high somatic seedling production handling costs. Therefore, it is still necessary to resolve several challenges so that SE could be used universally [33–35]. In this review, we summarize the current state of research on the problem of SE in *P. abies* and *P. sylvestris*. The efficiency of the SE process varies greatly from one stage to the next, so developing universal protocols for different tree species is a great challenge. Studies undertaken in recent years from both biochemical and molecular aspects to better understand the mechanisms controlling the SE process in conifers, as well as efforts aimed at improving the conditions for somatic embryo development, germination and acclimatization, give hope for the development of increasingly efficient protocols for these economically important forest tree species. We expect that it could result in the implementation of SE in European breeding programmes in the near future and will help the forest sector overcome the problems caused by climate change and market demands for wood products.

2. Explants and Initiation of SE

P. abies was the first conifer in which SE was successfully performed using mature zygotic embryos as explants [36]. However, for some coniferous species, including pines and common juniper (*Juniperus communis* L.), it is possible to induce SE only when zygotic embryos at the polyembryonic cleavage stage are used [37–39]. This is very inconvenient from a practical point of view because explants at this stage must be taken within a strictly defined, narrow time period. In contrast, mature explants may be used even from seeds stored in seed banks for many years, although the risk of lowering their SE induction potential increases over time [40].

In *Picea* spp., embryogenic cultures are usually induced from immature or mature zygotic embryos, but in most cases, the SE induction frequency is higher for the immature embryos [34]. The induction and proliferation media contain 2,4-dichlorophenoxyacetic acid (2,4-D) or rarely 1-naphthaleneacetic acid (NAA) or 4-amino-3,5,6-trichloropicolinic acid (Picloram) at 9.0–10.0 µM and benzyladenine (BA) at 4.5–5.0 µM [41–44]. In contrast, other explants derived from old trees are recalcitrant to SE induction. The oldest vegetative explants from which induction of embryogenic tissue has been achieved thus far were primordial shoots derived from 2- to 10-year-old white spruce (*Picea glauca* (Moench) Voss) [45] and 4- to 6-year-old trees obtained via SE [46]. Of the 17 clonal trees of *P. glauca*, five clones induced SE, while for, 5% of genotypes from 39 clonal trees tested had a positive response to this explant type [45,46]. For comparison, explants from trees of zygotic origin are much less susceptible to SE induction. Earlier, Ruaud et al. [47] demonstrated that in *P. abies*, explants of zygotic origin showed a significantly lower ability to induce SE (10%) than explants from SE-derived plantlets (80%) at the same age. The cause of these differences is not well understood. Klimaszewska et al. [45] hypothesized that during the first stages of SE, the capacity for this process undergoes permanent fixation, which is primarily associated with changes at the epigenetic level.

In many studies, it has been highlighted that the physiological state of the explant may be a major determinant of its ability to induce SE. Rutledge et al. [48] suggested that the underlying cause of this explant resistance is not a lack of activity of specific genes or other SE determinants but a potential activation of biotic defences that can work against this process. Based on the gene expression profile of shoot primordial explants originating from adult trees of *P. glauca*, for responsive and nonresponsive genotypes, high activity levels of the four candidate genes in the nonresponsive genotype were demonstrated. All these genes encoded proteins similar to angiosperm proteins, whose high activity over a prolonged period is associated with activation of the biotic defence response. At the same time, a more moderate response was obtained for the responsive genotype, indicating adaptation to stress conditions. This means a potential relationship between biotic defence and SE induction recalcitrance. Recently, Aronen et al. [49] drew attention to the possible connection between telomere length and SE induction, embryogenic tissue proliferation period and somatic embryo regeneration of *P. abies*. Telomeres are fragments of DNA with a repeating sequence that are found at the end of chromosomes. They have a specific structure; in plants, they are usually repeated DNA sequences of (TTTAGGG)n, and their role is to protect and stabilize DNA during cell division. However, telomeres may be shortened under the influence of various factors, such as exposure to stressors, chemicals or pollutants, pathogen attack, dietary patterns and weather harshness [50]. Under in vitro culture conditions, oxidative stress may be a significant contributor to telomere damage. The consequences of telomere shortening are changes in gene expression and cell cycle function, which ultimately lead to ageing and programmed cell death [49,50]. Analyses conducted for *P. abies* showed a clear correlation between maximum telomere length in explants and SE induction frequency [49]. It was found that the longer the telomeres are, the higher the ability of explants to regenerate embryogenic tissue. Changes in telomere length were probably the result of multiple stress factors that acted at the SE induction stage on the explants (wounding, harsh conditions and chemical treatments), inducing oxidative stress, which was a direct cause of telomere shortening. Interestingly, telomere shortening was not observed in either induced embryogenic tissues or somatic embryos that regenerated from these tissues [49]. However, prolonged in vitro multiplication of up to one year significantly affected telomere shortening, which as a consequence, could result in a decrease in the level of tissue productivity.

On the other hand, *P. sylvestris* is more recalcitrant to SE processes than the other pine species [19]. The first successful SE for *P. sylvestris* was reported in 1996 by Keinonen-Mettälä et al. [51]. In the beginning, it was assumed that the SE methods developed for spruce species would also apply to other coniferous species. However, it turned out that pines require significant modification. The most important differences result from the difficulty in obtaining SE initiation from mature zygotic embryos and from the quality of somatic embryos as determined by higher ABA concentration and higher osmolarity in the maturation medium [52–54]. Generally, in *Pinus* spp., SE induction is limited to the first weeks of zygotic embryo development, when the embryo is in a cleavage stage [39,51] or before it reaches the cotyledonary stage [55]; however, induction was also obtained from mature embryos [56]. According to Aronen and co-workers [57], the best explant type for this pine species is the intact gametophyte with immature zygotic embryo. The induction frequency was relatively low, reaching a maximum of 42%, and dependent on the mother tree [55,58]. Recently, Trontin et al. [59] induced embryogenic-looking tissues using slices from developing shoot buds. Only a low number of induced cell lines are able to proliferate and regenerate high-quality mature embryos. According to Abrahamsson et al. [60], this proliferation may be associated with the cleavage stage and the degeneration pattern of early and late embryos, which occur in a different way in normal and abnormal cell lines. In embryogenic cultures of *P. sylvestris*, a significant proportion of embryos at these stages degenerate. Studies have shown that in normal-cotyledon embryos regenerating cell lines, degenerating embryos are eliminated, as are subordinate embryos during zygotic embryogenesis. In contrast, in abnormal embryos regenerating cell lines, there is continuous degeneration and differentiation of new embryos [60]. According to the authors, this phenomenon already occurs during the initiation of the SE process and leads to an increased risk of producing abnormal embryos by lines induced from early zygotic embryos (at the cleavage polyembryony stage). To eliminate this risk, the authors suggest further exploration of other sources of explants for SE induction in pines. A recent analysis on the molecular regulation of somatic embryo development from these two types of *P. sylvestris* lines showed that there are differences in the expression patterns of selected transcripts. These were associated with phenomena such as the transition from morphogenesis to maturation, embryo degeneration or apical-basal polarization. Several genes probably related to the cleavage process were also differentially expressed during the development of somatic *P. sylvestris* and *P. abies* embryos. For example, upregulation of the SERK1 gene stimulated lobing of the embryonal mass, the first step of the cleavage process in *P. sylvestris*, which does not occur in *P. abies* [61].

Generally, embryogenic cultures both *P. abies* and *P. sylvestris* are induced and proliferated on the solid or semisolid media [51,52,62], although embryogenic tissues of *P. abies* had the ability to multiplication in liquid cultures [34,63]. For *P. sylvestris*, such attempts, to our knowledge, have not yet been undertaken. However, studies for black pine (*Pinus nigra* J.F. Arnold) conducted by Salaj et al. [64] showed such possibility in *Pinus* spp. Embryogenic cultures of *P. abies* are usually induced and maintained at 20–24 °C [62,63,65] and at 24–25 °C for *P. sylvestris* cultures [52,57]. For spruce SE, the most commonly used media are recommended by von Arnold and Eriksson [66] (LP) and by Litvay et al. [67] (LM); for pine SE, the most useful media are described by Gupta and Durzan [68] (DCR), Becwar et al. [69], Litvay et al. [67] (modified to contain half-strength macroelements; $\frac{1}{2}$ LM), Smith [70] and Teasdale et al. [71] (modified). According to Aronen [19], in the case of *P. sylvestris*, the most useful are modified Litvay's [67] and Gupta and Durzan's [68] media. In contrast to spruce species, induction of SE in pines usually requires a higher or lower concentration of auxin and cytokinin or a lack of these PGRs, plus supplementation of the induction medium with ABA and amino acids [42,72–74]. For example, Lelu-Walter et al. [52] applied 9.0 or 2.2 μM of 2,4-D and 4.4 or 2.3 BA to the induction medium with lower concentrations of PGRs in the proliferation medium. Induced ETs began to lose their potential for SE very quickly, even at 6 months of subculture [54]. For *P. nigra*, the potential for induction lasted slightly longer—2 or maximally 3 years [68]. To improve the multiplication rate of ETs filter paper discs were used [19,52].

Research conducted recently by García-Mendiguren et al. [74] for Monterey pine (*Pinus radiata* D. Don) showed that a lower initiation temperature (18 °C) promoted the SE process compared to that for the cultures incubated at 28 °C, indicating that the initiation stage has a long-term effect on embryogenesis in this species. The initiation rate was the highest at 18 °C (17%) and did not differ from the control variant 23 °C (13%), but was significantly lower at 28 °C (4%). The proliferation of embryonal masses initiated in 18 °C and 23 °C (54%) was higher as compared to 28 °C (15%). Moreover, the highest percentage of embryogenic cell lines was obtained at these lower temperatures as compared to 28 °C. Earlier, Montálban et al. [75] demonstrated that cold storage of cones for over 1 month increases SE initiation frequency. On the other hand, Gao et al. [76] proved that the success of the induction step in some pine species is also dependent on the family origin and collection date of explants. Open-pollinated cones of three families with explants were collected from June to July 2015. The age of each family was 28 years. The dates of seed collection were 23 June, 30 June, 6 July and 13 July (representing explants in four phases).

In chir pine (*Pinus roxburghii* Sargent), the role of salicylic acid (SA) in SE induction has been considered. SA could promote the inhibition of ethylene biosynthesis, which acts in plant differentiation, or by inhibiting enzyme detoxification of H_2O_2 [77]. ROS concentration and subcellular distribution in plants are carefully regulated, as imbalances cause redox state disturbances that have crucial effects on cell fate [78]. The steady state of ROS in cells is maintained through ROS-scavenging enzymes, including superoxide dismutases (SODs), ascorbate peroxidases (APXs), catalases (CATs), glutathione peroxidase (GPX), glutathione transferase (GST) and antioxidant molecules, such as glutathione and ascorbic acid [78].

An attempt to perform SE with explants derived from old trees for *P. sylvestris* was made by Trontin et al. [59]. Slices from developing shoot buds were used as explants, from which embryogenic tissues containing primordia were induced. However, only a few cell lines were able to proliferate and regenerate somatic embryos, but these were of poor quality and unable to germinate.

3. Maturation of Somatic Embryos

The lack of sufficient synchronization of the development of early somatic embryos (proembryos, PEMs) is still a serious factor reducing the efficiency of regeneration of somatic embryos in *Picea* spp. from embryogenic tissues. Only properly formed early somatic embryos develop into fully mature embryos, while others are eliminated. Synchronized embryo development is dependent on both the culture conditions and the inherent ability of the embryogenic cell lines to generate specific types of early somatic embryos [79]. During the maturation step, the multiplication process of early somatic embryos is stopped, and the accumulation of storage reserves (starch, proteins and lipids) in developing embryos starts. As a result, somatic embryos grow and go through successive, specific stages of development (globular, heart, torpedo, early cotyledonary and cotyledonary stages). The maturation of spruce somatic embryos is conditioned by the presence of abscisic acid (ABA) and their exposure to the osmotic stress provided by saccharides [80,81]. According to Varis et al. [62] the best ABA concentration for *P. abies* somatic embryos maturation was 30 μM. It was also proven that high-molecular-weight compounds such as polyethylene glycol (PEG) 4000 had a positive effect on somatic embryo quality and increased the number of mature spruce embryos [80,82,83]. However, their impact on post-maturation development is still unclear, and further investigations are needed [80,84]. An alternative to osmotic stress treatment during the maturation stage may be reduced glutathione (GSH) application to the medium [85] or low-temperature treatment [86,87]. Earlier, it was shown that *P. glauca* somatic embryos after treatment with GHS during maturation were characterized by improved shoot:root conversion and a higher frequency of embryos generating functional roots and shoots [85]. Additionally, they were able to develop into properly growing plants to a greater extent than the germinated control embryos. However, low-temperature treatment of mature somatic embryos has proven to be an effective approach in *P. abies*. Research results published by Varies et al. [88] indicated that treatment of mature somatic embryos of *P. abies* with a temperature of +4 °C did not interfere with their proper germination. Recent studies have shown that cold storage of mature embryos before their germination positively influences their further development during germination and conversion into plants, which could then be planted after a nursery period one year earlier than that of the control variant [87].

Despite the many protocols available, somatic embryo maturation in pines is not always successful [89]. The main reasons for this include asynchronous embryo production, abnormal morphology or poor root development. As in most pine species, the development of mature somatic embryos in *P. sylvestris* is dependent on the presence of high concentrations of ABA (80–90 µM) and osmoticum (9–10 g/L gellan gum) in the medium [52,57]. Recently, Salo et al. [90] demonstrated on the basis of changes in polyamine metabolism in this pine that ABA + PEG treatment may act in different ways on the cells in embryo-producing lines and in lines unable to produce somatic embryos. In the former, it is recognized as a signal to trigger the embryogenic pathway; in the latter, it is perceived as osmotic stress, which leads to the activation of stress defence mechanisms in the cells. Therefore, manipulation of the stress response pathways seems to be a promising solution for improving the somatic embryo production of recalcitrant *P. sylvestris* lines.

Some studies have revealed that when using improved protocols, up to 95% of established embryogenic cell lines of *P. sylvestris* are able to produce properly developed somatic embryos capable of germination [91]. Aronen [57] reported a germination rate greater than 90% if well-developed somatic embryos were used. The protocol developed at the Finnish Forest Research Institute allowed the production of high-quality *P. sylvestris* saplings, which were tested in the field [19]. Despite this, many genotypes remain resistant to propagation procedures through SE, and further research is needed in this area.

As previous studies have shown, SE is a good micropropagation system for *P. sylvestris* [52,57,91]. However, the need to develop efficient protocols and the lower demand for improved forest regeneration material in this species slow down the application of SE in breeding practice and for forestry purposes.

4. Growth Conditions

4.1. Nutrients

A prerequisite for the correct development of somatic embryos of *P. abies* (and other coniferous species) under in vitro culture is that they are provided with an adequate dose of both carbon (C) and nitrogen (N) in the medium. Under natural conditions, during the development of the zygotic embryo in the seed, these components are supplied by the surrounding megagametophyte [92], which do not occur during the SE process. Some studies have revealed that N and the form in which it is applied to the media have a significant influence on the maturation and germination of coniferous somatic embryos and further on the survival rate of somatic seedlings [92,93]. N is added to the nutrient solution either in inorganic (NH_4^+ and NO_3^-) or organic forms (Gln or casein hydrolysates) or as mixtures [92]. In *Picea* spp., enrichment of the medium with organic nitrogen additionally improved the frequency of initiation, multiplication of embryogenic tissue and the quality of mature somatic embryos [65,94,95]. Carlsson et al. [96] demonstrated the importance of Gln in *P. abies* PEM multiplication, suggesting that it is a significant source of N for germinating somatic embryos. This hypothesis was confirmed in recent studies, which showed that 50% of assimilated N was supplied by Gln [92]. Therefore, it must be assumed that manipulation with this medium component may be crucial for improving SE efficiency in some coniferous species. SE is a complex process affected by many factors, and over time, research has shown that stress factors play an important role during SE [97]. Several stress treatments, such as low or high temperature, osmotic stress, and heavy metals, may be key factors for inducing the SE process, even without the presence of PGRs in the medium [98]. In turn, studies on Japanese larch (*Larix kaempferi* (Lamb.) Carr.) SE have revealed that addition of ABA to the medium led to induction of this process not only in the presence of high H_2O_2 levels but also with increasing levels of CAT, SOD, and APX gene expression [99]. This indicates that oxidative stress conditions with high levels of ROS are needed for SE induction rather than low concentrations of ROS, as was previously considered in the context of the messenger role of ROS under physiological conditions, e.g., in seeds.

4.2. Light

Light is one of the main determinants of somatic embryo morphogenesis in conifers. Its presence or absence, intensity, spectrum and photoperiod are essential for the efficiency of embryogenic cultures. While SE induction, ET proliferation and/or somatic embryo maturation in coniferous species require darkness [41,100], the presence of light promotes germination and root growth [62,101]. Detailed research concerning light intensity and its spectra, using various light sources, on the particular stages of SE in *P. abies* was published recently by the research group from the LUKE Institute in Finland [62]. Low-intensity LED light was found to have little positive effect on the basic parameters of ET growth and somatic embryo production and survival. The proliferation of ET exposed to green light resulted in an increase in embryo productivity, but the quality of the germinating embryos was lower than that of embryos obtained from ET proliferated under far red light or in darkness. However, maturation under green light had a positive effect on root and shoot growth. The effect of blue light on the proliferation of ET was negative, similar to previous findings reported by Latkowska [102]. During maturation no or low intensity light was applied [41,63,65]. However, for the further growth of *P. abies* somatic seedlings the light intensity was increased for example from 5 μmol m^{-2} s^{-2} to 100–150 μmol m^{-2} s^{-2} at 20 °C during germination and plant regeneration [65]. However, different laboratories used various germinating conditions in this spruce somatic seedling development. In the case of *P. sylvestris* Lelu-Walter and coworkers [52] reported good results after treatment of mature cotyledonary embryos with darkness for 10–14 days at day/night temperatures of 24/21 °C and then with 10 μmol m^{-2} s^{-2} at the same temperatures scheme.

Additionally, germination performed under blue light resulted in shorter shoot and root development in this study. Generally, the spectral type and light intensity applied during germination had a significant influence on both shoot and root growth and somatic seedling survival. A beneficial effect on shoot growth was observed in the presence of red light. In previous investigations, it was reported that this light spectrum promoted somatic embryo maturation and germination in *P. abies* and in some *Pinus* spp. [101,103]. In turn, research results revealed that low-intensity red light favoured N storage during *P. abies* somatic embryo germination [92]. Based on the results presented by Varis et al. [62], it can be expected that maintaining an adequate balance between the different light spectra and the optimal light intensity will enable the effective control of the final stage of conifer somatic seedling production.

5. Germination and Acclimatization

At the germination and conversion stages PGRs are usually omitted in the media, both in *Picea* and *Pinus* spp. [41,52,65,87]. However, germination and acclimatization of saplings still remain major constraints for somatic embryos of spruces and pines due to low hypocotyl growth and subsequent abnormalities in growth [104]. These can lead to significant losses, as germination and acclimatization are the last stages of high-cost tissue culture. However, the successful production of high-quality saplings is highly dependent on previous steps of SE. Genotype, maturation method, duration of maturation and embryo quality have all been proven to influence the performance of somatic saplings [57]. Unbalanced root: shoot growth during germination may be a response to redox stress suffered by explants during the previous stages of SE [105]. The role of the redox state in the SE process was studied in white spruce and loblolly pine [106,107]. The addition of a low level of GSH during *P. glauca* somatic embryo formation led to higher production of embryos and root growth during the postembryonic phase. At higher concentrations, GSH inhibited these processes. When glutathione disulfide (GSSG) was added at a high concentration (1 mM), better-quality embryos were produced according to the control treatment. Supplementation of medium with GSSG led to a 20% increase in conversion frequency during postembryonic development [106]. A similar result was obtained for loblolly pine, where the addition of GSSG to germination medium led to an increase in SE germination. These results suggest that the level of oxidative stress during SE is high and that the process of oxidation of glutathione is very important [107].

In many spruce and pine species, desiccation of mature somatic embryos has been proven to considerably increase germination [108]. It is especially important in embryos matured in medium with the addition of PEG. Although PEG is beneficial to embryo maturation, it can inhibit later germination. For example, 2–3 weeks of desiccation was reported to be beneficial for egg-cone pine (*Pinus oocarpa* Schiede ex Schltdl.), increasing germination. Without this treatment, germinated embryos produce abnormal saplings incapable of further growth [109]. Apart from desiccation, additional cold stratification was proven to increase germination in slash pine (*Pinus elliottii* Engelm.) [110], Taiwan spruce (*Picea morrisonicola* Hayata) [111] and Fraser fir (*Abies fraseri* (Pursh.) Poir.) [112].

Saplings grown in vitro generally show altered morphology, physiology and anatomy, which result in their high mortality in field conditions. In vitro saplings have large stomata with changed shapes and structures, which result in increased transpiration. Their guard cells have thinner cell walls and contain more starch and chloroplasts [113]. In contrast, successfully acclimatized saplings generally have increased leaf thickness and decreased stomatal density. Subsequent development of cuticles, epicuticular waxes and effective stomatal regulation of transpiration leads to stabilization of transpiration in saplings [114]. Therefore, to ensure successful acclimatization, saplings should be exposed to ex vitro conditions slowly at high light intensity and low humidity. Additionally, carbohydrate accessibility plays a significant role in the acclimatization process as saplings change conditions from heterotrophic to autotrophic [115]. Biotization of micropropagated plants with either endophytic bacteria or mycorrhizal fungi promotes plant growth, survival and general persistence. Mycorrhizal inoculation has been proven to increase the survival rate and mean number of branches and roots, as well as the root length of stone pine (*Pinus pinea* L.) [116].

6. SE—The Future of Forestry?

6.1. Applied Biotechnology and SE

Elite trees with characteristics desired for forest management can be grown on plantations as part of breeding programs with strictly defined assumptions. For example, in New Zealand, the tree improvement program is based on the levels of genetic advantage gained by using previous generations of tree breeding [117,118]. Improved clones that allow for the commercial deployment of clones to plantations are an excellent opportunity for realizing higher genetic gains for breeding by applying genomic selection (GS). Benefits from the genomic methods implementing GS and its integration with SE are significant. They include reductions in identification errors, accurate estimates of relatedness and improved accuracy in genotyping for better control of breeding values. GS allows shortened breeding cycles and therefore results in greater flexibility than conventional conifer improvement, which is characterized by long breeding cycles necessary to make selections and reach the production stage [119]. It is likely that a drastic reduction in breeding cycle time while maximizing genetic gain is possible with the integration of GS selection and SE. This integration will lead to high synergies for implementation in the near future of multi-varietal forestry—the tested tree varieties in plantation forestry—and will be an effective way of maintaining the productivity and adaptation capacity of conifer plantations [119–121]. This solution to tree improvement will allow the deployment of genetically improved reforestation stock for long-living conifers, such as spruces. It is very important, especially in the face of climate change and the changing needs of wood product markets, to improve planting material for the various needs in a short time. In forest practice, tree breeding programs combine clonal testing with progeny testing to achieve the best selection accuracy [122]. In the case of plant material obtained via SE, such testing would be greatly facilitated by significant shortening of the breeding cycle and the possibility of long-term cryostorage of embryogenic cultures. The clonal populations will show the greatest gains for traits having low heritability, and the progeny will allow testing of large numbers of individuals for traits with higher heritability. Relatedly, a new strategy of breeding applied by New Zealand and New South Wales includes an elite population of *P. radiata* tested both as progeny and as clones [118].

The long-term breeding program of Sweden aims to adapt trees to different conditions (current and future), including changing climatic conditions, to conserve genetic diversity and to improve tree features important to timber production [123,124]. The main advantage of using vegetative reproduction in Sweden compared to seed orchards is the elimination of the 20–30 year time delay from breeding progress to the distribution of nursery material, which will increase breeding profits but also increase the stability of crops grown under different environmental conditions because of replication of the breeding program field test system [15]. In Finland, where the increased planting of *P. abies* and fluctuating seed yields caused shortages of improved regeneration material [125], SE may be a promising method for clonal propagation of conifers on a large scale to offset this lack of improved seeds [34].

Apart of GS, genetic engineering is also expected to be included on a larger scale in breeding programs in forestry management in the future [126]. It is considered that the combination of SE and genetic transformation will allow for the rapid and precise introduction of desired features into breeding material and, consequently, to improve production efficiency of species used in clonal forestry. However, the introduction of foreign DNA into the tree genome and subsequent regeneration of transformed plants is very difficult. A crucial factor determining successful genetic manipulation is the choice of explants competent for transformation and regeneration [127,128], such as embryogenic tissues, embryogenic suspensions and mature somatic [129–131]. The most convenient plant material is embryogenic tissue, which can multiply in vitro over a longer period of time and provide an excellent basis for experiments. In the case of somatic embryos, their availability may be limited [132].

Currently, several reports concerning the genetic transformation of *P. abies* embryogenic cultures are available [133–136]. Most studies were carried out on embryogenic tissues using *Agrobacterium tumefaciens*-mediated transformation. For example, Wenck et al. [133] obtained over 100 transformed embryogenic cell lines after their co-cultivation with *Agrobacterium* strains containing additional copies of virulence genes (either a constitutively active *virG* or extra copies of *virG* and *virB*, both from pTiBo542). Additionally, Briža et al. [136] reported 70 transgenic lines, which were transformed with *Agrobacterium tumefaciens*. In this experiment the modified versions of *Cry3A* gene of *Bacillus thuringiensis*, with the increased toxicity against spruce bark beetle, were introduced into embryogenic tissues. For comparison, when using the biolistic method, the authors obtained only 18 transformed cell lines. Earlier Walter et al. [134] obtained stable integration of *uidA* and *nptII* into the *P. abies* genome using also the biolistic transformation technology. Transgenic plants of *P. abies* were obtained after transformation of embryogenic cultures with an antisense construct of the *P. abies* gene encoding cinnamoyl CoA reductase (*CCR*), one of the genes that regulates lignin biosynthesis. The underexpression of the sense *CCR* gene in transgenic plants resulted in as much as an 8% decrease in lignin content, and stable expression was maintained in the plants for over five years [137]. Literature concerning the *P. abies* somatic embryos genetic transformation is very scarce due to difficulty in obtaining transgenic callus. In a study conducted by Pavingerová et al. [135], only one of the nine tested lines of cotyledonary somatic embryos were able to produce transgenic callus.

Until now, no studies were performed on the genetic transformation of *P. sylvestris* embryogenic cultures, although such attempts were carried out for its transformed pollen [138,139] and for other pine species [64,133,140,141]. Grace et al. [140] reported the regeneration of transgenic *P. radiata* plants transformed with the *cry1Ac* gene, which improved the level of resistance to larvae of the painted apple moth. For comparison, Salaj and coworkers [64] demonstrated the expression of the *uidA* reporter gene in five transgenic embryogenic tissue lines of *P. nigra* after one year of culture; however, poor or no transgenic plantlets were obtained.

Molecular studies have identified some genes regulating the process of SE. These include *SERK, LEA, LEC, YUC, AUX/IAA, BBM* or *WOX* genes. The activity of these genes was also detected during SE of trees. Introducing of these genes into explants (for example zygotic embryos, primordial shoots) or embryogenic tissues with low embryogenic potential would presumably improve induction and further embryo development of coniferous species somatic embryos. Thus, this research area is worth further exploration.

6.2. Possibility of Automation

The utilization of SE tissue culture protocols to produce millions of somatic seedlings for forestry with the aim of forest regeneration requires a long time because it is a multistage process. In most laboratories, conifer somatic embryo production is carried out on solid media. However, this system does not allow for the generation of large numbers of high-quality embryos and is expensive due to the considerable labour involved [33]. The solution to this problem may be to scale up by culturing in liquid media in suspension cultures or in bioreactors [142]. To achieve cost-efficient mass propagation, automation of SE is necessary, which will be possible via temporary immersion system bioreactor-based culture solutions. The choice of scale-up in bioreactors is possible thanks to the application of a multiplication step in liquid medium, which is feasible for most species [143]. However, although the use of suspension cultures improved the level of synchronization of the development of differentiating somatic embryos, the development of embryos at later stages in coniferous species remains problematic [33,64], and many more studies are needed in this area. To reduce the cost of somatic seedling production, investigations on bioreactors based on SE for some tree conifer species have been undertaken [144–147]. In the 1990s, perfusion bioreactors and regular immersion bioreactors were effective in the maturation of somatic embryos of several *Picea* spp. and Douglas fir (*Pseudotsuga menziesii* (Mirb.) Franco) [148]. However, in the last decade, TIS bioreactors have been successfully applied for Caucasian fir (*Abies nordmanniania* Spach.) [145] and *P. abies* [144,149]. The promising results of somatic embryo production after testing a TIS bioreactor platform for *P. abies* were demonstrated by Välimäki et al. [147]. It was found that both embryo production and survival were dependent on the frequency of bioreactor aeration and the support pad material used as a base. Moreover, the post-maturation desiccation of the mature embryos enabled the improvement of final somatic seedling survival rates. Embryo maturation in TIS was the most efficient on filter papers on plastic netting with 20 min/4 h aeration and when the mature embryos were desiccated at +2 °C for 5 wk on nested plates. However, the authors emphasize that to enable mass use of bioreactors for *P. abies* sapling production, it is still necessary to optimize the SE process in this spruce species.

The fluidics-based technology recently proposed by Swedish forestry companies is another promising solution for massive SE plant production [143]. In this approach, a special temporary immersion bioreactor model was designed for *P. abies* embryogenic cultures, allowing improvement of the quality and synchronization of the development of produced embryos. After reaching maturation, somatic embryos are harvested with the SE fluidics system; each embryo is documented individually and sorted by an image analysis system according to its morphological measurements [34,150]. It is important that the automation of maturation, germination and planting steps in this system enables the significant reduction of production time and costs of the SE process. Temporary immersion bioreactors may have advantages over other types of bioreactors based on total immersion and mixing of the culture. They provide support for the growth of culture by promoting the directional growth of PEM and the formation of polarized embryos at an early stage. In the case of SE coniferous cultures, reproduction and maturation can occur sequentially in the same bioreactor without disturbing the culture [143]. It is expected that the development of appropriate criteria for the selection of somatic embryos will allow for the rapid production of valuable material on a large scale. Therefore, it seems that *P. abies* SE process automation will be realized in the near future. If it can be carried out for one species, it will only be a matter of time before it can for others.

7. Concluding Remarks

In the last decade, significantly more research has been conducted to understand the mechanisms controlling tree embryogenesis and in vitro cultivation. SE still remains as a method with an unfulfilled potential. Despite being used commercially in some instances, application of SE in forestry is not well developed. Still, the main limiting factors appear to be the costs and difficulties in establishing cultures from some genotypes. However, in order to rapidly accelerate many tree improvement programs, an effective and practically reproducible technique must be developed to produce somatic plants from adult vegetative material (Figure 1). Especially in the era of climate change as well as biodiversity collapse possibility of combining SE with other biotechnological methods, cryopreservation, genome sequencing or genetic modification is an asset, which cannot be overestimated. It is especially helpful in trees that are generally characterized as having a long life span and therefore, slow reaction for changes.

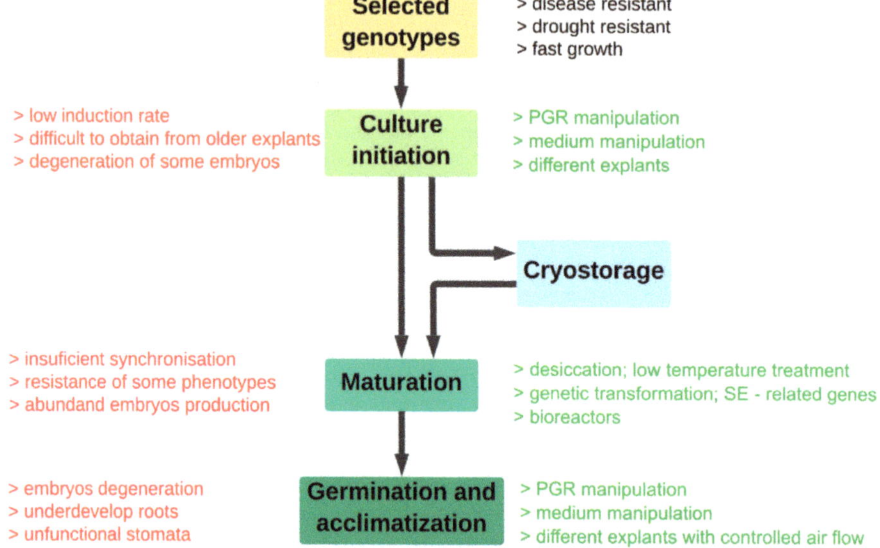

Figure 1. Steps of SE process of *Picea abies* and *Pinus sylvestris*. Major challenges are presented in red and potential solutions in green.

Amongst the challenges is to improve the general quality of mature somatic embryos and increase the conversion rate of somatic embryos and the survival rate of acclimatized saplings. This last part is especially significant, as losses in the number of received saplings increase the overall cost of their production. Improvements at each step of the SE process will contribute to the final success rate.

Author Contributions: Conceptualization, T.H.-P. and E.R. writing—original draft preparation, T.H.-P., M.K.W., J.K.-O., A.M.S. and E.R.; writing—review and editing, T.H.-P., M.K.W.; visualization, A.M.S.; supervision: T.H.-P., M.K.W.; All authors have read and agreed to the published version of the manuscript.

Funding: This work was supported by statutory research of the Institute of Dendrology, Polish Academy of Sciences.

Institutional Review Board Statement: Not applicable.

Informed Consent Statement: Not applicable.

Data Availability Statement: Not applicable.

Conflicts of Interest: The authors declare no conflict of interest.

References

1. Bruinsma, J. *World Agriculture: Towards 2015/2030: An FAO Study*; Earthscan Publications Ltd.: London, UK, 2017; p. 431.
2. Sutton, W.R.J. The Need for Planted Forests and the Example of *Radiata* Pine. *New For.* **1999**, *17*, 95–110. [CrossRef]
3. Pintilii, R.-D.; Andronache, I.; Diaconu, D.C.; Dobrea, R.C.; Zeleňáková, M.; Fensholt, R.; Peptenatu, D.; Drăghici, C.-C.; Ciobotaru, A.-M. Using Fractal Analysis in Modeling the Dynamics of Forest Areas and Economic Impact Assessment: Maramureș County, Romania, as a Case Study. *Forests* **2017**, *8*, 25. [CrossRef]
4. Keenan, R.J.; Reams, G.A.; Achard, F.; de Freitas, J.V.; Grainger, A.; Lindquist, E. Dynamics of Global Forest Area: Results from the FAO Global Forest Resources Assessment 2015. *For. Ecol. Manag.* **2015**, *352*, 9–20. [CrossRef]
5. Singha, K.; Sahariah, D.; Saikia, A. Shrinking Forest and Contested Frontiers: A Case of Changing Human-Forest Interface along the Protected Areas of Nagaon District, Assam, India. *Eur. J. Geogr.* **2019**, *10*, 120–136.
6. Andronache, I.; Marin, M.; Fischer, R.; Ahammer, H.; Radulovic, M.; Ciobotaru, A.-M.; Jelinek, H.F.; Di Ieva, A.; Pintilii, R.-D.; Drăghici, C.-C.; et al. Dynamics of Forest Fragmentation and Connectivity Using Particle and Fractal Analysis. *Sci. Rep.* **2019**, *9*, 12228. [CrossRef] [PubMed]
7. Kupfer, J.A.; Malanson, G.P.; Franklin, S.B. Not Seeing the Ocean for the Islands: The Mediating Influence of Matrix-Based Processes on Forest Fragmentation Effects. *Glob. Ecol. Biogeogr.* **2006**, *15*, 8–20. [CrossRef]
8. Cao, Q.; Liu, Y.; Georgescu, M.; Wu, J. Impacts of Landscape Changes on Local and Regional Climate: A Systematic Review. *Landsc. Ecol.* **2020**, *35*, 1269–1290. [CrossRef]
9. Davin, E.L.; de Noblet-Ducoudre, N. Climatic Impact of Global-Scale Deforestation: Radiative versus Nonradiative Processes. *J. Clim.* **2010**, *23*, 97–112. [CrossRef]
10. Strandberg, G.; Kjellström, E. Climate Impacts from Afforestation and Deforestation in Europe. *Earth Interact.* **2019**, *23*, 1–27. [CrossRef]
11. Mendes, C.B.; Prevedello, J.A. Does Habitat Fragmentation Affect Landscape-Level Temperatures? A Global Analysis. *Landsc. Ecol.* **2020**, *35*, 1743–1756. [CrossRef]
12. Wu, J. Key Concepts and Research Topics in Landscape Ecology Revisited: 30 Years after the Allerton Park workshop. *Landsc. Ecol.* **2013**, *28*, 1–11. [CrossRef]
13. Acharya, K.P.; Paudel, P.K.; Jnawali, S.R.; Neupane, P.R.; Köhl, M. Can Forest Fragmentation and Configuration Work as Indicators of Human–Wildlife Conflict? Evidences from Human Death and Injury by Wildlife Attacks in Nepal. *Ecol. Indic.* **2017**, *80*, 74–83. [CrossRef]
14. Hazubska-Przybyl, T.; Bojarczuk, K. Tree Somatic Embryogenesis in Science and Forestry. *Dendrobiology* **2016**, *76*, 105–116. [CrossRef]
15. Rosvall, O. Using Norway Spruce Clones in Swedish Forestry: General Overview and Concepts. *Scand. J. For. Res.* **2019**, *34*, 336–341. [CrossRef]
16. Park, Y.-S.; Beaulieu, J.; Bousquet, J. Multi-Varietal Forestry Integrating Genomic Selection and Somatic Embryogenesis. In *Vegetative Propagation of Forest Trees*; Park, Y.S., Bonga, J.M., Moon, H.-K., Eds.; National Institute of Forest Science (Nifos): Seoul, Korea, 2016; pp. 302–318.
17. Montwé, D.; Spiecker, H.; Hamann, A. An Experimentally Controlled Extreme Drought in a Norway Spruce Forest Reveals Fast Hydraulic Response and Subsequent Recovery of Growth Rates. *Trees Struct. Funct.* **2014**, *28*, 891–900. [CrossRef]
18. Skrøppa, T. *EUFORGEN Technical Guidelines for Genetic Conservation and Use for Norway Spruce (Picea abies)*; International Plant Genetic Resources Institute: Rome, Italy, 2003.
19. Aronen, T. From Lab to Field-Current State of Somatic Embryogenesis in Scots Pine. In *Vegetative Propagation of Forest Trees*; Park, Y.S., Bonga, J.M., Moon, H.-K., Eds.; National Institute of Forest Science (Nifos): Seoul, Korea, 2016; pp. 515–527.
20. Hytönen, J.; Jylhä, P. Long-Term Response of Weed Control Intensity on Scots Pine Survival, Growth and Nutrition on Former Arable Land. *Eur. J. For. Res.* **2011**, *130*, 91–98. [CrossRef]
21. O'Reilly-Wapstra, J.M.; Moore, B.D.; Brewer, M.; Beaton, J.; Sim, D.; Wiggins, N.L.; Iason, G.R. *Pinus sylvestris* Sapling Growth and Recovery from Mammalian Browsing. *For. Ecol. Manag.* **2014**, *325*, 18–25. [CrossRef]

22. Mátyás, C.; Ackzell, L.; Samuel, C.J.A. *EUFORGEN Technical Guidelines for Genetic Conservation and Use for Scots Pine (Pinus sylvestris)*; International Plant Genetic Resources Institute: Rome, Italy, 2004.

23. Kellomäki, S.; Peltola, H.; Nuutinen, T.; Korhonen, K.T.; Strandman, H. Sensitivity of Managed Boreal Forests in Finland to Climate Change, with Implications for Adaptive Management. *Philos. Trans. R. Soc. B Biol. Sci.* **2008**, *363*, 2341–2351. [CrossRef] [PubMed]

24. Subramanian, N.; Bergh, J.; Johansson, U.; Nilsson, U.; Sallnäs, O. Adaptation of Forest Management Regimes in Southern Sweden to Increased Risks Associated with Climate Change. *Forests* **2016**, *7*, 8. [CrossRef]

25. Sierota, Z.; Grodzki, W.; Szczepkowski, A. Abiotic and Biotic Disturbances Affecting Forest Health in Poland over the Past 30 Years: Impacts of Climate and Forest Management. *Forests* **2019**, *10*, 75. [CrossRef]

26. Dyderski, M.K.; Paź, S.; Frelich, L.E.; Jagodziński, A.M. How Much Does Climate Change Threaten European Forest Tree Species Distributions? *Glob. Chang. Biol.* **2018**, *24*, 1150–1163. [CrossRef] [PubMed]

27. Högberg, K.; Varis, S. Vegetative Propagation of Norway Spruce: Experiences and Present Situation in Sweden and Finland. In *Vegetative Propagation of Forest Trees*; Park, Y.S., Bonga, J.M., Moon, H.-K., Eds.; National Institute of Forest Science (Nifos): Seoul, Korea, 2016; pp. 528–550.

28. Cyr, D.R.; Attree, S.M.; El-Kassaby, Y.A.; Ellis, D.D.; Polonenko, D.R.; Sutton, B.C.S. Application of Somatic Embryogenesis to Tree Improvement in Conifers. *Prog. Biotechnol.* **2001**, *18*, 305–312. [CrossRef]

29. Lelu-Walter, M.-A.; Thompson, D.; Harvengt, L.; Sanchez, L.; Toribio, M.; Pâques, L.E. Somatic Embryogenesis in Forestry with a Focus on Europe: State-of-the-Art, Benefits, Challenges and Future Direction. *Tree Genet. Genomes* **2013**, *9*, 883–899. [CrossRef]

30. Egertsdotter, U. Plant Physiological and Genetical Aspects of the Somatic Embryogenesis Process in Conifers. *Scand. J. For. Res.* **2019**, *34*, 360–369. [CrossRef]

31. Isah, T. Induction of Somatic Embryogenesis in Woody Plants. *Acta Physiol. Plant.* **2016**, *38*, 118. [CrossRef]

32. Högberg, K.-A.; Hajek, J.; Gailis, A.; Stenvall, N.; Zarina, I.; Teivonen, S.; Aronen, T. *Practical Testing of Scots Pine Cutting Propagation—A Joint Metla-Skogforsk-Silava Project*; Finnish Forest Research Institute: Vantaa, Finland, 2011; p. 20.

33. Thompson, D. Challenges for the Large-Scale Propagation of Forest Trees by Somatic Embryogenesis–a Review. In Proceedings of the Third International Conference of the IUFRO Unit 2.09.02 on Woody Plant Production Integrating Genetic and Vegetative Propagation Technologies, Vitoria-Gasteiz, Spain, 8–12 September 2014; pp. 81–91.

34. Bonga, J. Conifer Clonal Propagation in Tree Improvement Programs. In *Vegetative Propagation of Forest Trees*; Park, Y.S., Bonga, J.M., Moon, H.-K., Eds.; National Institute of Forest Science (NIFoS): Seoul, Korea, 2016; pp. 3–31.

35. Lelu-Walter, M.A.; Klimaszewska, K.; Miguel, C.; Aronen, T.; Hargreaves, C.; Teyssier, C.; Trontin, J.F. Somatic Embryogenesis for More Effective Breeding and Deployment of Improved Varieties in *Pinus* spp.: Bottlenecks and Recent Advances. *Somat. Embryog. Fundam. AsP. Appl.* **2016**, 319–365. [CrossRef]

36. Toon, S.M. Somatic Embryogenesis of Lodgepole Pine (*Pinus contorta*) and Subalpine Larch (*Larix Lyallii*) for Conservation Purposes. Bachelor's Thesis, University of British Columbia, Vancouver, BC, Canada, 12 April 2010.

37. Park, Y.S.; Lelu-Walter, M.A.; Harvenght, L.; Trontin, J.F.; MacEacheron, I.; Klimaszewska, K.; Bonga, J.M. Initiation of Somatic Embryogenesis in *Pinus* Banksiana, *P. strobus*, *P. pinaster*, and *P. sylvestris* at Three Laboratories in Canada and France. *Plant Cell Tissue Organ Cult.* **2006**, *86*, 87–101. [CrossRef]

38. Helmersson, A.; von Arnold, S. Embryogenic Cell Lines of *Juniperus Communis*; Easy Establishment and Embryo Maturation, Limited Germination. *Plant Cell Tissue Organ Cult. PCTOC* **2009**, *96*, 211–217. [CrossRef]

39. Abrahamsson, M.; Valladares, S.; Larsson, E.; Clapham, D.; von Arnold, S. Patterning during Somatic Embryogenesis in Scots Pine in Relation to Polar Auxin Transport and Programmed Cell Death. *Plant Cell Tissue Organ Cult.* **2012**, *109*, 391–400. [CrossRef]

40. Park, Y.S.; Barrett, J.D.; Bonga, J.M. Application of Somatic Embryogenesis in High-Value Clonal Forestry: Deployment, Genetic Control, and Stability of Cryopreserved Clones. *Vitr. Cell. Dev. Biol. Plant* **1998**, *34*, 231–239. [CrossRef]

41. Hazubska-Przybył, T.; Ratajczak, E.; Obarska, A.; Pers-Kamczyc, E. Different Roles of Auxins in Somatic Embryogenesis Efficiency in Two *Picea* Species. *Int. J. Mol. Sci.* **2020**, *21*, 3394. [CrossRef]

42. Klimaszewska, K.; Cyr, D.R. Conifer Somatic Embryogenesis: I. Development. *Dendrobiology* **2002**, *48*, 31–39.

43. Lu, C.-Y.; Thorpe, T.A. Somatic Embryogenesis and Plantlet Regeneration in Cultured Immature Embryos of *Picea glauca*. *J. Plant Physiol.* **1987**, *128*, 297–302. [CrossRef]

44. Ramarosandratana, A.V.; Van Staden, J. Tissue Position, Explant Orientation and Naphthaleneacetic Acid (NAA) Affect Initiation of Somatic Embryos and Callus Proliferation in Norway Spruce (*Picea abies*). *Plant Cell Tissue Organ Cult.* **2003**, *74*, 249–255. [CrossRef]

45. Klimaszewska, K.; Overton, C.; Stewart, D.; Rutledge, R.G. Initiation of Somatic Embryos and Regeneration of Plants from Primordial Shoots of 10-Year-Old Somatic White Spruce and Expression Profiles of 11 Genes Followed during the Tissue Culture Process. *Planta* **2011**, *233*, 635–647. [CrossRef]

46. Varis, S.; Klimaszewska, K.; Aronen, T. Somatic Embryogenesis and Plant Regeneration from Primordial Shoot Explants of *Picea abies* (L.) H. Karst. Somatic Trees. *Front. Plant Sci.* **2018**, *9*, 1551. [CrossRef]

47. Ruaud, J.-N.; Bercetche, J.; Pâques, M. First Evidence of Somatic Embryogenesis from Needles of 1-Year-Old *Picea Abies* Plants. *Plant Cell ReP.* **1992**, *11*, 563–566. [CrossRef]

48. Rutledge, R.G.; Stewart, D.; Caron, S.; Overton, C.; Boyle, B.; MacKay, J.; Klimaszewska, K. Potential Link between Biotic Defense Activation and Recalcitrance to Induction of Somatic Embryogenesis in Shoot Primordia from Adult Trees of White Spruce (*Picea glauca*). *BMC Plant Biol.* **2013**, *13*, 116. [CrossRef] [PubMed]

49. Aronen, T.; Virta, S.; Varis, S. Telomere Length in Norway Spruce during Somatic Embryogenesis and Cryopreservation. *Plants* **2021**, *10*, 416. [CrossRef] [PubMed]

50. Chatelain, M.; Drobniak, S.M.; Szulkin, M. The Association between Stressors and Telomeres in Non-Human Vertebrates: A Meta-Analysis. *Ecol. Lett.* **2020**, *23*, 381–398. [CrossRef] [PubMed]

51. Keinonen–Mettälä, K.; Jalonen, P.; Eurola, P.; von Arnold, S.; von Weissenberg, K. Somatic Embryogenesis of *Pinus sylvestris*. *Scand. J. For. Res.* **1996**, *11*, 242–250. [CrossRef]

52. Lelu-Walter, M.-A.; Bernier-Cardou, M.; Klimaszewska, K. Clonal Plant Production from Self- and Cross-Pollinated Seed Families of *Pinus sylvestris* (L.) through Somatic Embryogenesis. *Plant Cell Tissue Organ Cult.* **2008**, *92*, 31–45. [CrossRef]

53. Yang, F.; Xia, X.-R.; Ke, X.; Ye, J.; Zhu, L.-H. Somatic Embryogenesis in Slash Pine (*Pinus elliottii* Engelm): Improving Initiation of Embryogenic Tissues and Maturation of Somatic Embryos. *Plant Cell Tissue Organ Cult.* **2020**, *143*, 159–171. [CrossRef]

54. Peng, C.; Gao, F.; Wang, H.; Shen, H.; Yang, L. Optimization of Maturation Process for Somatic Embryo Production and Cryopreservation of Embryogenic Tissue in *Pinus koraiensis*. *Plant Cell Tissue Organ Cult.* **2021**, *144*, 185–194. [CrossRef]

55. Lelu, M.-A.; Bastien, C.; Drugeault, A.; Gouez, M.-L.; Klimaszewska, K. Somatic Embryogenesis and Plantlet Development in *Pinus sylvestris* and *Pinus pinaster* on Medium with and without Growth Regulators. *Physiol. Plant.* **1999**, *105*, 719–728. [CrossRef]

56. Hohtola, A. Somatic Embryogenesis in Scots Pine (*Pinus sylvestris* L.). In *Somatic Embryogenesis in Woody Plants. Forestry Sciences*; Jain, S.M., Gupta, P.K., Newton, R.J., Eds.; Springer: Berlin/Heidelberg, Germany, 1995; Volume 44–46, pp. 269–285.

57. Aronen, T.; Pehkonen, T.; Ryynänen, L. Enhancement of Somatic Embryogenesis from Immature Zygotic Embryos of *Pinus sylvestris*. *Scand. J. For. Res.* **2009**, *24*, 372–383. [CrossRef]

58. Niskanen, A.-M.; Lu, J.; Seitz, S.; Keinonen, K.; Von Weissenberg, K.; Pappinen, A. Effect of Parent Genotype on Somatic Embryogenesis in Scots Pine (*Pinus sylvestris*). *Tree Physiol.* **2004**, *24*, 1259–1265. [CrossRef]

59. Trontin, J.F.; Aronen, T.; Hargreaves, C.; Montalbán, I.A.; Reeves, C.; Quoniou, S.; Lelu-Walter, M.A.; Klimaszewska, K. International Effort to Induce Somatic Embryogenesis in Adult Pine Trees. In *Vegetative Propagation of Forest Trees*; Park, Y.S., Bonga, J.M., Moon, H.-K., Eds.; National Institute of Forest Science (Nifos): Seoul, Korea, 2016.

60. Abrahamsson, M.; Valladares, S.; Merino, I.; Larsson, E.; von Arnold, S. Degeneration Pattern in Somatic Embryos of *Pinus sylvestris* L. *Vitr. Cell. Dev. Biol. Plant* **2017**, *53*, 86–96. [CrossRef] [PubMed]

61. Merino, I.; Abrahamsson, M.; Larsson, E.; von Arnold, S. Identification of Molecular Processes That Differ among Scots Pine Somatic Embryogenic Cell Lines Leading to the Development of Normal or Abnormal Cotyledonary Embryos. *Tree Genet. Genomes* **2018**, *14*, 1–17. [CrossRef]

62. Varis, S.; Tikkinen, M.; Välimäki, S.; Aronen, T. Light Spectra during Somatic Embryogenesis of Norway Spruce—Impact on Growth, Embryo Productivity, and Embling Survival. *Forests* **2021**, *12*, 301. [CrossRef]

63. Filonova, L.H.; Bozhkov, P.V.; Brukhin, V.B.; Daniel, G.; Zhivotovsky, B.; von Arnold, S. Two waves of programmed cell death occur during formation and development of somatic embryos in the gymnosperm, Norway spruce. *J. Cell Sci.* **2000**, *113*, 4399–4411. [CrossRef] [PubMed]

64. Salaj, T.; Blehová, A.; Salaj, J. Embryogenic Suspension Cultures of *Pinus* Nigra Arn.: Growth Parameters and Maturation Ability. *Acta Physiol. Plant.* **2007**, *29*, 225–231. [CrossRef]

65. Dahrendorf, J.; Clapham, D.; Egertsdotter, U. Analysis of Nitrogen Utilization Capability during the Proliferation and Maturation Phases of Norway Spruce (*Picea abies* (L.) H. Karst.) Somatic Embryogenesis. *Forests* **2018**, *9*, 288. [CrossRef]

66. Arnold, S.V.; Eriksson, T. In vitro Studies of Adventitious Shoot Formation in *Pinus contorta*. *Can. J. Bot.* **1981**, *59*, 870–874. [CrossRef]

67. Litvay, J.D.; Verma, D.C.; Johnson, M.A. Influence of a Loblolly Pine (*Pinus taeda* L.). Culture Medium and Its Components on Growth and Somatic Embryogenesis of the Wild Carrot (*Daucus carota* L.). *Plant Cell Rep.* **1985**, *4*, 325–328. [CrossRef] [PubMed]

68. Gupta, P.K.; Durzan, D.J. Shoot Multiplication from Mature Trees of Douglas-Fir (*Pseudotsuga menziesii*) and Sugar Pine (*Pinus lambertiana*). *Plant Cell ReP.* **1985**, *4*, 177–179. [CrossRef]

69. Becwar, M.R.; Nagmani, R.; Wann, S.R. Initiation of Embryogenic Cultures and Somatic Embryo Development in Loblolly Pine (*Pinus taeda*). *Can. J. For. Res.* **1990**, *20*, 810–817. [CrossRef]

70. Smith. Growth Medium. U.S. Patent 5,565,355, 15 October 1996.

71. Teasdale, R.; Buxton, P.A. Culture of *Pinus radiata* embryos with reference to artificial seed production. *NZJ For. Sci.* **1986**, *16*, 387–391.

72. Montalbán, I.A.; Setién-Olarra, A.; Hargreaves, C.L.; Moncaleán, P. Somatic Embryogenesis in *Pinus halepensis* Mill.: An Important Ecological Species from the Mediterranean Forest. *Trees* **2013**, *27*, 1339–1351. [CrossRef]

73. Salaj, T.; Klubicová, K.; Matusova, R.; Salaj, J. Somatic Embryogenesis in Selected Conifer Trees *Pinus nigra* Arn. And *Abies* Hybrids. *Front. Plant Sci.* **2019**, *10*, 13. [CrossRef]

74. García-Mendiguren, O.; Montalbán, I.A.; Goicoa, T.; Ugarte, M.D.; Moncaleán, P. Environmental Conditions at the Initial Stages of *Pinus radiata* Somatic Embryogenesis Affect the Production of Somatic Embryos. *Trees Struct. Funct.* **2016**, *30*, 949–958. [CrossRef]

75. Montalbán, I.A.; García-Mendiguren, O.; Goicoa, T.; Ugarte, M.D.; Moncaleán, P. Cold Storage of Initial Plant Material Affects Positively Somatic Embryogenesis in *Pinus radiata*. *New For.* **2015**, *46*, 309–317. [CrossRef]

76. Gao, F.; Peng, C.; Wang, H.; Tretyakova, I.N.; Nosov, A.M.; Shen, H.; Yang, L. Key Techniques for Somatic Embryogenesis and Plant Regeneration of *Pinus koraiensis*. *Forests* **2020**, *11*, 912. [CrossRef]

77. Malabadi, R.B.; Teixeira da Silva, J.A.; Nataraja, K. Salic Acid Induces Somatic Embryogenesis from Mature Trees of *Pinus roxburghii* (Chir Pine) Using TCL Technology. In *Tree Forest Science and Biotechnology*; Global Science Books Ltd: East Sussex, UK; Ikanobe: Takamatsu, Japan, 2008; pp. 34–39.

78. Kocsy, G.; Tari, I.; Vanková, R.; Zechmann, B.; Gulyás, Z.; Poór, P.; Galiba, G. Redox Control of Plant Growth and Development. *Plant Sci.* **2013**, *211*, 77–91. [CrossRef]

79. Filonova, L.H.; Bozhkov, P.V.; Von Arnold, S. Developmental Pathway of Somatic Embryogenesis in *Picea Abies* as Revealed by Time-Lapse Tracking. *J. Exp. Bot.* **2000**, *51*, 249–264. [CrossRef] [PubMed]

80. Hudec, L.; Konrádová, H.; Hašková, A.; Lipavská, H. Norway Spruce Embryogenesis: Changes in Carbohydrate Profile, Structural Development and Response to Polyethylene Glycol. *Tree Physiol.* **2016**, *36*, 548–561. [CrossRef] [PubMed]

81. Tremblay, L.; Tremblay, F.M. Carbohydrate Requirements for the Development of Black Spruce (*Picea* Mariana (Mill.) B.S.P.) and Red Spruce (*P. rubens* Sarg.) Somatic Embryos. *Plant Cell Tissue Organ Cult.* **1991**, *27*, 95–103. [CrossRef]

82. Chen, S.; Chen, S.; Chen, F.; Wu, T.; Wang, Y.; Yi, S. Somatic Embryogenesis in Mature Zygotic Embryos of *Picea likiangensis*. *Biologia* **2010**, *65*, 853–858. [CrossRef]

83. Stasolla, C.; Yeung, E.C. Recent Advances in Conifer Somatic Embryogenesis: Improving Somatic Embryo Quality. *Plant Cell Tissue Organ Cult.* **2003**, *74*, 15–35. [CrossRef]

84. Hazubska-Przybył, T.; Wawrzyniak, M. Stimulation of Somatic Embryo Growth and Development in *Picea* spp. by Polyethylene Glycol. *Dendrobiology* **2017**, *78*, 168–178. [CrossRef]

85. Stasolla, C. The Effect of Reduced Glutathione on Morphology and Gene Expression of White Spruce (*Picea glauca*) Somatic Embryos. *J. ExP. Bot.* **2004**, *55*, 695–709. [CrossRef]

86. Pond, S.E. The Effect of Temperature on Conversion of White Spruce Somatic Embryos. *Propag Ornam Plants* **2005**, *5*, 35–44.

87. Tikkinen, M.; Varis, S.; Peltola, H.; Aronen, T. Improved Germination Conditions for Norway Spruce Somatic Cotyledonary Embryos Increased Survival and Height Growth of Emblings. *Trees* **2018**, *32*, 1489–1504. [CrossRef]

88. Varis, S.; Ahola, S.; Jaakola, L.; Aronen, T. Reliable and Practical Methods for Cryopreservation of Embryogenic Cultures and Cold Storage of Somatic Embryos of Norway Spruce. *Cryobiology* **2017**, *76*, 8–17. [CrossRef]

89. Montalbán, I.A.; De Diego, N.; Moncaleán, P. Bottlenecks in *Pinus radiata* Somatic Embryogenesis: Improving Maturation and Germination. *Trees Struct. Funct.* **2010**, *24*, 1061–1071. [CrossRef]

90. Salo, H.M.; Sarjala, T.; Jokela, A.; Häggman, H.; Vuosku, J. Moderate Stress Responses and Specific Changes in Polyamine Metabolism Characterize Scots Pine Somatic Embryogenesis. *Tree Physiol.* **2016**, *36*, 392–402. [CrossRef] [PubMed]

91. Krakau, U.-K.; Liesebach, M.; Aronen, T.; Lelu-Walter, M.-A.; Schneck, V. Scots Pine (*Pinus Sylvestris* L.). In *Forest Tree Breeding in Europe: Current State-of-the-Art and Perspectives*; Pâques, L.E., Ed.; Managing Forest Ecosystems; Springer: Dordrecht, The Netherlands, 2013; pp. 267–323, ISBN 978-94-007-6146-9.

92. Carlsson, J.; Egertsdotter, U.; Ganeteg, U.; Svennerstam, H. Nitrogen Utilization during Germination of Somatic Embryos of Norway Spruce: Revealing the Importance of Supplied Glutamine for Nitrogen Metabolism. *Trees Struct. Funct.* **2019**, *33*, 383–394. [CrossRef]

93. Llebrés, M.T.; Avila, C.; Cánovas, F.M.; Klimaszewska, K. Root Growth of Somatic Plants of Hybrid *Pinus strobus* (L.) and *P. wallichiana* (A. B. Jacks.) Is Affected by the Nitrogen Composition of the Somatic Embryo Germination Medium. *Trees Struct. Funct.* **2018**, *32*, 371–381. [CrossRef]

94. Bozhkov, P.V.; Mikhlina, S.B.; Shiryaeva, G.A.; Lebedenko, L.A. Influence of Nitrogen Balance of Culture Medium on Norway Spruce [*Picea abies* (L.) Karst] Somatic Polyembryogenesis: High Frequency Establishment of Embryonal-Suspensor Mass Lines from Mature Zygotic Embryos. *J. Plant Physiol.* **1993**, *142*, 735–741. [CrossRef]

95. Barrett, J.D.; Park, Y.S.; Bonga, J.M. The Effectiveness of Various Nitrogen Sources in White Spruce [*Picea glauca* (Moench) Voss] Somatic Embryogenesis. *Plant Cell ReP.* **1997**, *16*, 411–415. [CrossRef] [PubMed]

96. Carlsson, J.; Svennerstam, H.; Moritz, T.; Egertsdotter, U.; Ganeteg, U. Nitrogen Uptake and Assimilation in Proliferating Embryogenic Cultures of Norway Spruce—Investigating the Specific Role of Glutamine. *PLoS ONE* **2017**, *12*, e0181785. [CrossRef] [PubMed]

97. Jin, F.; Hu, L.; Yuan, D.; Xu, J.; Gao, W.; He, L.; Yang, X.; Zhang, X. Comparative Transcriptome Analysis between Somatic Embryos (SEs) and Zygotic Embryos in Cotton: Evidence for Stress Response Functions in SE Development. *Plant Biotechnol. J.* **2014**, *12*, 161–173. [CrossRef]

98. Nic-Can, G.I.; Avilez-Montalvo, J.R.; Aviles-Montalvo, R.N.; Márquez-López, R.E.; Mellado-Mojica, E.; Galaz-Ávalos, R.M.; Loyola-Vargas, V.M. The Relationship between Stress and Somatic Embryogenesis. In *Somatic Embryogenesis: Fundamental Aspects and Applications*; Springer International Publishing: Cham, Switzerland, 2016; pp. 121–170.

99. Zhang, S.-G.; Han, S.-Y.; Yang, W.-H.; Wei, H.-L.; Zhang, M.; Qi, L.-W. Changes in H_2O_2 Content and Antioxidant Enzyme Gene Expression during the Somatic Embryogenesis of *Larix leptolepis*. *Plant Cell Tissue Organ Cult.* **2010**, *100*, 21–29. [CrossRef]

100. Salajova, T.; Salaj, J.; Kormutak, A. Initiation of Embryogenic Tissues and Plantlet Regeneration from Somatic Embryos of *Pinus nigra* Arn. *Plant Sci.* **1999**, *145*, 33–40. [CrossRef]

101. Kvaalen, H.; Appelgren, M. Light Quality Influences Germination, Root Growth and Hypocotyl Elongation in Somatic Embryos but Not in Seedlings of Norway Spruce. *Vitr. Cell. Dev. Biol. Plant* **1999**, *35*, 437–441. [CrossRef]

102. Latkowska, M.J.; Kvaalen, H.; Appelgren, M. Genotype Dependent Blue and Red Light Inhibition of the Proliferation of the Embryogenic Tissue of Norway Spruce. *Vitr. Cell. Dev. Biol. Plant* **2000**, *36*, 57–60. [CrossRef]
103. Merkle, S.A.; Montello, P.M.; Xia, X.; Upchurch, B.L.; Smith, D.R. Light Quality Treatments Enhance Somatic Seedling Production in Three Southern Pine Species. *Tree Physiol.* **2006**, *26*, 187–194. [CrossRef] [PubMed]
104. Montalbán, I.A.; Moncaleán, P. Rooting of *Pinus radiata* Somatic Embryos: Factors Involved in the Success of the Process. *J. For. Res.* **2019**, *1*, 65–71. [CrossRef]
105. Bonga, J.M.; Klimaszewska, K.K.; von Aderkas, P. Recalcitrance in Clonal Propagation, in Particular of Conifers. *Plant Cell Tissue Organ Cult. PCTOC* **2010**, *100*, 241–254. [CrossRef]
106. Belmonte, M.F.; Yeung, E.C. The Effects of Reduced and Oxidized Glutathione on White Spruce Somatic Embryogenesis. *Vitr. Cell. Dev. Biol. Plant* **2004**, *40*, 61–66. [CrossRef]
107. Pullman, G.S.; Zeng, X.; Copeland-Kamp, B.; Crockett, J.; Lucrezi, J.; May, S.W.; Bucalo, K. Conifer Somatic Embryogenesis: Improvements by Supplementation of Medium with Oxidation-Reduction Agents. *Tree Physiol.* **2015**, *35*, 209–224. [CrossRef] [PubMed]
108. Maruyama, T.E.; Hosoi, Y. Progress in Somatic Embryogenesis of Japanese Pines. *Front. Plant Sci.* **2019**, *10*, 31. [CrossRef]
109. Lara-Chavez, A.; Flinn, B.S.; Egertsdotter, U. Initiation of Somatic Embryogenesis from Immature Zygotic Embryos of Oocarpa Pine (*Pinus oocarpa* Schiede Ex Schlectendal). *Tree Physiol.* **2011**, *31*, 539–554. [CrossRef] [PubMed]
110. Liao, Y.K.; Amerson, H.V. Slash Pine (*Pinus elliottii* Engelm.) Somatic Embryogenesis I. Initiation of Embryogenic Cultures from Immature Zygotic Embryos. *New For.* **1995**, *10*, 145–163. [CrossRef]
111. Liao, Y.K.; Juan, I.-P. Improving the Germination of Somatic Embryos of *Picea morrisonicola* Hayata: Effects of Cold Storage and Partial Drying. *J. For. Res.* **2015**, *20*, 114–124. [CrossRef]
112. Pullman, G.S.; Olson, K.; Fischer, T.; Egertsdotter, U.; Frampton, J.; Bucalo, K. Fraser Fir Somatic Embryogenesis: High Frequency Initiation, Maintenance, Embryo Development, Germination and Cryopreservation. *New For.* **2016**, *47*, 453–480. [CrossRef]
113. Marin, J.A.; Gella, R.; Herrero, M. Stomatal Structure and Functioning as a Response to Environmental Changes in Acclimatized Micropropagated *Prunus cerasus* L. *Ann. Bot.* **1988**, *62*, 663–670.
114. Pospóšilová, J.; Tichá, I.; Kadleček, P.; Haisel, D.; Plzáková, Š. Acclimatization of Micropropagated Plants to Ex Vitro Conditions. *Biol. Plant.* **1999**, *42*, 481–497. [CrossRef]
115. Chandra, S.; Bandopadhyay, R.; Kumar, V.; Chandra, R. Acclimatization of Tissue Cultured Plantlets: From Laboratory to Land. *Biotechnol. Lett.* **2010**, *32*, 1199–1205. [CrossRef] [PubMed]
116. Ragonezi, C.; Caldeira, A.T.; Martins, M.R.; Dias, L.S.; Santos-Silva, C.; Ganhão, E.; Miralto, O.; Pereira, I.; Louro, R.; Klimaszewska, K.; et al. *Pisolithus arhizus* (Scop.) Rauschert Improves Growth of Adventitious Roots and Acclimatization of in vitro Regenerated Plantlets of *Pinus pinea* L. *Propag. Ornam. Plants* **2012**, *12*, 139–147.
117. Jayawickrama, K.J.S.; Carson, M.J. A Breeding Strategy for the New Zealand *Radiata* Pine Breeding Cooperative. *Silvae Genet. Ger.* **2000**, *49*, 82–89.
118. Dungey, H.S.; Brawner, J.T.; Burger, F.; Carson, M.; Henson, M.; Jefferson, P.; Matheson, A.C. A New Breeding Strategy for *Pinus Radiata* in New Zealand and New South Wales. *Silvae Genet.* **2009**, *58*, 28–38. [CrossRef]
119. Weng, Y.H.; Park, Y.S.; Krasowski, M.J.; Mullin, T.J. Allocation of Varietal Testing Efforst for Implementing Conifer Multi-Varietal Forestry Using White Spruce as a Model Species. *Ann. For. Sci.* **2011**, *68*, 129–138. [CrossRef]
120. Park, Y.-S. Implementation of Conifer Somatic Embryogenesis in Clonal Forestry: Technical Requirements and Deployment Considerations. *Ann. For. Sci.* **2002**, *59*, 651–656. [CrossRef]
121. Plomion, C.; Bastien, C.; Bogeat-Triboulot, M.-B.; Bouffier, L.; Déjardin, A.; Duplessis, S.; Fady, B.; Heuertz, M.; Le Gac, A.-L.; Le Provost, G.; et al. Forest Tree Genomics: 10 Achievements from the Past 10 Years and Future Prospects. *Ann. For. Sci.* **2016**, *73*, 77–103. [CrossRef]
122. Baltunis, B.S.; Wu, H.X.; Dungey, H.S.; Mullin, T.J.; Brawner, J.T. Comparisons of Genetic Parameters and Clonal Value Predictions from Clonal Trials and Seedling Base Population Trials of *Radiata* Pine. *Tree Genet. Genomes* **2009**, *5*, 269–278. [CrossRef]
123. Danell, Ö. Breeding Programmes in Sweden. General Approach. In Proceedings of the Progeny Testing and Breeding Strategies. Proceedings from a Meeting with the Nordic Group for Tree Breeding, Edinburg, UK, 6–10 October 1993.
124. Rosvall, O. Review of the Swedish Tree Breeding Program. *Skogforsk Upps. Swed.* **2011**, 4–37.
125. Jansson, G.; Hansen, J.K.; Haapanen, M.; Kvaalen, H.; Steffenrem, A. The Genetic and Economic Gains from Forest Tree Breeding Programmes in Scandinavia and Finland. *Scand. J. For. Res.* **2017**, *32*, 273–286. [CrossRef]
126. Find, J.I.; Charity, J.A.; Grace, L.J.; Kristensen, M.M.M.H.; Krogstrup, P.; Walter, C. Stable genetic transformation of embryogenic cultures of *Abies nordmanniana* (Nordmann fir) and regeneration of transgenic plants. *In vitro Cell. Dev. Biol. Plant* **2005**, *41*, 725–730. [CrossRef]
127. Giri, C.C.; Shyamkumar, B.; Anjaneyulu, C. Progress in tissue culture, genetic transformation and applications of biotechnology to trees: An overview. *Trees Struct. Funct.* **2004**, *18*, 115–135. [CrossRef]
128. Sun, Z.-L.; Li, X.; Zhou, W.; Yan, J.-D.; Gao, Y.-R.; Li, X.-W.; Sun, J.-C.; Fang, K.-F.; Zhang, Q.; Xing, Y.; et al. *Agrobacterium*-mediated genetic transformation of Chinese chestnut (*Castanea mollissima* Blume). *Plant Cell Tissue Organ Cult.* **2020**, *140*, 95–103. [CrossRef]
129. Tereso, S.; Miguel, C.; Zoglauer, K.; Valle-Piquera, C.; Oliveira, M.M. Stable *Agrobacterium*-mediated transformation of embryogenic tissues from *Pinus pinaster* Portuguese genotypes. *Plant Growth Regul.* **2006**, *50*, 57–68. [CrossRef]

130. Teixeira da Silva, J.A.; Kher, M.M.; Soner, D.; Page, T.; Zhang, X.; Nataraj, M.; Ma, G. Sandalwood: Basic biology, tissue culture, and genetic transformation. *Planta* **2016**, *243*, 847–887. [CrossRef] [PubMed]

131. McGuigan, L.; Fernandes, P.; Oakes, A.; Stewart, K.; Powell, W. Transformation of American chestnut (*Castanea dentata* (marsh.) borkh) using Rita® temporary immersion bioreactors and in vitro containers. *Forests* **2020**, *11*, 1196. [CrossRef]

132. Klimaszewska, K.; Lachance, D.; Pelletier, G.; Lelu, M.-A.; Séguin, A. Regeneration of transgenic *Picea glauca*, *P. mariana*, and *P. abies* after cocultivation of embryogenic tissue with *Agrobacterium tumefaciens*. *In Vitro Cell.-Dev. Biol.-Plant* **2001**, *37*, 748–755. [CrossRef]

133. Wenck, R.A.; Quinn, M.; Whetten, R.W.; Pullman, G.; Sederoff, R. High-efficiency *Agrobacterium*-mediated transformation of Norway spruce (*Picea abies*) and loblolly pine (*Pinus taeda*). *Plant Mol. Biol.* **1999**, *39*, 407–416. [CrossRef]

134. Walter, C.; Grace, L.J.; Donaldson, S.S.; Moody, J.; Gemmell, J.E.; van der Maas, S.; Kvaalen, H.; Lönneborg, A. An efficient Biolistic® transformation protocol for *Picea abies* embryogenic tissue and regeneration of transgenic plants. *Can. J. For. Res.* **1999**, *29*, 1539–1546. [CrossRef]

135. Pavingerová, D.; Bříza, J.; Niedermeierová, H.; Vlasák, J. Stable *Agrobacterium*-mediated transformation of Norway spruce embryogenic tissues using somatic embryo explants. *J. For. Sci.* **2011**, *57*, 277–280. [CrossRef]

136. Bříza, J.; Pavingerová, D.; Vlasák, J.; Niedermeierová, H. Norway spruce (*Picea abies*) genetic transformation with modified Cry3A gene of *Bacillus thuringiensis*. *Acta Biochim. Pol.* **2013**, *60*, 395–400. [CrossRef]

137. Wadenbäck, J.; Von Arnold, S.; Egertsdotter, U.; Walter, M.H.; Grima-Pettenati, J.; Goffner, D.; Gellerstedt, G.; Gullion, T.; Clapham, D. Lignin biosynthesis in transgenic Norway spruce plants harboring an antisense construct for cinnamoyl CoA reductase (CCR). *Transgenic Res.* **2008**, *17*, 379–392. [CrossRef]

138. Häggman, H.M.; Aronen, T.S.; Nikkanen, T.O. Gene transfer by particle bombardment to Norway spruce and Scots pine pollen. *Can. J. For. Res.* **1997**, *27*, 928–935. [CrossRef]

139. Aronen, T.S.; Nikkanen, T.O.; Häggman, H.M. The Production of Transgenic Scots Pine (*Pinus sylvestris* L.) via the Application of Transformed Pollen in Controlled Crossings. *Transgenic Res.* **2003**, *12*, 375–378. [CrossRef] [PubMed]

140. Grace, L.J.; Charity, J.A.; Gresham, B.; Kay, N.; Walter, C. Insect-resistant transgenic *Pinus radiata*. *Plant Cell Rep.* **2005**, *24*, 103–111. [CrossRef]

141. Alvarez, J.M.; Ordás, R.J. Stable *Agrobacterium* -mediated transformation of maritime pine based on kanamycin selection. *Sci. World J.* **2013**, *2013*, 681792. [CrossRef] [PubMed]

142. Mamun, N.H.A.; Egertsdotter, U.; Aidun, C.K. Bioreactor Technology for Clonal Propagation of Plants and Metabolite Production. *Front. Biol.* **2015**, *10*, 177–193. [CrossRef]

143. Egertsdotter, U.; Ahmad, I.; Clapham, D. Automation and Scale Up of Somatic Embryogenesis for Commercial Plant Production, With Emphasis on Conifers. *Front. Plant Sci.* **2019**, *10*, 109. [CrossRef] [PubMed]

144. Aidun, C.K.; Egertsdotter, U. SE Fluidics System. In *Step Wise Protocols for Somatic Embryogenesis of Important Woody Plants*.; Jain, S.M., Gupta, P., Eds.; Springer International Publishing: Berlin/Heidelberg, Germany, 2018; Volume I, pp. 211–227.

145. Businge, E.; Trifonova, A.; Schneider, C.; Rödel, P.; Egertsdotter, U. Evaluation of a New Temporary Immersion Bioreactor System for Micropropagation of Cultivars of Eucalyptus, Birch and Fir. *Forests* **2017**, *8*, 196. [CrossRef]

146. Salonen, F.; Varis, S.; Aronen, T.S. From Petri Dishes to Bioreactors—First Experiences on Optimization of Norway Spruce SE-Process for Bioreactors. In Proceedings of the 4th International Conference of the IUFRO Unit 2.09.02. on Development and Application of Vegetative Propagation Technologies in Plantation Forestry to Cope with a Changing Climate and Environment, La Plata, Argentina, 19–23 September 2016; pp. 293–297.

147. Välimäki, S.; Paavilainen, L.; Tikkinen, M.; Salonen, F.; Varis, S.; Aronen, T. Production of Norway Spruce Embryos in a Temporary Immersion System (TIS). *Vitr. Cell. Dev. Biol. Plant* **2020**, *56*, 430–439. [CrossRef]

148. Yoeup, P.K.; Chakrabarty, D. Micropropagation of Woody Plants Using Bioreactor. In *Micropropagation of Woody Trees and Fruits*; Jain, S.M., Ishii, K., Eds.; Forestry Sciences; Springer Netherlands: Dordrecht, The Netherlands, 2003; pp. 735–755, ISBN 978-94-010-0125-0.

149. Mamun, N.H.A.; Aidun, C.K.; Egertsdotter, U. Improved and Synchronized Maturation of Norway Spruce (*Picea abies* (L.) H. Karst.) Somatic Embryos in Temporary Immersion Bioreactors. *Vitr. Cell. Dev. Biol. Plant* **2018**, *54*, 612–620. [CrossRef]

150. Le, K.-C.; Weerasekara, A.B.; Ranade, S.S.; Egertsdotter, E.-M.U. Evaluation of Parameters to Characterize Germination-Competent Mature Somatic Embryos of Norway Spruce (*Picea abies*). *Biosyst. Eng.* **2021**, *203*, 55–59. [CrossRef]

Article

How to Capture Thousands of Genotypes—Initiation of Somatic Embryogenesis in Norway Spruce

Saila Varis *, Mikko Tikkinen, Jaanika Edesi and Tuija Aronen

Natural Resources Institute Finland (Luke), Vipusenkuja 5, FI-57200 Savonlinna, Finland;
mikko.tikkinen@luke.fi (M.T.); jaanika.edesi@luke.fi (J.E.)
* Correspondence: saila.varis@luke.fi

Abstract: Somatic embryogenesis (SE) is considered the most effective method for vegetative propagation of Norway spruce (*Picea abies* L. Karst). When the aim is commercial production, the process needs scaling up. This includes many initiations to increase the number of available genotypes in the cryo-bank. Numerous genotypes are needed to maintain genetic diversity in reforestation and, at the same time, are a prerequisite for the efficient improvement of breeding traits. Norway spruce is also highly susceptible to Heterobasidion root rot. We analysed the data from the SE initiations of Norway spruce from six different years, including a total of 126 families and almost 13,000 initiations, and used several genetic (including allele PaLAR3B improving Heterobasidion resistance), environmental, and operational variables to explain the initiation success and the number of cryopreserved embryogenic tissue (ET). Overall, the cone collection date was the best and most comprehensive single variable for predicting the initiation success and the number of cryopreserved ET in the logistic regression models. PaLAR3B allele did not interfere with SE initiation or the cryopreservation. In the optimal scenario, according to the current data, Norway spruce cones would be collected in southern Finland during the first two weeks of July (in approximately 800 d.d. accumulation) from the seed orchard or greenhouse and delivered quickly to the laboratory, and the cones would be cold-stored for five days or less before initiations on mLM media. Lower initiation frequencies in some families can be compensated by increasing the number of explants—however, taking operational limitations into account.

Keywords: cryo-bank; embryogenic tissue; *Picea abies*; vegetative propagation

Citation: Varis, S.; Tikkinen, M.; Edesi, J.; Aronen, T. How to Capture Thousands of Genotypes—Initiation of Somatic Embryogenesis in Norway Spruce. *Forests* **2023**, *14*, 810. https://doi.org/10.3390/f14040810

Academic Editors: Ricardo Javier Ordás and José Manuel Álvarez Díaz

Received: 14 March 2023
Revised: 11 April 2023
Accepted: 13 April 2023
Published: 14 April 2023

1. Introduction

To increase forest growth and, thus, increase the possibility of their sustainable use, the best possible regeneration material, i.e., superior tree genotypes, should be used [1]. To ensure the availability of good-quality forest regeneration material, effective vegetative propagation methods such as somatic embryogenesis (SE) were introduced [2]. SE must provide improved genetic gain to be utilised in forestry and breeding, i.e., SE plants must be proven to be superior to unimproved seedlings (at least) to support the costs of production [3]. Genetic gain can be proven by field testing [4].

From the practical perspective, establishing field trials with SE plants requires numerous genotypes per family to ensure the possibility of genotype selection and to avoid reduced genetic gain [4–7]. To obtain the required number of plants for field testing, well-performing cell lines should be initiated, i.e., lines that have vigorous proliferation, survive cryopreservation, and produce a large number of somatic embryos after cryopreservation [5,6]. Moreover, embryos must germinate and perform well in the nursery. For this, many initiations, 60 to 80 explants per family, from several families must be undertaken [7]. These preceding requirements regarding field testing combined with irregular flowering of Norway spruce in Finland lay out the framework for the magnitude of annual initiations for thousands rather than hundreds [8]. Labour and financial resources are often limited and to optimise successful SE propagation, variables affecting initiation should be carefully analysed.

Genetic variation in the outcome of SE initiations was addressed in white spruce (*Picea glauca*) [9,10], loblolly pine (*Pinus taeda*) [11], and black spruce (*Picea mariana*) [12,13]. Park et al. [9] found that the initiation phase of SE was under strong genetic control, but genetic influence declined steadily through the proliferation, maturation, and germination phases [2]. In Norway spruce, differences in the initiation rates and establishment of embryogenic cultures were reported among families [14]. However, to estimate genetic parameters for the initiation capacity of a population, much larger samples are needed than those reported above [13].

SE propagation enables the efficient propagation of selected genotypes with desired traits. Norway spruce is highly susceptible to Heterobasidion root rot, and the pressure of disease is predicted to increase with ongoing climate change [15]. Therefore, an effort was made to identify more root-rot-tolerant genotypes through genome-based tools as marker-assisted selection [16,17]. It was shown that in the presence of leucoanthocyanidin reductase, PaLAR3B allele root rot spread slowed down by 27% [16]. In a previous study, we showed that the resistance allele Palar3B is successfully delivered through the SE process [18]. The scale of the present study enables us to study whether this allele interferes with the SE initiation and cryopreservation process.

The developmental stage of explants had a significant effect on the initiation of Norway spruce [19], black spruce [20], and white spruce [9]. The optimal stage depends on the species. For example, for maritime pine (*Pinus pinaster*), the precotyledonary stage of embryo development was the most responsive [21], and for jack pine (*Pinus banksiana*), only embryos at the polyembryonic cleavage stage will respond [22]. The initiation rate gradually diminishes as the zygotic embryo matures, although at the earliest stage, immature ZEs may not be capable of forming embryogenic tissue (ET) [2,9,19]. Various environmental factors and genetic variability affect ZE development, making it difficult to predict the right moment for cone collection.

Due to limited labour resources, immature cones are usually cold-stored before preparations. However, cold storage may reduce initiation potential remarkably, as observed in Park's study [9]. Another storage option is to keep sterilised immature seeds in sterile water in the cold prior to initiation, but, to our knowledge, there are no previous studies about the effect of this.

Many studies were conducted on optimising SE initiation media for different species [23,24]. In spruce, the media for SE are usually based either on those described by von Arnold and Eriksson (LP) [25], Litvay et al. (LM) [26], or Murashige-Skoog medium (MS) [27]. Modifications were made, for example, by reducing the amount of macroelements [14,28]. Media are supplemented with organic nitrogen (L-glutamine and casein hydrolysate), and typically contain 2,4-dichlorophenoxyacetic acid (2,4-D) at 9.0–10 M, benzyl adenine (BA) at 4.5–5.0 M, sucrose at 1 or 2%, and are solidified with agar (0.8%) or gellan gum (0.4%) [28]. In some cases, proliferation media may need changes especially in the concentrations of plant growth regulators [23,29], but media of the same composition are usually used [28].

Proliferating ET should be cryopreserved when enough culture is grown to avoid losing ET's embryogenic potential and save resources [14]. The criteria for cryopreservation should be the growth rate to avoid the delay caused by waiting for the results of embryo productivity testing [14,30].

In this study, we utilise the extensive data from SE initiations of Norway spruce made in several years to identify the genetic, environmental, and operational variables affecting initiation and proliferation up to cryostorage. Several studies were previously conducted on factors affecting the SE success of Norway spruce and other coniferous species, but this study covers families and initiations among families more than previous ones. Compared to other studies, this unique dataset is also extended over a decade. To our knowledge, this is the first study in which such extensive SE initiation and cryopreservation data are analysed for root rot tolerance related to the genetic component for resistance breeding. Our findings will help optimise the successful initiation and cryopreservation of Norway spruce using the SE technique.

2. Materials and Methods

2.1. Plant Material

Embryogenic lines of Norway spruce were initiated from immature and mature seed embryos originating from the controlled crossings of a Finnish tree breeding programme (Supplementary Table S1). Crossings were made in 2011, 2012, 2014, 2015, 2019, and 2022 using seed orchard grafts in southern Finland (60°55′ N, 26°13′ E, 80 m, 60°41′ N, 24°02′ E, 130 m) or grafted trees in a greenhouse. Grafted trees and pollen donors originated from different locations in southern and central Finland. From 2011 to 2015, the plus trees used in the SE were chosen based on the flowering of the grafts and the sufficiency of cones. In 2019, the measured progeny testing results (unpublished data) were available for selection [31]. In 2022, the plus trees were genotyped and chosen based on the root rot resistance marker PaLAR3 alleles [16,18]. Over the years, a total of 126 crossing combinations (families) were made, using 69 mother and 85 father genotypes. Three combinations were repeated in two different years, and 14 mothers and 13 fathers were used in more than one year.

Controlled crossings were conducted on 7 May at the earliest in 2019, and on 17 May at the latest in 2011 and 2014. Immature cones were collected at the earliest on 24 June in the summer (2019) and at the latest on 9 August (2012). Mature cones were collected in November or December (2011 and 2012).

Days between the first pollination day and cone collection were used to describe the development of ZE. In addition, the temperature sum accumulation in degree days (d.d.) was used to describe the timing of cone collection. The d.d. was calculated by summing the daily mean temperatures exceeding the threshold value of 5 °C from 1 January of that year. Temperature sums were available from 2011, 2012, 2014 (seed orchards), and 2019 (greenhouse).

2.2. Initiation, Proliferation, and Cryopreservation

Embryogenic lines were initiated and proliferated by applying the methods developed by Klimaszewska et al. [32] and Lelu-Walter et al. [33] and described in Varis et al. [34]. In short, immature cones were wiped with 70% ethanol, seeds were dissected in sterile water, and washing-up liquid (15%–30% anionic surfactants, 5%–15% non-ionic surfactants, benzisothiazolinone, methylisothiazolinone, phenoxyethanol, perfumes) was added to clean the seeds. After one rinse in sterile water, the seeds were surface-sterilised in 70% ethanol for 5 min and rinsed three times in sterilised water. The ZEs were dissected from megagametophytes and placed on modified Litvay's medium (mLM) [26,32] containing 10 µM 2,4-dicholophenoxyacetic acid (2,4-D) and 5 µM 6-benzyladenine (BA) as plant growth regulators (Supplementary Table S2). The sucrose concentration of the medium was 0.03 M. The pH of the medium was adjusted to 5.8 prior to adding gelling agent (Phytagel 4 g/L) and sterilisation in the autoclave. After autoclaving at 120 degrees for 20 min, the medium was cooled to 60 °C, and 500 mg/L of L-glutamine was added using filter sterilising.

In 2011, half the initiations were placed on 1/2LP medium [25], modified according to [14], containing 9 µM 2.4-D and 4.4 µM BA (Supplementary Table S2). In both cases, Petri dishes (9 cm in diameter) were filled with 20 mL of the medium. Ten ZEs were placed on the same Petri dish, and they were kept in the dark and at 24 °C. The ZEs were kept on the same media until ET started to grow and then transferred to a fresh medium, each genotype on a separate dish. The established ETs were subcultured bi-weekly. The proliferation media was the same as that used for culture initiation.

Embryogenic lines were chosen for cryopreservation based on their embryo productivity (2011 and 2012) or good growth in proliferation (2014 and since). In 2011, the two-day pre-treatment method in liquid mLM media described by Kartha [35] and modified by

Find et al. [36] was applied to cryopreserve six samples from every 260 cell lines. From 2012 to 2019, a two-day pre-treatment method in semi-solid media was used as described in Varis et al. [34], and in 2022, the pre-treatment method was simplified to a one-day dehydration in semi-solid media [37]. After pre-treatment, 1.5 mL suspended ET was placed in a 2 mL sterile cryovial, or 200 mg of ET from semi-solid media was placed in sterile cryovials containing 400 mL of liquid mLM medium with 0.4 M of sucrose without plant growth regulators (PGR) or glutamine, after which 400 mL of preservative mixture containing polyethylene glycol 6000, glucose, and DMSO 10% w/v each were added. The cryovials were cooled at the rate of 0.17 °C/min to −38 °C in a programmable cooling device (Planer, Kryo 10 Series III, Planer Products, Middlesex, UK), followed by immersion in LN.

2.3. Cone and Seed Cold Storage

Because of the limited human resources or for experimental reasons, some of the cones were stored in a cold room at +2 °C. The cones were in paper bags and stored from 1 to 38 days. For the same reason, sterilised seeds were left in the last sterile rinsing water and stored in sterile water in a refrigerator at +4 °C from 1 to 22 days.

When the effect of storing cones in the cold room was analysed, initiations in which the seed was also stored in cold water were not used. Because there was an unbalanced number of families between cone storing days, the data were merged into groups of five days. When the effect of seed storage in cold water was described, only data in which cones were cold-stored for one to seven days were used.

2.4. Statistical Analyses

The initiation was defined as successful when the ET was growing and could be separated from the ZE and moved to its own medium. Each initiation was counted as successful (1) or unsuccessful (0). Cryopreserved ET is the number of initiations which ended up in cryopreservation. The variables which were investigated reflecting initiation and cryopreservation success were categorised in three groups of variables: genetic; environmental; and operational. The genetic variables investigated were: full-sib family, father tree, mother tree, PaLAR3 cross type PaLAR3 father, PaLAR3 mother. Because mother and father PaLAR3 genotypes AA, AB, and BB were available in different years (Supplementary Table S1), the PaLAR3 cross types of the mother × father were AA × AA, AA × AB, AB × AA, AB × AB, AB × BB, i.e., BB × BB was missing.

The environmental variables were year of initiation, pollination date, cone collection d.d. and date, and time period from pollination to cone collection and to initiation. The operational variables tested were location of the mother graft (in a seed orchard or in a greenhouse), mature or immature ZEs, duration of cone or seed cold storage before initiation, initiation date, and initiation media lot.

The data in which $\frac{1}{2}$LP was used as an initiation medium were included only when the initiation rate between different media was analysed but excluded from further analysis. Cryopreservation data from 2011 and 2012 was excluded from the analysis because ETs for cryopreservation were selected based on embryo production capacity, i.e., following maturation experiments, and the material was, therefore, not comparable with the ETs selected for their vigorous proliferation as in later years.

Differences in initiation and cryopreservation success between seed orchard and greenhouse grafts, initiation success between mature and immature seeds, and the effect of the PaLAR3 genotype were analysed by the Pearson chi-square test. The Pearson correlation test was used to analyse the correlation between the number of initiations in the full-sib family and initiation success %.

The cryopreservation percentages were calculated as cryopreserved ET/initiated explants × 100 if not mentioned otherwise. Binary logistic regression ($log(p/1 - p)$) modelling was used to estimate the effects of genetic, environmental, and operational factors on the initiation success and cryopreserved ET, because they are 0/1 data. Variables were added to the model if values were available throughout the data (although several were included with missing values, which was considered in the selection of final models and in comparison of variables).

In the final models, only cone collection date was used as an experimental variable, as it was the best (and most comprehensive) single variable predicting initiation and cryopreservation success in the overall data. When more than one variable was accepted for the model, the best model response was achieved including full-sib family as a genetic variable and the initiation date as an operational variable for the models (Supplementary Table S3).

All the results were analysed using the IBM SPSS Statistics 28 software package (International Business Machines Corporation, Ammonk, NY, USA). The level of confidence used was 5%. A Sankey diagram was plotted by ggplot2 [38] and using the ggsankey package by David Sjöberg in R [39].

3. Results

3.1. Genetic Variables in Initiation and Cryopreserved ET

Embryogenic cell lines were obtained from 118 full-sib families out of the 126 (Supplementary Table S1). ET initiation % varied from 0 to 100 among full-sib families (Supplementary Table S1), and the mean ET initiation % of all 12,907 initiations was 59.8% (Figure 1a). The ET initiation % of the full-sib family was positively correlated with the number of initiations made (Pearson correlation 0.233, $p < 0.01$). Not all ET continued growing vigorously, and in 2014–2022, 46.9% of the original initiations ended up in cryopreservation (=81.2% of the succeeded initiations, i.e., growing cell lines) (Figure 1b).

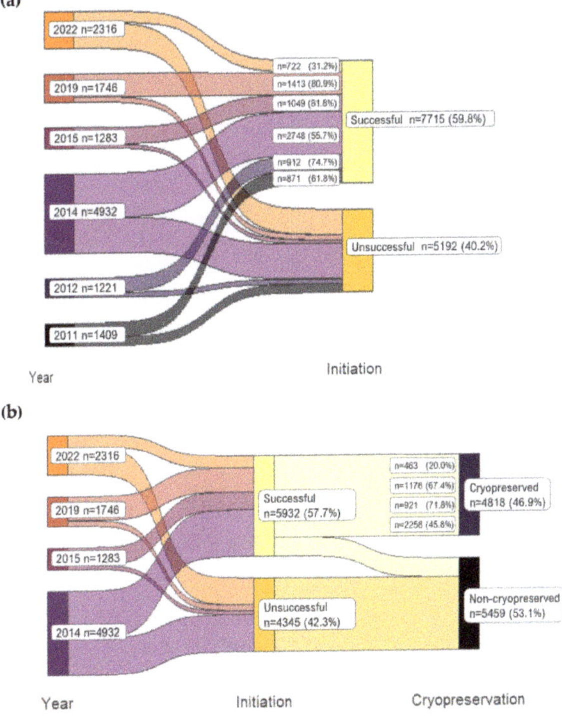

Figure 1. Sankey diagram of Norway spruce initiation success (**a**) and cryopreserved ET (**b**) percentages in different years. n = the number of initiations.

In the logistic regression model, the full-sib family, father tree, and mother tree were significant ($p < 0.01$, in all). The full-sib family improved the model most by increasing the overall % of cases predicted correctly to 69.3% (improvement in the model 9.5 units of %). The father tree and mother tree resulted in 68% (improvement in the model 8.2 units of %) and 65.2% (improvement in the model 5.4 units of %) of cases predicted correctly, respectively.

The effect of the PaLAR3 cross type, PaLAR3 father and PaLAR3 mother, was significant ($p < 0.01$, in all both in initiation and cryopreservation). However, the improvements in the initiation model were only 1.5, 0.0, and 0.2 units of %, respectively. The effect of PaLAR3 variables was similar in the model of cryopreserved ET. The PaLAR3 cross type, PaLAR3 father and PaLAR3 mother, improved the model by 8.4, 4.8, and 6.1 units of %. Additionally, when PaLAR3 variables were included in the model with the full-sib family, the PaLAR3 variables did not improve the model at all.

The effect of PaLAR3 genotype was further studied using the chi-squared test of independence. The PaLAR3 cross type and allele frequency of root rot resistance-related B-allele had a significant effect on the initiation and cryopreserved ET when the data from all years were pooled. However, when these variables were tested for each year separately, the PaLAR3 genotype had no effect on the initiation and cryopreserved ET result.

3.2. Environmental Variables in Initiation and Cryopreserved ET

The highest percentage of cases predicted correctly in initiation was found with cone collection d.d. (71.6%, improving the cases predicted correctly by 7.8 units of %) (Figure 2). However, there were missing data in cone collection d.d. (4556 cases of the overall 12,907), making the variable not comprehensive and not fully comparable with other variables. In cryopreserved ET, the situation was similar in the cone collection d.d. (64.7% of cases predicted correctly, improving the cases predicted correctly by 13.6 units of %).

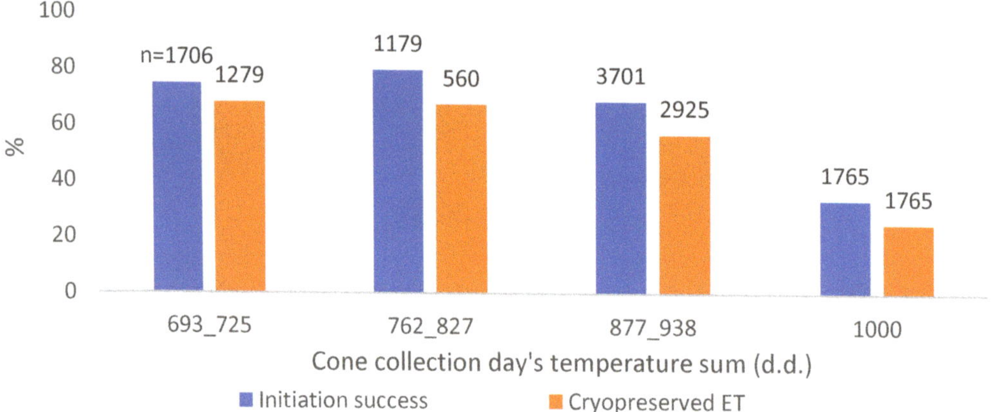

Figure 2. Norway spruce SE initiation success percentages from 2011 to 2022, and the percentage of initiations which led to cryopreservation from 2014 to 2022 in different cone collection days' temperature sums (d.d.).

Cone collection date was the best and most comprehensive single variable for predicting the initiation and cryopreserved ET in the logistic regression models (71.0% and 69.1%, respectively) and predicted correctly by 11.2 and 16.0 units of %, respectively (Table 1, Figure 3).

Table 1. Logistic regression models used to analyse binary response in the initiation (proliferating ET or not) and cryopreservation (cryopreserved or not). In the models, a1 to a15 are design variables for cone collection date.

Dependent Variable	Model, $log(p/1-p)$	Variable	p-Value	Odds Ratio (95% CI)	Date	% of Cases Predicted Correctly by Model
Initiation	$log(p/1-p) = 1.160 + 0.321a_1 - 0.045a_2 - 1.038a_3 + 0.340a_4 - 1.349a_5 - 0.349a_6 + 159a_7 + 0.287a_8 - 0.566a_9 - 2.183a_{10} - 2.474a_{11} - 0.426a_{12} - 1.816a_{13} - 0.049a_{14}$					71.0
		Collection date	<0.001	1	24 June	
				1.378 (1.043–1.821)	1 July	
				0.956 (0.760–1.204)	8 July	
				0.354 (0.275–0.457)	12 July	
				1.406 (1.116–1.77)	13 July	
				0.259 (0.207–0.325)	15 July	
				0.674 (0.519–0.876)	18 July	
				1.172 (0.915–1.501)	21 July	
				1.332 (0.916–1.939)	24 July	
				0.568 (0.433–0.744)	25 July	
				0.113 (0.085–0.15)	26 July	
				0.084 (0.067–0.106)	28 July	
				0.653 (0.535–0.797)	31 July	
				0.163 (0.132–0.200)	5 August	
				0.952 (0.594–1.525)	9 August	
Cryopreservation	$log(p/1-p) = 0.553 + 0.209a_1 + 0.347a_2 + 0.380a_4 - 1.240a_5 + 0.250a_7 - 2.125a_{10} - 2.743a_{11} - 0.408a_{12} - 1.621a_{13}$					69.1
		Collection date	<0.001	1	24 June	
				1.232 (0.971–1.564)	1 July	
				1.415 (1.105–1.811)	8 July	
				1.463 (1.196–1.789)	13 July	
				0.289 (0.234–0.359)	15 July	
				1.284 (1.032–1.599)	21 July	
				0.119 (0.088–0.162)	26 July	
				0.064 (0.050–0.083)	28 July	
				0.665 (0.555–0.796)	31 July	
				0.198 (0.163–0.240)	5 August	

The pollination date was found to have a significant effect on the initiation and cryopreserved ET ($p < 0.01$ for both in the logistic regression model). Including pollination date in the model resulted in the logistic regression model correctly predicting 63.9 and 64.9% percent of cases in initiation and cryopreserved ET. The improvement in the model was 5.7 and 11.8 units of % in predicting the initiation and cryopreservation success, respectively.

The duration between pollination and cone collection, i.e., cone development duration, had a significant effect on initiation and cryopreserved ET ($p < 0.01$ for both in the logistic regression model) (Figure 4). It improved the cases predicted correctly by the model by 8.9 and 12.7 units of % to 66.9 and 66.8% in initiation and cryopreserved ET, respectively.

The period from the pollination of flowers to initiation had a significant effect on the initiation and cryopreserved ET ($p < 0.01$ for both in the logistic regression model). It improved the cases predicted correctly by the model by 10.3 and 12.7 units of % to 68.4 and 66.8% in initiation and cryopreserved ET, respectively.

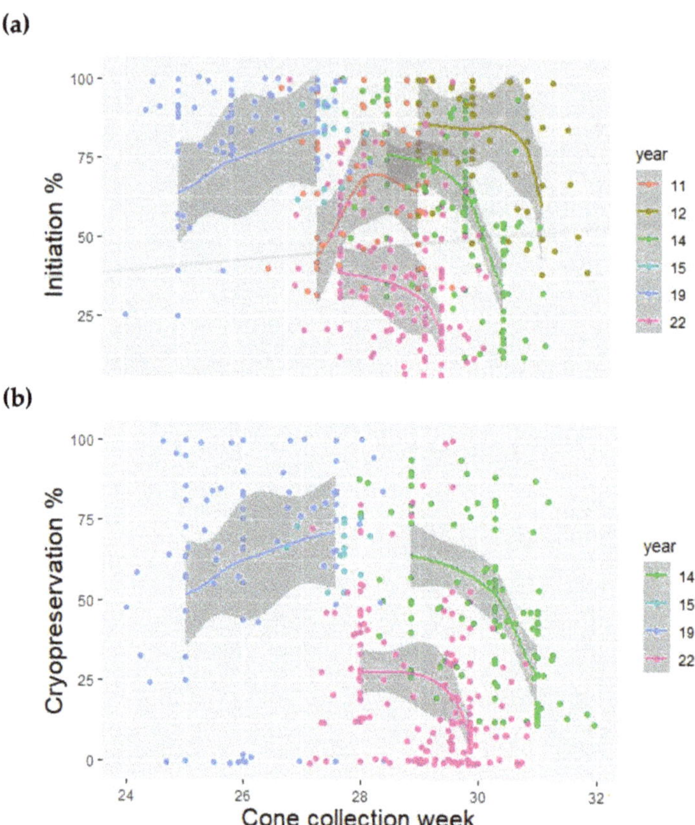

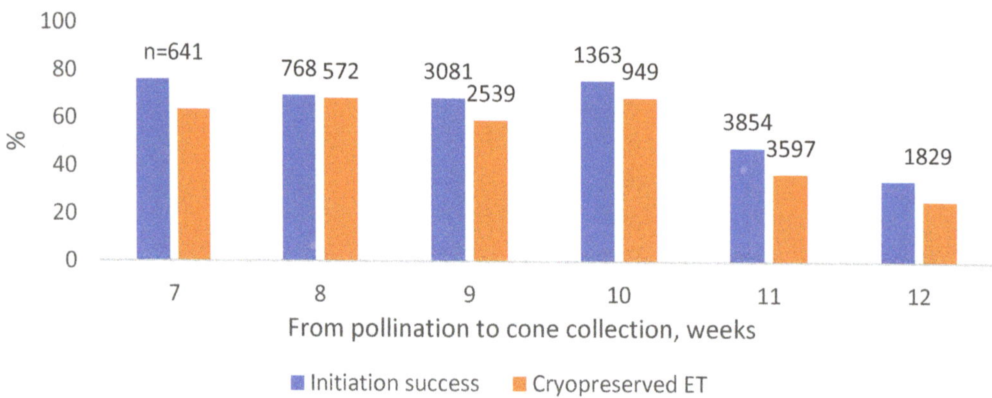

Figure 3. The effect of cone collection date on initiation (**a**) and cryopreserved ET (**b**) of Norway spruce SE cultures. Each dot represents the mean initiation or cryopreserved ET percentage of each family according to the cone collection date.

Figure 4. Norway spruce SE initiation success percentages from 2011 to 2022, and the percentage of initiations which led to cryopreservation from 2014 to 2022 in different cone development duration.

In 2022, the mean initiation success % was lowest, 31.2%, and in 2015, it was highest, 81.8% (Figure 1a). In 2022, the cryopreserved ET was also low, when only 20.2% of initiations led to a cryopreserved cell line (Figure 1b), which was 64.1% of the initiations that started to grow in the first place. In the best year, 2015, 71.8% of the initiations led to cryopreservation, which was 87.8% of successful initiations. The initiation year had a significant effect on initiation and cryopreserved ET in the logistic regression model ($p < 0.01$ for both). In the model explaining initiation success, initiation year improved the cases predicted correctly by 6.7 units of % to 66.5, and in the cryopreservation model, 11.4 units to 64.5%.

3.3. Operational Variables in Initiation and Cryopreserved ET

When mother trees were grafts in the seed orchard, 59.2% of the initiations started to grow ET (n = 6605 initiations), which is the same as when grafts were potted in the greenhouse (n = 6302 initiations). From seed orchards, 45.2% of the original explants, i.e., 81.9% of the initiated ET lines were cryopreserved (in 2014, n = 4783 initiations), and from the greenhouse grafts, the corresponding cryopreserved ET % was 65.8 of the original explants, i.e., 87.5% of the initiated ETs (n = 149 initiations). The difference in cryopreserved ET % was significant: $\chi^2 = 10.641, p = 0.001$.

Mature (468 initiations) and immature (1891 initiations) seeds were used as a source of ZEs from 17 families in 2011 and 2012. When mature ZEs were used, 47.4% started to grow ET, and when immature ZEs were used, the ET initiation % was 69.1 ($\chi^2 = 76.909, p < 0.01$).

The initiation date had a significant effect on initiation and cryopreserved ET in the logistic regression model ($p < 0.01$ in both). When the initiation date was included in the model, the cases predicted correctly rose to 68.0 (8.2 units of % increment) and 68.5% (15.4 units of % increment).

In 2011, the ET initiation % from 1/2LP media was 16.2%, and from mLM media 61.8%. After the first year, only mLM media was used in the SE initiations of Norway spruce. The initiation media (mLM) lot also had a significant effect on the initiation and cryopreserved ET ($p < 0.01$ in both). When the initiation media lot was added to the model, the cases predicted correctly rose to 69.4 (6 units of % increment) and 66.0 (14.7 units of % increment).

When cones were cold-stored for one to seven days, and initiations were made immediately afterwards, the ET initiation % was 73.8, and 70.4% of the original initiations led to cryopreservation between 2014 and 2022 (Figure 5a). When cones were stored for more than one week, the ET initiation % was 50.7, and 39.1% of the original initiations led to cryopreservation (2014 to 2022). When both cones and seeds were stored for one to seven days, the ET initiation % was 73.6, and 65.5% of the original initiations led to cryopreservation (2014 to 2022), and if seeds were stored for more than one week, 55.3% of the initiations started to grow ET (Figure 5b).

When the cold storage of cones or seeds was added to the logistic regression model, it was found to have a significant effect on initiation and cryopreserved ET ($p < 0.01$ for all in the logistic regression model). However, the cold storage of cones (65.0 and 64.3% cases predicted correctly with an increment of 4.8 and 11.9 units of % in initiation and cryopreserved ET, respectively) improved the model much more than the cold storage of seeds (54.7 and 57.4% cases predicted correctly with an increment of 0.4 and 1 units of % in initiation and cryopreserved ET, respectively).

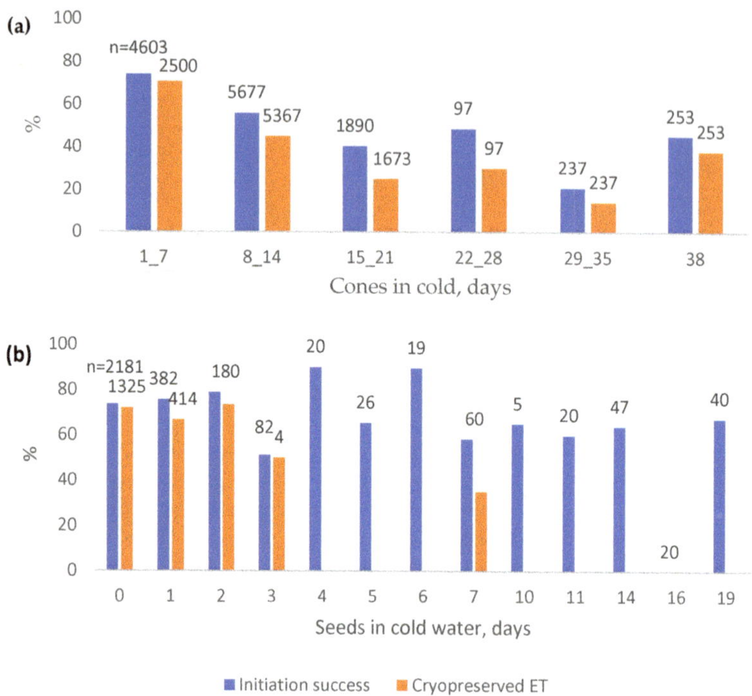

Figure 5. Norway spruce initiation success percentages from 2011 to 2022 and the percentage of initiations which led to cryopreservation from 2014 to 2022 when (**a**) cones were stored at +2 °C, and (**b**) sterilised seeds were stored in cold water at +4 °C. n = number of initiations.

4. Discussion

4.1. Relations of the Different Variables and Limitations of the Study

When more than one variable was included in the model, the highest percentage of cases predicted correctly was found with full-sib family and initiation date in initiation (73.9%, improving the cases predicted correctly by 14.1 units of %) and cryopreserved ET (72.0% success, improving the cases predicted correctly by 18.9 units of %) (Supplementary Table S3).

In the current data, several of the studied variables were partially overlapping among different variable categories. For example, the genetic full-sib family (and other genetic variables) also reflects the variation between the year of initiation and the initiation date (as full-sib families were not replicated every year), and the variables were also derived from the aforementioned ones.

The cone collection date variable, which was used in the final models, reflects the variation caused partly by the different full-sib families and other genetic variables, cone collection d.d., and the time related variables resulting from biological development. The cone collection day also had a significant effect on the initiation day. The cone collection day variable worked well in the model because it was the single variable which most comprehensively predicted the effects caused by other variables. The cone collection day will probably work well as an operational indicator when carrying out the collections. However, it must be noted that unexpected change in the other variables may have a drastic effect on the outcome if the whole process from flowering to initiation is not well known and monitored.

Although the data were very extensive, covering dozens of full-sib families and thousands of initiations, initiated, and cryopreserved cell lines, there were weaknesses that complicated the interpretation of results. First, there were a fair amount of missing data in several variables, which weakened their potential in the statistical modelling. Second, although the data were vast, there were very few replicates with the same genetic background in different experimental variables. Third, in the large dataset, very small differences became significant in the models, which needs to be considered when making conclusions.

4.2. The Effects of Genetic Variables on Initiation Success and the Number of Cryopreserved ET

The overall initiation percentage, 59.8%, in the current study was higher than the 15% in Egertsdotter et al. [13], who used mature ZEs as explants. In Högberg et al.'s [14] study with initiations using immature ZEs, 53.5% of the initiations were growing vigorously after four months. Small steps were taken since Hakman et al. started in 1985 and reported that about 50% of the cultures obtained from ZEs were embryogenic [40].

In the present study, ETs were obtained from 94% of families, which was similar to the previous report by Egertsdotter et al. [13] in which 1 to 3% of Norway spruce families did not generate any SE initiations at all. In our study, the unsuccessful 6% consisted of eight families, of which seven were initiated in 2022. When they were further investigated, it was obvious that there were difficulties in seed and ZE development, because only one to six ZEs were found from two dissected cones per family. It must also be noted that the cones from the aforementioned seven families were cold-stored for 14 days on average (varying from 8 to 18). Judging those families as recalcitrant to SE may, therefore, be misleading. Additionally, five of the mother trees from those eight families were also crossed with different fathers (the number of ZEs was higher in these cases). From these families, successful initiations and cryopreserved ETs were achieved, which can be related to an earlier cone collection date and the greater number of ZEs or shorter cone cold storage (mean 12 days, varying from three to 38 days).

Of the genetic variables related to the parent trees, the full-sib family predicted the outcome best in the logistic regression model. This is understandable, as it covers the variation related to both the mother and father trees. Of the parent trees, the mother tree was slightly more effective in the model, probably because mother tree as a variable also covers some environmental variation due to its growing location and the physiological effects of the mother tree on the development of ZE.

Another tested genetic variable was the effect of PaLAR3B allele on root rot tolerance. The fact that all parents in the present study were genotyped for the B allele enabled it to be confirmed that the allele did not interfere with SE initiation or ET growth that led to cryopreservation, based on this very large dataset extending over different years. In the present study, only the parent genotype was known, whereas the progeny genotype was not analysed. However, we assume that the allele frequency among progeny of different PaLAR3 cross types follows the Mendelian inheritance rules as shown in a previous study [18]. In that study, the progeny from seven different PaLAR3 crosses from 2014 initiations was shown to follow the inheritance rule as expected.

In 2022, the initiation and cryopreserved ET percentages were lower than in previous years. One difference from previous years is that the proportion of heterozygous crosses for the PaLAR3 allele was more numerous than in other years. However, there were no differences in initiation or cryopreserved ET percentages between different PaLAR3 cross types or the amount of B-allele frequency in 2022. The same result was obtained for all previous years in the present study. The reason for the lower initiation percentages is, therefore, probably related to factors other than the PaLAR3 genotype, such as the long storage of cones.

4.3. The Effects of Environmental Variables on Initiation Success and the Number of Cryopreserved ET

The time period variables, duration from pollination to cone collection, and initiation were included to explain the developmental stage of the cone and ZE better than just the date variables (date of collection, cone collection date d.d., and initiation date). Surprisingly, the date variables were more effective in the model, probably because they also reflected variation caused by other factors. It was shown that maritime pine explant's ability to initiate embryogenic masses was significantly influenced by the collection date of the cones [21,41], and Kvaalen et al. [42] state that the most important factor that determined the initiation success in subalpine fir (Abies lasiocarpa) was the time of cone collection.

The cone collection d.d. and cone collection date are both variables that reflect the developmental stage of the cones and ZEs within the cones. In addition to this, cone collection d.d. reflects the annual variation in temperature and the relationship of the growing season with the calendar, which the cone collection date does not [43]. This probably explains why cone collection d.d. worked better in the part of the data where it was available. On the other hand, the cone collection date affects and partly determines the following operational variables.

Pollination date is highly dependent on flower development, which is affected by the timing of the spring, i.e., temperature sum accumulation. In addition to the d.d. of the year, Norway spruce flower development, especially the number of flowers, is affected by the environmental conditions a year before [44]. Lindgren et al. [44] states that the weather conditions after flowering certainly influence the production of good seeds. If the summer and autumn are cool, the seeds may not mature properly, and this will be reflected by a high proportion of embryos that have not fully developed [44]. In our model, pollination date only covers the running years' environmental variation before actual pollination, so it does not explain the initiation success and the number of cryopreserved ET very well.

Overall initiation success varied remarkably in the years of our study, although the initiation year as a variable did not work well in the models. This is probably related to genetic material and operational factors varying from year to year, with different goals set for initiations.

4.4. The Effects of Operational Variables on Initiation Success and the Number of Cryopreserved ET

Labour resources are limited in the laboratory, which affects the scheduling of the initiations, as well as the operational workflow later when the ETs are picked and placed in subculture. Moreover, the cryopreservation of hundreds of cell lines with more than one sample is itself a massive process. Previously, e.g., Häggman et al. [45] proposed cold storage of the collected cones to even the laboratory workload, and reported that cold storage could be applied for up to two months without an effect on SE initiation in Scots pine (*Pinus sylvestris*). However, in the present study, the initiation success decreased when cones were stored in the cold for more than one week, either reflecting a difference between *Pinus* and *Picea* or suggesting that the overall low initiation frequency in Scots pine in Häggman's study masked the cold storage effect. The present result is confirmed in the study of Park [9], in which a significantly higher percentage of the white spruce explants excised from fresh material immediately after cone collection resulted as ETs than of explants excised from cold-stored cones. It seems that sterilised seeds could be stored longer than cones, but the data from long storage are limited in the present study, and the good outcome may be due to a genetic effect.

Due to the limited resources or varying objectives, different approaches to initiation may be applied. For example, in 2019, the aim was to initiate and cryopreserve enough cell lines to serve as future forest regeneration material in Finland, with enough genotypes per family to achieve improved genetic gain after field testing. When the first forest regeneration material was compiled from a collection of families initiated in 2014, we were able to estimate an initiation threshold for 2019 families. The initiation threshold was achieved when the number of initiations that was estimated to produce enough cell lines was achieved from a family [7]. When this threshold was applied, the initiations were carried out in a short period (no excess cold storage), and cryopreservation was carried out in a larger number per cell line than was possible in 2014. Applying this strict threshold probably affected the higher initiation and cryopreserved ET rates in 2019 than in previous years.

Another approach to the conduct of initiations was applied in 2022, when families with a different PaLAR3 background were initiated. The aim was to obtain as many cell lines as possible from the 54 genotyped families. This resulted in a prolonged initiation period, which, in turn, resulted in a lower initiation and cryopreservation rate than in previous years. However, the sheer number of the obtained and cryopreserved cell lines was vast, and the aim was achieved.

Being able to collect material for SE initiations from small potted grafts in the greenhouse near you also saves resources compared to collections from seed orchards or outdoor clonal archives with bigger trees and potentially located at a distance. Based on our results, both options are equally good from the SE initiation perspective, with factors other than the location of donor trees having a greater effect. However, when planning initiations, one should remember that the temperature in the greenhouse is higher than the temperature outdoors, and this is related to the accumulation of the d.d. sum and development of explants.

In 2011, the first year of our Norway spruce SE studies, we tested two different media for initiations, 1/2LP and mLM (Supplementary Table S2), with a drastic superiority of mLM. There were remarkable differences in the quantities of inorganic elements, vitamins, and amino acids, and in addition, $\frac{1}{2}$LP contains more different amino acids and sugars. Interestingly, the quantities of plant growth regulators were similar, suggesting that Norway spruce SE benefits from a richer medium, and that additional amino acids and sugars present in $\frac{1}{2}$LP were not threshold components for SE initiation or proliferation in this species. mLM is used in conifer SE, being preferable for initiation not only in spruce species [28,46] but also, e.g., in *Pinus banksiana*, *P. strobus*, *P. pinaster*, and *P. sylvestris* [22,47]. However, it should be noted that there are also interactions among basal media and plant growth regulators and trace elements, as described by Park et al. [22].

As previously elaborated with other variables, the effect of the media lot probably describes variation caused by the combination of other factors, i.e., variation among years, cone collection, and initiation dates, as well as genetic variables. Explants prepared previously normally form ETs earlier and are cryopreserved earlier than those managed later in the initiation season, when different media lots were used. In the models, the media lot did not cover any variation which could not be covered with other variables.

4.5. Conclusions and Future Remarks

When trying to capture thousands of genotypes, it is critical to make a strategic plan keeping in mind what the purpose of the material will be. Operational decisions (such as the use of either immature or mature ZE as explants, cold-storing explants awaiting preparation versus collecting them later, etc.) will also affect the outcome, but knowing these effects helps in the optimisation of resources.

According to the current data, the best overall crossings subjected to SE were E799 × E1366 (carried out in 2014), E9990 × 9902, and K1181 × E318 (both carried out in 2019). In the two first mentioned crossings, few ZEs (45 and 5, respectively) were found, and the rates in initiation (93.3 and 100%, respectively) and cryopreserved ET (80 and 100% of ZEs) were high. Thus, the focus was shifted to the K1181 × E318 full-sib family, where 184 ZEs were found, and the initiation (92.4%) and cryopreserved ET (84.8% of ZE's) rates were excellent. In this crossing, the mother tree was a potted greenhouse graft, and there were two collection dates (24 June in 718 d.d. and 8 July in 883 d.d.), which were both rather early. From the first collection, initiations were made on 1 July (30 ZEs with 96.7% initiation and 93.3% cryopreservation from ZEs) and 5 July (40 ZEs with 87.5% initiation and 80.0% cryopreservation from ZEs) as from the latter collection initiations were made on 11 July (114 ZEs 93.0% initiation and 84.8% cryopreservation from ZEs).

In the optimal scenario, according to the current data, the Norway spruce cones would be collected in southern Finland during the first two weeks of July (in approximately 800 d.d. accumulation) from the seed orchard or greenhouse and delivered quickly to the laboratory, and the cones would be cold-stored for five days or less before initiations on mLM. In the current data, this happened in practice in two years (2015 and 2019) for four full-sib families, with an initiation rate of 94%, and ET from 87% of the original ZEs were cryopreserved. Our results show that with the right timing of explant collection and an efficient SE initiation procedure, the genotype capture from amenable full-sib families of Norway spruce can be very high. Furthermore, almost all families are responsive, and lower initiation frequencies in some families can be compensated by increasing the number of explants—however, considering operational limitations. In practice, biological and societal constraints may have a conflict here because the optimal initiation time with the highest labour intensity coincides with the well-established holiday season in our latitudes.

Supplementary Materials: The following are available online at https://www.mdpi.com/article/10.3390/f14040810/s1, Table S1: year of the crossing, mother tree and its location, pollen donor tree, number of initiations, cell lines picked up, and the percentage of picked up cell lines; Table S2: media used in the initiations and proliferation of Norway spruce somatic embryogenesis; Table S3: logistic regression models used for analysing a binary response (living or dead) in the initiation and cryopreservation success. In the models, a1 to a125 are design variables for full-sib families, and b1 to b57 are design variables for the cone collection date.

Author Contributions: Conceptualisation, M.T. and S.V.; methodology, T.A., M.T., S.V. and J.E.; validation, T.A., M.T., S.V. and J.E.; formal analysis, M.T., J.E. and S.V.; investigation, M.T., J.E. and S.V.; resources, T.A.; writing—original draft preparation, S.V., M.T. and J.E.; writing—review and editing, T.A.; visualisation, J.E. and S.V.; supervision, T.A.; project administration, T.A., M.T. and S.V.; funding acquisition, T.A., M.T. and S.V. All authors have read and agreed to the published version of the manuscript.

Funding: This research was funded by European Regional Development Fund (A31652, A70606, A72814), and Savonlinna municipality, the ForestValue Research Program, which is a transnational research, development, and innovation program jointly funded by national funding organizations within the framework of the ERA-NET Cofund 'ForestValue—Innovating forest-based bioeconomy', grant agreement No. 773324, Public Private Partnership—project in Norway spruce vegetative propagation, and The Royal Swedish Academy of Agriculture and Forestry (KSLA) under Tandem Forest Values research program, grant number TFV 2018-0027.

Data Availability Statement: The data presented in this study are available on request from the corresponding author.

Acknowledgments: The authors wish to thank the technical personnel of Natural Resources Institute Finland for their skilful contributions.

Conflicts of Interest: The authors declare no conflict of interest. The funders had no role in the design of the study; the collection, analyses, or interpretation of data; the writing of the manuscript; or in the decision to publish the results.

References

1. Bonga, J. Conifer Clonal Propagation in Tree Improvement Programs. In *Vegetative Propagation of Forest Trees*; Springer: Berlin/Heidelberg, Germany, 2016; pp. 3–31.
2. Park, Y.-S. Implementation of Conifer Somatic Embryogenesis in Clonal Forestry: Technical Requirements and Deployment Considerations. *Ann. For. Sci.* **2002**, *59*, 651–656. [CrossRef]
3. Sutton, B. Commercial Delivery of Genetic Improvement to Conifer Plantations Using Somatic Embryogenesis. *Ann. For. Sci.* **2002**, *59*, 657–661. [CrossRef]
4. Haapanen, M.; Mikola, J. *Metsänjalostus 2050—Pitkän Aikavälin Metsänjalostusohjelma*; Working Papers of the Finnish Forest Research Institute; METLA: Helsinki, Finland, 2008; 50p. Available online: http://jukuri.luke.fi/handle/10024/535985 (accessed on 8 July 2016).
5. Haines, R.J.; Woolaston, R.R. The Influence of Reproductive Traits on the Capture of Genetic Gain. *Can. J. For. Res.* **1991**, *21*, 272–275. [CrossRef]
6. Högberg, K.-A. *Possibilities and Limitations of Vegetative Propagation in Breeding and Mass Propagation of Norway Spruce*; Dept. of Plant Biology and Forest Genetics, Swedish Univ. of Agricultural Sciences: Uppsala, Sweden, 2003; Volume 294, ISBN 91-576-6528-1.
7. Tikkinen, M. Improved Propagation Efficiency in a Laboratory–Nursery Interface for Somatic Embryogenesis in Norway Spruce. *Diss. For.* **2018**, *265*, 35. [CrossRef]

8. Haapanen, M.; Leinonen, H.; Leinonen, K. Männyn ja kuusen siemenviljelyssiemenen taimitarhakäytön kehitys 2006–2016: Alueellinen tarkastelu. *Metsätieteen Aikakauskirja* **2017**, *2017*, 7716. [CrossRef]
9. Park, Y.S.; Pond, S.E.; Bonga, J.M. Initiation of Somatic Embryogenesis in White Spruce (*Picea glauca*): Genetic Control, Culture Treatment Effects, and Implications for Tree Breeding. *Theor. Appl. Genet.* **1993**, *86*, 427–436. [CrossRef]
10. Park, Y.S.; Pond, S.E.; Bonga, J.M. Somatic Embryogenesis in White Spruce (*Picea glauca*): Genetic Control in Somatic Embryos Exposed to Storage, Maturation Treatments, Germination, and Cryopreservation. *Theor. Appl. Genet.* **1994**, *89*, 742–750. [CrossRef]
11. MacKay, J.J.; Becwar, M.R.; Park, Y.-S.; Corderro, J.P.; Pullman, G.S. Genetic Control of Somatic Embryogenesis Initiation in Loblolly Pine and Implications for Breeding. *Tree Genet. Genomes* **2006**, *2*, 1–9. [CrossRef]
12. Cheliak, W.M.; Klimaszewska, K. Genetic Variation in Somatic Embryogenic Response in Open-Pollinated Families of Black Spruce. *Theor. Appl. Genet.* **1991**, *82*, 185–190. [CrossRef]
13. Egertsdotter, U. Plant Physiological and Genetical Aspects of the Somatic Embryogenesis Process in Conifers. *Scand. J. For. Res.* **2019**, *34*, 360–369. [CrossRef]
14. Högberg, K.-A.; Ekberg, I.; Norell, L.; Von Arnold, S. Integration of Somatic Embryogenesis in a Tree Breeding Programme: A Case Study with *Picea abies*. *Can. J. For. Res.* **1998**, *28*, 1536–1545. [CrossRef]
15. Thor, M.; Moykkynen, T.; Pratt, J.E.; Pukkala, T.; Ronnberg, J.; Shaw, C.G.; Stenlid, J.; Stahl, G.; Woodward, S. Modeling Infection and Spread of *Heterobasidion annosum* in Coniferous Forests in Europe. *U. S. Dep. Agric. For. Serv. Gen. Tech. Rep. PNW* **2005**, *656*, 105.
16. Nemesio-Gorriz, M.; Hammerbacher, A.; Ihrmark, K.; Källman, T.; Olson, Å.; Lascoux, M.; Stenlid, J.; Gershenzon, J.; Elfstrand, M. Different Alleles of a Gene Encoding Leucoanthocyanidin Reductase (PaLAR3) Influence Resistance against the Fungus *Heterobasidion parviporum* in *Picea abies*. *Plant Physiol.* **2016**, *171*, 2671–2681. [CrossRef] [PubMed]
17. Lind, M.; Källman, T.; Chen, J.; Ma, X.-F.; Bousquet, J.; Morgante, M.; Zaina, G.; Karlsson, B.; Elfstrand, M.; Lascoux, M. A *Picea abies* Linkage Map Based on SNP Markers Identifies QTLs for Four Aspects of Resistance to *Heterobasidion parviporum* Infection. *PLoS ONE* **2014**, *9*, e101049. [CrossRef] [PubMed]
18. Edesi, J.; Tikkinen, M.; Elfstrand, M.; Olson, Å.; Varis, S.; Egertsdotter, U.; Aronen, T. Root Rot Resistance Locus PaLAR3 Is Delivered by Somatic Embryogenesis (SE) Pipeline in Norway Spruce (*Picea abies* (L.) Karst.). *Forests* **2021**, *12*, 193. [CrossRef]
19. Hakman, I.; Fowke, L.C.; Von Arnold, S.; Eriksson, T. The Development of Somatic Embryos in Tissue Cultures Initiated from Immature Embryos of *Picea abies* (*Norway spruce*). *Plant Sci.* **1985**, *38*, 53–59. [CrossRef]
20. Tautorus, T.E.; Attree, S.M.; Fowke, L.C.; Dunstan, D.I. Somatic Embryogenesis from Immature and Mature Zygotic Embryos, and Embryo Regeneration from Protoplasts in Black Spruce (*Picea mariana* Mill.). *Plant Sci.* **1990**, *67*, 115–124. [CrossRef]
21. Bercetche, J.; Pâques, M. Somatic Embryogenesis in Maritime Pine (*Pinus pinaster*). In *Somat. Embryogenesis Woody Plants Vol. 3 Gymnosperms*; Springer: Berlin/Heidelberg, Germany, 1995; Volume 3, p. 221.
22. Park, Y.S.; Lelu-Walter, M.A.; Harvengt, L.; Trontin, J.F.; MacEacheron, I.; Klimaszewska, K.; Bonga, J.M. Initiation of Somatic Embryogenesis in *Pinus banksiana*, *P. strobus*, *P. pinaster*, and *P. sylvestris* at Three Laboratoriesin Canada and France. *Plant Cell Tissue Organ Cult.* **2006**, *86*, 87–101. [CrossRef]
23. Gupta, P.K.; Durzan, D.J. Plantlet Regeneration via Somatic Embryogenesis from Subcultured Callus of Mature Embryos of *Picea abies* (*Norway spruce*). *Vitro Cell Dev. Biol.* **1986**, *22*, 685–688. [CrossRef]
24. Phillips, G.C.; Garda, M. Plant Tissue Culture Media and Practices: An Overview. *Vitro Cell Dev. Biol.-Plant* **2019**, *55*, 242–257. [CrossRef]
25. Arnold, S.V.; Eriksson, T. In Vitro Studies of Adventitious Shoot Formation in *Pinus contorta*. *Can. J. Bot.* **1981**, *59*, 870–874. [CrossRef]
26. Litvay, J.D.; Verma, D.C.; Johnson, M.A. Influence of a Loblolly Pine (*Pinus taeda* L.). Culture Medium and Its Components on Growth and Somatic Embryogenesis of the Wild Carrot (*Daucus carota* L.). *Plant Cell Rep.* **1985**, *4*, 325–328. [CrossRef]
27. Murashige, T.; Skoog, F. A Revised Medium for Rapid Growth and Bio Assays with Tobacco Tissue Cultures. *Physiol. Plant.* **1962**, *15*, 473–497. [CrossRef]
28. Klimaszewska, K.; Cyr, D.R. Conifer Somatic Embryogenesis: I. Development. *Dendrobiology* **2002**, *74*, 69–76.
29. Salaj, T.; Matusova, R.; Salaj, J. Conifer Somatic Embryogenesis-an Efficient Plant Regeneration System for Theoretical Studies and Mass Propagation. *Dendrobiology* **2015**, *35*, 209–224. [CrossRef]
30. Nørgaard, J.V.; Duran, V.; Johnsen, Ø.; Krogstrup, P.; Baldursson, S.; Arnold, S.V. Variations in Cryotolerance of Embryogenic *Picea abies* Cell Lines and the Association to Genetic, Morphological, and Physiological Factors. *Can. J. For. Res.* **1993**, *23*, 2560–2567. [CrossRef]
31. Haapanen, M. Performance of Genetically Improved Norway Spruce in One-Third Rotation-Aged Progeny Trials in Southern Finland. *Scand. J. For. Res.* **2020**, *35*, 221–226. [CrossRef]
32. Klimaszewska, K.; Lachance, D.; Pelletier, G.; Lelu, M.-A.; Séguin, A. Regeneration of Transgenic *Picea glauca*, *P. mariana*, and *P. abies* after Cocultivation of Embryogenic Tissue with Agrobacterium Tumefaciens. *Vitro Cell Dev. Biol.-Plant* **2001**, *37*, 748–755. [CrossRef]
33. Lelu-Walter, M.-A.; Bernier-Cardou, M.; Klimaszewska, K. Clonal Plant Production from Self- and Cross-Pollinated Seed Families of *Pinus sylvestris* (L.) through Somatic Embryogenesis. *Plant Cell Tissue Organ Cult.* **2008**, *92*, 31–45. [CrossRef]
34. Varis, S.; Ahola, S.; Jaakola, L.; Aronen, T. Reliable and Practical Methods for Cryopreservation of Embryogenic Cultures and Cold Storage of Somatic Embryos of Norway Spruce. *Cryobiology* **2017**, *76*, 8–17. [CrossRef]

35. Kartha, K.K.; Fowke, L.C.; Leung, N.L.; Caswell, K.L.; Hakman, I. Induction of Somatic Embryos and Plantlets from Cryopreserved Cell Cultures of White Spruce (*Picea glauca*). *J. Plant Physiol.* **1988**, *132*, 529–539. [CrossRef]
36. Find, J.I.; Floto, F.; Krogstrup, P.; Møller, J.D.; Nørgaard, J.V.; Kristensen, M.M.H. Cryopreservation of an Embryogenic Suspension Culture of *Picea sitchensis* and Subsequent Plant Regeneration. *Scand. J. For. Res.* **1993**, *8*, 156–162. [CrossRef]
37. Varis, S.A.; Virta, S.; Montalbán, I.A.; Aronen, T. Reducing Pre-and Post-Treatments in Cryopreservation Protocol and Testing Storage at −80 °C for Norway Spruce Embryogenic Cultures. *Int. J. Mol. Sci.* **2022**, *23*, 15516. [CrossRef] [PubMed]
38. Wickham, H. Data Analysis. In *Ggplot2: Elegant Graphics for Data Analysis*; Wickham, H., Ed.; Springer International Publishing: Cham, Switzerland, 2016; pp. 189–201. ISBN 978-3-319-24277-4.
39. R Core Team. *R: A Language and Environment for Statistical Computing*; R Foundation for Statistical Computing: Vienna, Austria, 2008; Available online: https://www.R-project.org/ (accessed on 15 March 2023).
40. Hakman, I.; Arnold, S.V. Plantlet Regeneration through Somatic Embryogenesis in *Picea abies* (*Norway spruce*). *J. Plant Physiol.* **1985**, *121*, 149–158. [CrossRef]
41. Miguel, C.; Gonçalves, S.; Tereso, S.; Marum, L.; Maroco, J.; Margarida Oliveira, M. Somatic Embryogenesis from 20 Open-Pollinated Families of Portuguese plus Trees of Maritime Pine. *Plant Cell Tissue Organ Cult.* **2004**, *76*, 121–130. [CrossRef]
42. Kvaalen, H.; Daehlen, O.G.; Rognstad, A.T.; Grønstad, B.; Egertsdotter, U. Somatic Embryogenesis for Plant Production of Abies Lasiocarpa. *Can. J. For. Res.* **2005**, *35*, 1053–1060. [CrossRef]
43. Almqvist, C.; Bergsten, U.; Bondesson, L.; Eriksson, U. Predicting Germination Capacity of *Pinus sylvestris* and *Picea abies* Seeds Using Temperature Data from Weather Stations. *Can. J. For. Res.* **1998**, *28*, 1530–1535. [CrossRef]
44. Lindgren, K.; Ekberg, I.; Eriksson, G. *External Factors Influencing Female Flowering in Picea abies (L.) Karst.*; Community Organized Relief Effort: Los Angeles, CA, USA, 1977; ISBN 91-38-03762-9.
45. Häggman, H.; Jokela, A.; Krajnakova, J.; Kauppi, A.; Niemi, K.; Aronen, T. Somatic Embryogenesis of Scots Pine: Cold Treatment and Characteristics of Explants Affecting Induction. *J. Exp. Bot.* **1999**, *50*, 1769–1778. [CrossRef]
46. Hazubska-Przybyl, T.; Bojarczuk, K. Somatic Embryogenesis of Selected Spruce Species [*Picea abies, P. omorika, P. pungens* 'Glauca' and *P. breweriana*]. *Acta Soc. Bot. Pol.* **2008**, *77*, 189–199. [CrossRef]
47. Harju, A.; Heiska, S.; Julkunen-Tiitto, R.; Venäläinen, M.; Aronen, T. Somatic Embryogenesis of *Pinus sylvestris* L. from Parent Genotypes with High-and Low Stilbene Content in Their Heartwood. *Forests* **2022**, *13*, 557. [CrossRef]

Article

Efficient Procedure for Induction Somatic Embryogenesis in Holm Oak: Roles of Explant Type, Auxin Type, and Exposure Duration to Auxin

María Teresa Martínez and Elena Corredoira *

Misión Biológica de Galicia (MBG-CSIC), Sede Santiago de Compostela, Avd de Vigo s/n, Campus Sur, 15705 Santiago de Compostela, Spain
* Correspondence: elenac@mbg.csic.es

Abstract: Holm oak is the dominant tree species in the Mediterranean climate. Currently, worrisome degradation of its ecosystems has been observed, produced, among other factors, by changes in land use, extreme weather events, forest fires, climate change, and especially the increasingly frequent episodes of high tree mortality caused by "oak decline", which has brought with it a social concern that transcends the productive interest. Breeding and conservation programs for this species are necessary to ensure the prevalence of these ecosystems for future generations. Biotechnological tools such as somatic embryogenesis (SE) have great potential value for tree improvement and have been shown to be highly efficient in the propagation and conservation of woody species. One challenge to this approach is that SE induction in holm oak has not yet been optimized. Here, we present a new reproducible procedure to induce SE in holm oak; we evaluated the responsiveness of different initial explants exposed to different types, concentrations, and durations of auxin. SE rates were significantly improved (37%) by culturing nodal segments for two weeks in induction medium. In addition, a significant auxin–genotype interaction was observed.

Keywords: auxin; indole-3-acetic acid; indole-3-butyric acid; leaves; 1-naphthaleneacetic acid; node explants; *Quercus ilex*; somatic embryogenesis induction; shoot apex

Citation: Martínez, M.T.; Corredoira, E. Efficient Procedure for Induction Somatic Embryogenesis in Holm Oak: Roles of Explant Type, Auxin Type, and Exposure Duration to Auxin. *Forests* **2023**, *14*, 430. https://doi.org/10.3390/f14020430

Academic Editors: Ricardo Javier Ordás and José Manuel Álvarez Díaz

Received: 30 January 2023
Revised: 15 February 2023
Accepted: 17 February 2023
Published: 19 February 2023

1. Introduction

The evergreen oak *Quercus ilex* L. (holm oak) is one of the most important forest species in arid and semi-arid Mediterranean environments [1]. Approximately 90% of its worldwide distribution is in Morocco and the Iberian Peninsula [2]. In Spain, holm oak can form forest woodlands, and it is the predominant species in dehesas, which are the largest forest arrangements in the country and are equivalent to 27% of the Spanish forest area [3]. Dehesas are representative agroforestry systems of European agricultural systems and have high natural and cultural values [4,5]; they are models of compatibility between an efficient, diversified, extensive production system and the generation and conservation of high levels of biological diversity [6,7]. In addition to the productive role of dehesas (livestock, agriculture, forestry, hunting products), the traditional management of dehesas provides a wide variety of services, contributes to in the regulation of important natural cycles, and contributes to mitigating climate change and enhancing the conservation of biodiversity [4,5]. Finally, dehesas are an important part of the historical–cultural heritage of Spain and are increasingly exploited for tourism and recreational uses [8].

In recent years, the sustainability of these important ecosystems has been at high risk due to the lack of natural regeneration, extreme weather events, forest fires, climate change, and especially the presence of a severe disease named "oak decline syndrome" that has caused the loss of tens of thousands of hectares of dehesas and holm oak woodlands [9]. Oak decline is a complex syndrome, which causes the gradual and general deterioration of affected trees until their death and is produced by the joint action of silvicultural practices and abiotic (episodes of drought or floods, and air or soil pollution) and biotic (pests and diseases) factors. *Phytophthora cinnamomi* is the biotic agent most related to oak decline, although other species such as *P. gonapodyides*, *P. quercina*, *P. psychrophila*, and *P. pseudocryptea* have also been identified as causative agents of this syndrome [10,11]. Despite the high economic and ecological importance of holm oak, currently there are no effective methods to control oak decline syndrome, and the vegetative propagation of tolerant genotypes of holm oak and their progenies may be one of the most realistic ways to address this problem.

Biotechnological tools such as somatic embryogenesis (SE) have great potential for tree improvement, and its high efficiency has been shown in many hardwood species [12]. Somatic embryogenesis in combination with genetic modification has enormous potential for improving forest species, but several bottlenecks must first be investigated and solved. Among these limitations, induction from adult tissues remains a challenge, as in many woody species, SE has only been reported from juvenile tissues. According to previous research on the topic, SE in oak species from very juvenile tissues (e.g., immature zygotic embryos) is relatively feasible, and induction rates of up to 100% have been achieved in some instances [13]. In contrast, the induction of somatic embryos from non-zygotic tissues, especially when derived from adult trees, remains problematic. To date, only three oak species, i.e., *Q. suber* [14], *Q. robur* [15,16], and *Q. alba* [17], have shown acceptable induction rates. In the case of holm oak, SE has been developed from zygotic embryos [18,19], floral tissues [20,21], and shoot and leaf explants [22–24], but induction frequencies were low, ranging from 0.2 to 11% [23–25]. Therefore, more efficient procedures for SE in holm oak need to be developed to apply this micropropagation technique for mass propagation of this species.

Substantial effort has been expended in recent decades to determine the factors that control SE. It is accepted that selection of the appropriate initial explant and the choice of plant growth regulators (PGRs) incorporated into the induction medium, as well as the exposure duration, are the most important factors for the successful induction of SE [12,26,27]. It is generally accepted that there are two stimuli that induce the reprogramming of differentiated plant cells to convert them into competent cells: (i) strong stress and (ii) changes in the internal and/or external cellular levels of PGRs [28,29]. There is also a consensus that among PGRs, auxins play a key role during SE induction, especially when the initial explants are non-zygotic tissues [12,30]. It is recognized that high doses of auxin with or without a cytokinin at low concentrations are crucial as an initial trigger in the acquisition of cellular competence, firstly promoting dedifferentiation followed by embryogenic differentiation [31]. The addition of exogenous auxins seems to act as a stressing agent and/or induces endogenous indole-3-acetic acid (IAA) production, which regulates the expression of a great number of transcription factors, several of them related to stress, and provokes changes in chromatin status [32,33]. Among the different types of auxin, the most used in SE induction, in order of frequency, are as follows: 2,4-dichlorophenoxyacetic (2,4-D) acid, 1-naphthaleneacetic acid (NAA), IAA, picloram, and dicamba [34]. Usually, NAA is applied when a strong auxin is not required to induce SE or because of its specificity for a given species. In the *Fagaceae* family, and specifically in the case of the *Quercus* species, to date NAA has been the most widely employed auxin to induce SE in non-zygotic tissues, whereas 2,4-D has been used to induce SE in zygotic tissues [13]. By contrast, IAA and indole-3-butyric acid (IBA) have hardly been used in these species [13,35]. Until now, most of the papers published on the effect of auxin on the induction of SE focus on the type and concentration of auxin used. However, less attention has been paid to the determination of the necessary exposure duration to trigger SE once the auxin type and concentration are set. Usually, a long exposure duration is routinely applied, whereas short periods and pulses have rarely been mentioned.

In addition to auxins, the type of explant and its well-defined developmental stage seem to be the most important factors that determine embryogenesis [12,36,37]. For SE induction in explants derived from adult trees, the general approach is to select explants that retain juvenile characteristics. Initially, maternal tissues (e.g., nucellus or inner teguments) and floral tissues (e.g., immature inflorescences, petals, floral staminodes, pistils, stamens, or anther teguments) were the most used explants [12]. There was a general opinion that these tissues could contain dedifferentiated cells due to their proximity to the sites of fertilization and formation of zygotic embryos which facilitated the return to the embryogenic state [36]. However, in the last two decades, the use of other explant types such as shoot tips, nodes, internodes, and especially leaves has gained relevance in SE induction, mainly due to the fact that they are more abundant, which enables the evaluation of more factors, and, above all, they are easier to manage than flower tissues [25]. These explants can be isolated directly from a tree, although the strategy that has offered better results involves excision from shoots derived from forced flushing of branch segments or from axillary shoot cultures established from them [12,25,37]. In addition, the embryogenic ability of the explant type shows great variability in the function of the species. For instance, in *Eucalyptus globulus*, shoot apices presented a greater embryogenic response than leaves [38], while in *Q. alba*, the explants that best responded were the leaves [17]. In *Vitis vinifera*, nodes with a single axillary bud showed the highest rate of embryogenic induction [39].

In order to optimize the frequency of SE induction in holm oak, our objective was to identify the factors that can improve this step using axillary shoot cultures as the source of initial explants. In this study, we investigated the effects of (i) the explant type (leaves, nodes, and shoot tips), (ii) three auxins (NAA, IAA, and IBA), (iii) the auxin exposure duration, and (iv) the genotype role.

2. Materials and Methods

2.1. Plant Material

We used axillary shoot cultures of three holm oak genotypes, Q3-SE, Q10-SE, and E00, as the explant source. Genotype E00 was established from epicormic shoots of a 30-year-old tree as previously described [40]. Axillary shoot cultures of Q3-SE and Q10-SE were established by axillary budding from shoots derived from germinated somatic embryos induced from centenary trees as previously described [22]. Axillary shoot cultures of the three genotypes were maintained by subculture on Woody Plant Medium [41] (Duchefa Biochemie, The Netherlands) supplemented with sucrose (30 g/L), silver thiosulphate (20 μM), and Sigma agar (8 g/L), with a sequence of transfers performed every 2 weeks over a 6-week multiplication cycle as follows: 0.1 mg/L 6-benzyadenine (BA) for the first 2 weeks, 0.05 mg/L BA for the next 2 weeks, and 0.01 mg/L BA for the last 2 weeks. All culture media were brought to pH 5.6–5.7 before autoclaving at 115 °C for 20 min. Stock cultures were cultivated in a growth chamber with a 16 h photoperiod, provided by cool white fluorescent lamps (photon flux density of 50–60 μmol m^{-2} s^{-1}) at 25 °C light/20 °C dark (i.e., standard culture conditions).

2.2. Somatic Embryogenesis Initiation

To determine the optimal explant type for SE initiation, 3 types of explants were tested in this study: shoot apex (2 to 2.5 mm long, comprising the apical meristem and 2 to 3 pairs of leaf primordia), the most apical expanding leaf, and the node below the shoot apex of the three genotypes (Figure 1). Then, to determine the best combination of growth regulators, explants were cultured on a basal induction medium consisting of Murashige and Skoog medium (MS) [42] (Duchefa Biochemie, The Netherlands) added with casein hydrolysate (500 mg/L), sucrose (30 g/L), and Plant Propagation Agar (6 g/L; Pronadisa, Spain). This medium was added with three different PGR combinations: IAA (4 mg/L) plus BA (0.5 mg/L) (IAA treatment), NAA (4 mg/L) plus BA (0.5 mg/L) (NAA treatment), and IBA (3 mg/L) plus NAA (0.1 mg/L) (IBA treatment). The IAA and NAA treatments were chosen based on previous studies of SE induction in holm oak [22]. The IBA treatment was chosen following the appearance of somatic embryos when this treatment was used to induce adventitious root formation on E00 axillary shoots (Supplementary information S1). Nodal segments and shoot tips were cultured on auxin medium in the dark at 25 °C for 2, 4, and 8 weeks, whereas leaves were cultured only for 2 and 8 weeks. Then, explants were transferred to basal induction medium without PGRs and cultured in light conditions (standard culture conditions) without transfers to fresh medium for at least 24 weeks. After this period, SE induction efficiency was estimated.

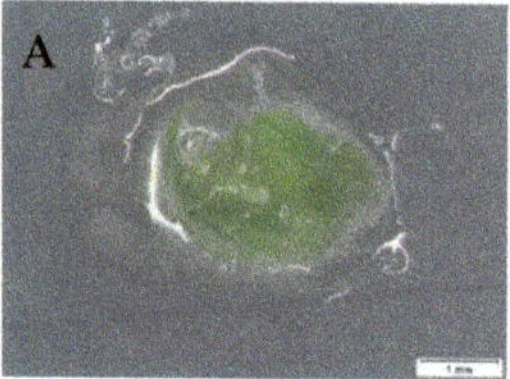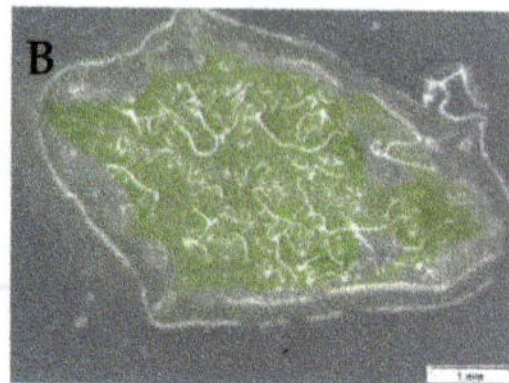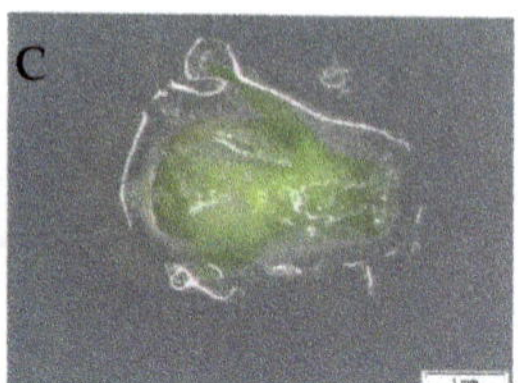

Figure 1. Initial explants used to induce somatic embryogenesis on holm oak at the excision day. (**A**) Shoot apex explant. (**B**) Leaf explant. (**C**) Node explant. Bar: 1 mm.

Ten leaves (abaxial side down) and ten shoot apices and nodal segments (horizontally orientated) (Figure 1) were cultured in 90×15 mm Petri dishes containing 25 mL of auxin treatment. For each genotype, explant type, auxin treatment, and auxin exposure duration, 50 explants were used, and each experiment was repeated twice.

At the end of the culture period, the following data were recorded: the percentage of explants that formed calli, the percentage of explants that formed roots, the percentage of explants that showed an embryogenic response, and the number of somatic embryos or nodular embryogenic structures per initial explant. An embryogenic response was described as the presence of nodular embryogenic structures and/or somatic embryos (torpedo/cotyledonary stage) on the initial explants. These parameters were determined by periodically examining the explants under a stereomicroscope (Olympus SZX9, Japan). Photographs were taken with an Olympus SC100 digital camera (Japan).

2.3. Statistical Analysis

The influence of the main experimental factors on the embryogenic response (measured as a percentage) and their interactions were statistically evaluated using analysis of variance (ANOVA). Prior to analysis, an arcsine square root transformation was applied to proportional data. A Levene test for normality and homogeneity of variance was performed prior to ANOVA. In the tables, non-transformed data are presented. SPSS for Windows (version 26.0, Chicago, IL, USA) was utilized to perform the statistical analysis.

3. Results

For the three genotypes evaluated, the first response observed was callus formation. The percentage, callus size, and its appearance mainly depended on the auxin treatment applied. The highest callus percentage (between 83 and 100%) was obtained when IBA treatment was used regardless of the genotype, exposure duration, and explant type (see Supplementary information S2). In this treatment, before and after the explants were transferred to media without PGRs, callus growth was disorderly, so the original explant became unrecognizable. With NAA treatment, the percentage of calli formed was also high, with the exception of the apex of genotype Q3-SE and the leaves of genotype E00. In this treatment, calli appeared only around wounded areas, and consequently, the initial explants were perfectly recognizable. Finally, IAA treatment produced only very small calli and usually in very low proportions. In addition to callus formation, after culture on induction media without PGRs, adventitious root formation was also observed. Among the three treatments, IBA treatment showed the highest percentage of roots, especially when the initial explants were leaves (e.g., about 95% in Q3-SE) (Supplementary information S3).

The embryogenic response was always indirect through callus formation on the original explant regardless of the type of explant, exposure duration, and auxin treatment; however, the size and percentage of the calli formed did not seem to be directly related to SE induction. In all genotypes, somatic embryos or nodular embryogenic structures arose after the initial explants were transferred into PGR-free medium, becoming visible between 3 and 8 months after culture initiation (Figure 2A). Shoot tips and node explants produced the highest number of somatic embryos and/or nodular structures per explant, ranging mostly from one to three, while in foliar explants, the number of somatic embryos was always one per reactive explant (Figure 2B–D).

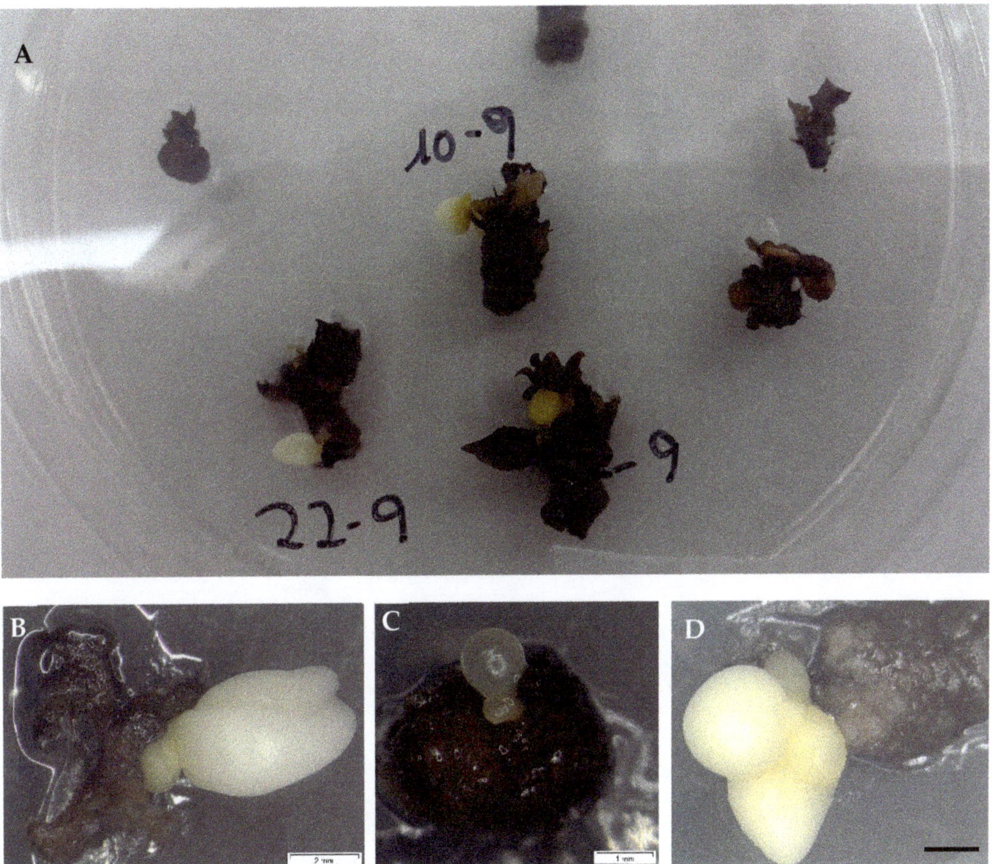

Figure 2. Somatic embryogenesis induction on different explants excised from axillary shoot cultures established from adult holm oak trees. (**A**) Somatic embryos and nodular embryogenic structures generated on different apex explants of genotype Q10-SE cultured on induction medium with IAA. (**B–D**) Embryogenic response on an apex (**B**), node (**C**), and leaf (**D**) explant of genotype Q3-SE. (**A**): diameter dish 90 mm. (**D**): bar 1 mm.

3.1. Embryogenic Response in Apex Explants

Somatic embryogenesis from apex explants was significantly affected by genotype ($p = 0.001$), auxin treatment ($p = 0.001$), and auxin exposure duration ($p = 0.049$), as well as by the genotype–auxin treatment interaction ($p = 0.001$) and the genotype–auxin treatment exposure-duration interaction ($p = 0.001$) (Table 1).

Table 1. Embryogenic response in apex tips excised from axillary shoot cultures of three adult genotypes of holm oak under the effects of different treatments and exposure duration to auxins.

Treatment (mg/L)	Somatic Embryogenesis (%)		
	Q10-SE	Q3-SE	E00
IAA 4 + BA 0.5			
2 wks	15.0 ± 2.9	2.0 ± 1.9	12.0 ± 3.1
4 wks	11.0 ± 3.6	7.0 ± 2.5	2.0 ± 1.3
8 wks	12.0 ± 2.8	15.0 ± 5.0	1.0 ± 0.9
NAA 4 + BA 0.5			
2 wks	3.0 ± 2.9	33.0 ± 4.3	0.0 ± 0.0
4 wks	0.0 ± 0.0	20.0 ± 7.2	0.0 ± 0.0
8 wks	0.0 ± 0.0	22.0 ± 4.2	0.0 ± 0.0
IBA 3 + NAA 0.1			
2 wks	0.0 ± 0.0	12.0 ± 4.2	0.0 ± 0.0
8 wks	0.0 ± 0.0	14.0 ± 2.9	0.0 ± 0.0
ANOVA			
Genotype (A)		$p = 0.001$ ***	
Treatment (B)		$p = 0.001$ ***	
Exposure duration (C)		$p = 0.049$ *	
A × B		$p = 0.001$ ***	
A × C		0.111 ns	
B × C		0.495 ns	
A × B × C		$p = 0.001$ ***	

BA: 6-Benzylaminopurine; IAA: indole-3-acetic acid; IBA: indole-3-butyric acid; NAA: 1-naphthaleneacetic acid; wks: weeks. Each value is the mean ± standard error of ten replicate dishes with ten explants per dish. ANOVA significance values are shown for each parameter. ns: not significant; * significant difference at 95.0% ($p \leq 0.05$); *** significant difference at 99.9% ($p \leq 0.001$).

The induction rates were higher with the shoot tips of Q3-SE cultured with NAA (33%) for two weeks. By contrast, in genotypes Q10-SE and E00, the best percentages were obtained with IAA also applied for 2 weeks (15% and 12%, respectively), but the values were lower than those in Q3-SE (Table 1). IAA treatment induced SE in all three genotypes, whereas NAA induced SE in genotypes Q10-SE and Q3-SE, although marked differences in the induction rates were observed between both genotypes (33% in Q3-SE versus 3% in Q10-SE). IBA induced SE only in genotype Q3-SE but without differences between the two auxin exposure periods evaluated (12% with two weeks versus 14% with eight weeks). Regarding the auxin application regime, a clear interaction between the genotype and the type of auxin was observed (Table 1). In the IAA treatment, 2 weeks of auxin exposure produced the best results in genotypes Q10-SE and E00, whereas in genotype Q3-SE, the highest induction percentages were obtained with an induction-medium culture period of 8 weeks. By contrast, in this genotype and in the NAA treatment, the best values were achieved with a 2-week culture period (Table 1).

3.2. Embryogenic Response in Node Explants

As occurred with the apex explants, SE from nodes was significantly influenced by genotype ($p = 0.001$), auxin treatment ($p = 0.001$), and auxin exposure duration ($p = 0.001$) (Table 2). Also similar to what was found with the shoot apex explants, three of the four possible interactions between treatments had significant effects on SE induction (i.e., genotype–auxin treatment, genotype–exposure duration, and genotype–auxin treatment–exposure duration).

Table 2. Embryogenic response in node explants excised from axillary shoot cultures of three adult genotypes of holm oak under the effects of different treatments and exposure duration to auxins.

Treatment (mg/L)	Somatic Embryogenesis (%)		
	Q10-SE	Q3-SE	E00
IAA 4 + BA 0.5			
2 wks	37.0 ± 5.3	2.0 ± 1.9	4.0 ± 2.1
4 wks	21.0 ± 5.6	2.0 ± 1.3	4.0 ± 2.1
8 wks	3.0 ± 2.0	4.0 ± 1.6	3.0 ± 2.0
NAA 4 + BA 0.5			
2 wks	0.0 ± 0.0	29.0 ± 5.0	1.0 ± 0.9
4 wks	0.0 ± 0.0	4.0 ± 2.1	0.0 ± 0.0
8 wks	0.0 ± 0.0	3.0 ± 1.5	0.0 ± 0.0
IBA 3 + NAA 0.1			
2 wks	4.0 ± 2.9	24.0 ± 3.2	1.0 ± 0.0
8 wks	0.0 ± 0.0	1.0 ± 0.95	0.0 ± 0.0
ANOVA			
Genotype (A)	$p = 0.001$ ***		
Treatment (B)	$p = 0.001$ ***		
Exposure duration (C)	$p = 0.001$ ***		
A × B	$p = 0.001$ ***		
A × C	$p = 0.001$ ***		
B × C	0.262 ns		
A × B × C	$p = 0.001$ ***		

BA: 6-benzylaminopurine; IAA: indole-3-acetic acid; IBA: indole-3-butyric acid; NAA: 1-naphthaleneacetic acid; wks: weeks. Each value is the mean $\pm$ standard error of ten replicate dishes with ten explants per dish. ANOVA significance values are shown for each parameter. ns: not significant; *** significant differences at 99.9% ($p \leq 0.001$).

Regarding auxin treatment, IAA induced SE in the three genotypes, but the best rate was observed in genotype Q10-SE (37%) (Table 2). Conversely, this treatment was less effective with the Q3-SE genotype (2%–4%), which responded with high SE induction frequencies when nodes were treated with NAA (29%) or IBA (24%) (Table 2). Treatment with IBA also produced an embryogenic response in the nodes of genotypes Q10-SE (4%) and E00 (1%), but the values were significantly lower than those obtained with IAA. With respect to the auxin exposure duration, the auxin treatment applied to the three genotypes for 2 weeks on induction medium produced the best results, with values of 4% for E00 in the IAA treatment, 29% for Q3-SE in the NAA treatment, and 37% for Q10-SE in the IAA treatment (Table 2). In addition, in genotypes Q10-SE and E00, for nodes cultured on NAA or IBA treatments, embryogenic responses were observed only when explants were cultured for 2 weeks.

3.3. Embryogenic Response in Leaf Explants

For the three genotypes, regardless of the auxin treatment and exposure duration, leaf explants were the least responsive explants, with values ranging between 1 and 9% (Table 3). Somatic embryogenesis in leaves was significantly influenced by genotype ($p = 0.003$) and auxin treatment ($p = 0.019$), and by the genotype–auxin treatment interaction ($p = 0.001$), the auxin treatment–exposure duration interaction ($p = 0.019$), and the interaction among the three factors ($p = 0.001$) (Table 3). The highest embryogenic response (9%) was obtained for genotype Q3-SE with leaves cultured on medium with NAA for 2 weeks (Table 3). The results provide strong support for the argument that the leaves are not the best explant source by which to induce SE in holm oak if we continue to use these conditions. This low response is probably related to the high degree of necrosis shown by the leaves in comparison with the apex and node explants once they are excised from the shoots.

Table 3. Embryogenic response in leaf explants excised from axillary shoot cultures of three adult genotypes of holm oak under the effects of different treatments and exposure duration to auxins.

Treatment (mg/L)	Somatic Embryogenesis (%)		
	Q10-SE	**Q3-SE**	**E00**
IAA 4 + BA 0.5			
2 wks	1.0 ± 0.9	0.0 ± 0.0	0.0 ± 0.0
8 wks	0.0 ± 0.0	2.0 ± 1.3	0.0 ± 0.0
NAA 4 + BA 0.5			
2 wks	0.0 ± 0.0	9.0 ± 3.0	0.0 ± 0.0
8 wks	0.0 ± 0.0	1.0 ± 0.9	1.0 ± 0.95
IBA 3 + NAA 0.1			
2 wks	0.0 ± 0.0	0.0 ± 0.0	0.0 ± 0.0
8 wks	1.0 ± 0.9	1.0 ± 0.95	1.0 ± 0.9
ANOVA			
Genotype (A)	$p = 0.003$ **		
Treatment (B)	$p = 0.019$ *		
Exposure duration (C)	0.506 ns		
A × B	$p = 0.001$ ***		
A × C	0.071 ns		
B × C	$p = 0.019$ *		
A × B × C	$p = 0.001$ ***		

BA: 6-Benzylaminopurine; IAA: indole-3-acetic acid; IBA: indole-3-butyric acid; NAA: 1-naphthaleneacetic acid; wks: weeks. Each value is the mean ± standard error of ten replicate dishes with ten explants per dish. ANOVA significance values are shown for each parameter. ns: not significant; * significant differences at 95.0% ($p \leq 0.05$); ** significant differences at 99% ($p \leq 0.01$); *** significant differences at 99.9% ($p \leq 0.001$).

3.4. Overview of Results

To summarize, the results obtained indicate that all genotypes exhibited embryogenic response, but the induction rates varied substantially, with genotypes Q10-SE and Q3-SE being those with the highest embryogenic responses (37% and 33%, respectively). However, a very strong auxin–genotype interaction was observed regardless of the type of explant used and the auxin exposure duration; E00 and Q10-SE showed the highest induction rates when the IAA treatment was applied (12% and 15%, respectively, for apices and 4.0% and 37.0%, respectively, for nodal segments), while for genotype Q3-SE, the best response was obtained with the NAA and IBA treatments (33.0% and 14% for apex tips and 29% and 24% for nodes).

Among all three genotypes, regardless of the treatment and auxin exposure duration, leaves were the least reactive explants to induce SE. The shoot tip and node response rates were conditioned by the genotype, so that in Q3-SE (33%) and E00 (12%), the highest induction rates were recorded when the apices were used, and in Q10-SE (37%), the most reactive explants were the nodal segments.

The induction of embryos was possible for the three auxin exposure periods evaluated, although the results show that the percentage of explants with somatic embryos decreased when a 4-week exposure duration was applied. Regardless of the genotype, the best rates were recorded when the explants remained in the medium with auxin for 2 weeks.

In all three genotypes, embryogenic capacity was maintained by secondary embryogenesis subculturing proembryogenic masses according to the procedure described by [22] (see Supplementary Information S4). Likewise, plant regeneration was achieved by somatic embryo germination following the procedure developed by [22] (see Supplementary Information S4).

4. Discussion

The highly recalcitrant nature of holm oak represented by low rates of induction via somatic embryogenesis, together with the difficulty in obtaining the desired response in explants derived from mature trees, has slowed the application of this micropropagation pathway in breeding programs. One of the most important issues arising in SE induction of forest species, but also in the other two micropropagation pathways, is the difficulty in obtaining the desired response in explants derived from mature trees since as the tree age increases, the regeneration ability decreases [37,43,44]. In the present paper, regeneration through SE from adult tissues in the recalcitrant species holm oak was considerably improved by manipulating factors such as the explant type, auxin type, and auxin exposure duration for the three genotypes.

The initiation of the SE process may be conditioned by the explant source and by the condition of the donor plant [12,45]. A first key point in the present paper was the use of axillary shoot cultures as a source of initial explants. This type of explant source presents a great advantage over ex vitro explants, as it allows better control of the growing conditions of the stock material and avoids differences caused by the time of the collection of plant material from trees growing in the field. In addition, it enables the production of physiologically uniform explants while ensuring the supply of an unlimited number of explants throughout the year, which makes it possible to simultaneously evaluate multiple factors [12,37]. In particular, in holm oak, axillary shoot cultures have been shown be a good alternative by improving the embryogenic efficiency of previous attempts to induce SE from leaves or apex tips collected directly from selected field-growing trees [20,25]. In the same way, the induction of somatic embryos in *Arbutus unedo* was possible only when explants excised from axillary shoot cultures were cultured [46]. It is known that repetitive subcultures of shoots on medium supplemented with cytokinins exhibit certain rejuvenating effects in the shoots, which facilitates the induction of the embryogenic process [47]. Additionally, the embryogenic competence of genotypes Q3 and Q10 was enhanced in this research when axillary shoot cultures of Q3-SE and Q10-SE, established from somatic plants, were used as a source of explants. The induction rates obtained for Q3-SE and Q10-SE were significantly higher than the values previously published for the same genotypes when explants derived from axillary shoots established from forced shoots of the trees were used [22]. These results agree with those obtained by [48] for pedunculate oak; it was reported that for the same genotype, shoot multiplication and rooting rates were significantly higher in isolated shoots from cultures derived from germinated somatic embryos than those established from forced shoots of crown branches [48]. Both results confirm the idea that some rejuvenation occurred during the process of somatic embryogenesis, and axillary shoots derived from germinated embryos have more morphogenic ability than shoots derived directly from the tree.

The role of auxins in SE induction has been widely investigated on the basis of 'one-factor-at-a-time' and 'trial-and-error' assays, and in most of these tests, the evaluation of the auxin effect was insufficient due to the low availability of explants from which to study different experimental conditions. Moreover, the optimal PGR contents have been determined only for certain cultivars or genotypes. In the present study, the use of axillary shoot cultures as a source of initial explants allowed us to analyze three induction treatments and three auxin exposure periods in three explant types collected from three different genotypes. Until now, NAA was the most widely used auxin to initiate somatic embryos in oak species when non-zygotic explants were employed [13,35]. However, in holm oak, IAA induced SE in the three genotypes, whereas NAA was effective only in genotype Q3-SE. There are some examples in which other less conventional auxins also had a successful effect on SE induction [12]. For example, IBA was applied in the SE of petioles of olive [49], leaves of *Camellia japonica* [50], and nodal segments of *C. sinensis* [51], whereas picloram was more effective in inducing somatic embryos in the apex of *Eucalyptus* [38]. We believe that this different embryogenic behavior of each genotype in the functioning of auxin types could be due to a different content of endogenous auxins. However, at present, very little is known about how exogenous auxin applied during the induction step interacts with the endogenous auxin of the initial explant used to induce SE.

While it has been suggested that a high concentration of auxin during a long period is required for the acquisition of embryogenic capacity, our results in holm oak show that the 2-week pulse treatment, together with the removal of auxin for the subsequent differentiation of somatic embryos, was sufficient to provide the necessary stimuli to induce somatic embryogenesis, although SE was also achieved when the period was extended to 8 weeks. Likewise, in *Cercis canadensis*, a greater number of somatic embryos was obtained using auxin pulse treatments compared to a long exposure to auxin [52]. A similar finding was also obtained for nucellar tissues of mango, where a short exposure time of 4-week culture was the optimal induction period [53]. In addition to added exogenous auxin, our results again emphasize the important role played by endogenous auxin levels, as well as the stress caused by the excision of the explant to generate an embryogenic response. Both factors combined with the 2-week pulse with a high auxin concentration seem sufficient to induce an embryogenic response for explants derived from adult trees of holm oak. It is well recognized that wounding is the first event that provides signals triggering the embryogenic process, as explant excision produces hormonal balance changes [54,55]. This is in accordance with our previous findings in holm oak, as we were able to induce somatic embryos in apex explants cultured on induction medium devoid of PGRs, concluding that in this species, wounding has a clear effect on the embryo induction process, but the addition of an auxin improves the induction rate [22,23]. Although two weeks of culture on induction medium increased the values, this approach did not reduce the time period required for the whole process of somatic embryo generation.

The novel achievement of this study is that the application of a two-step procedure, in which after culture on auxin medium explants are directly transferred to medium without PGRs, is effective. Our protocol makes it possible to induce somatic embryos or nodular embryogenic structures faster and at a lower cost than the three-step procedure previously published for holm oak [22] and other oak species such as cork oak [14], bicolor oak [56], or pedunculate oak [57], for which it has been observed that the embryogenic response occurs with a three-step procedure in which somatic embryos are induced in the presence of a high auxin concentration, following transfer to a second medium with a lower PGR concentration, and finally transfer to a third medium without PGRs. By contrast, as in the present work, in *Q. alba* [17], the application of a two-step culture induction process enhanced the embryogenic frequencies.

With regard to the choice of explant type, the aim is to identify those tissues that contain competent dedifferentiated cells, i.e., with the capacity to generate somatic embryos as a response to external or internal stimuli and/or signals [58]. To induce SE on adult genotypes of hardwoods, one of the most commonly used explants in recent decades is leaves [12]; however, in the case of holm oak, the morphogenetic ability of leaves was much lower than that of shoot tips and nodes, confirming previous published results for this species [22–24]. Holm oak leaves have the ability to form calli and roots, especially in induction medium supplemented with IBA, but their embryogenic competence is low. By contrast, the shoot apices of the three genotypes and node explants of Q3-SE and Q10-SE showed a strong ability to generate somatic embryos. Similar results have been reported by [59], who found that shoot tips were more appropriate than petioles and entire leaves to obtain SE in two adult genotypes of olive. In the same way, the shoot apex of *Phoenix dactylifera* was the most reactive explant to generate somatic embryos in this species [60]. Equally, nodal segments taken from new sprouting branches of mature trees of *Santalum album* [61] or in vitro grown plantlets of *Vitis vinifera* [39] were able to generate somatic embryos. There is evidence in the literature supporting the hypothesis that the presence of meristematic tissues is strongly involved in the embryogenic response as they can be considered stem cell niches (i.e., pluripotent and totipotent cells) [37,39,62]. Moreover, it is important to highlight the effect of the timing of explant excision that determines the induction of SE; shoot apices should be collected with two or three primordial leaves and very young nodes (i.e., 2–3 days after formation).

In addition to auxin treatment and explant type and its developmental stage, the genotype of the mother tree used as the source of explants has a strong influence on embryogenic competence, and it is one of the main factors limiting SE induction [12,30,37]. There are many reports that have highlighted the influence of genotype on SE induction; however, there is little information regarding why the embryogenic response varies in different genotypes of the same species when the same induction treatments are used. In the present study, SE was achieved in the three genotypes evaluated, but large differences in embryogenic response were observed among them. Thus, for genotypes Q10-SE and E00, the best embryogenic responses were observed on induction medium supplemented with IAA, but very low or no somatic embryo formation was obtained in the medium with IBA or NAA. Conversely, in genotype Q3-SE, high embryogenic responsiveness was yielded with the latter two auxin types. These results are consistent with previous reports on SE induction in different oak species, which revealed a substantial participation of the genotype in the embryogenic response capacity. For example, for pedunculate oak, the authors of [57] obtained somatic embryos in three out of the five adult genotypes evaluated, with embryogenic rates ranging from 1.7 to 5.6%, while the authors of [63] obtained embryogenic lines in 12 out of 19 mature cork oak trees from four provenances with values ranging from 1.3 to 28.5%. The results obtained in the present research show the strong effect of the genotype–auxin interaction on SE induction but also indicate that the differences in embryogenic ability between genotypes can be significantly reduced by altering and optimizing culture conditions. Similar conclusions were drawn in the literature, demonstrating that SE induction in recalcitrant genotypes of coffee [64] or grapevine [45] was possible when the composition of the media was optimized.

5. Conclusions

The higher SE induction rates described in this work represent a considerable improvement with respect to what was previously published for this species. In addition to significantly increasing the induction percentages, the procedure was simplified by reducing the exposure duration to auxins to 6 weeks and eliminating the intermediate step to an expression medium with reduced PGRs (4 weeks), with consequent savings in time. The proposed experimental model might also be useful for developing studies on the anatomical, physiological, and epigenetic facets implicated in the embryogenic process of this species. This protocol has the potential to be used for mass-scale propagation of holm oak; however, several bottlenecks including scaled-up SE production, high-frequency germination and conversion, and improving somatic seedling quality should be overcome before SE could be employed for mass propagation or integration with breeding programs.

Supplementary Materials: The following supporting information can be downloaded at https://www.mdpi.com/article/10.3390/f14020430/s1, Supplementary information S1: Somatic embryo induced on the callus formed at the base of the axillary shoot subjected to rooting treatment consisting of IBA (3 mg/L) plus NAA (0.1 mg/L) for two weeks, Supplementary information S2: Callus response of three different explant types excised from axillary shoot cultures of three adult genotypes of holm oak and cultured on three different treatments, Supplementary information S3: Adventitious root formation in leaves cultured on induction medium with 3 mg/L IBA plus 0.1 mg/L NAA,

Supplementary information S4: A. Somatic embryos originated from proembryogenic masses after 6 weeks of culture on proliferation medium; B. Plant recovery after 8 weeks of culture of somatic embryos in germination medium; A: diameter dish 90 mm; B: diameter jar 90 cm.

Author Contributions: Conceptualization, methodology, investigation, writing—review and editing, E.C. and M.T.M. All authors have read and agreed to the published version of the manuscript.

Funding: This research was funded in part by MINECO (Spain), MICINN (Spain), and MITECO-TRAGSA (Spain) through the projects AGL2016-76143-C4-4-R and PID2020-112627RB-C33 and through the contract TSA0068372, respectively.

Data Availability Statement: Not applicable.

Acknowledgments: We wish to thank Fátima Mosteiro and Sol Campañó for their technical assistance, enthusiasm, and involvement in the work.

Conflicts of Interest: The authors declare no conflict of interest.

References

1. De Rigo, D.; Cadullo, G. *Quercus ilex* in Europe: Distribution, habitat, usage and threats. In *European Atlas of Forest Tree Species*; Publication Office of the European Union: Luxembourg, 2016; pp. 152–153.
2. Villar-Salvador, P.; Nicolás, J.L.; Heredia, N.; Uscola, M. *Quercus ilex* L. In *Producción y Manejo de Semillas y Plantas Forestales. Tomo II. Organismo Autónomo Parques Nacionales*; Pemán, J., Navarro-Cerrillo, R.M., Nicolás, J.L., Prada, M.A., Serrada, R., Eds.; Serie Forestal; Ministerio de Agricultura, Alimentación y Medio Ambiente: Madrid, Spain, 2013; pp. 226–250.
3. Agricultura, Alimentación y Medio Ambiente en España. 2015. Available online: https://www.mapa.gob.es/es/ministerio/servicios/publicaciones/memoria_magrama_2015_completo_tcm30-83965.pdf (accessed on 12 January 2023).
4. Oppermann, R.; Beaufoy, G.; Jones, G. High nature value farming in Europe. In *35 European Countries—Experiences and Perspectives*; Verlag Regionalkultur: Ubstadt-Weiher, Germany, 2012; p. 544.
5. Plieninger, T.; Hartel, T.; Martín-López, B.; Beaufoy, G.; Bergmeier, E.; Kirby, K.; Montero, M.J.; Moreno, G.; Oteros-Rozas, E.; Van Uytvanck, J. Wood-pastures of Europe: Geographic coverage, social-ecological values, conservation management, and policy implications. *Biol. Conserv.* **2015**, *190*, 70–79. [CrossRef]
6. Roig, S.; San Miguel, A. ¿Cómo se mide el estado de conservación de la dehesa? Los pastos: Nuevos retos, nuevas oportunidades. In Proceedings of the Reunión Científica de la Sociedad Española para el Estudio de los Pastos, Badajoz, Spain, 8–11 April 2013.
7. Plieninger, T.; Flinzberger, L.; Hetman, M.; Horstmannshoff, I.; Reinhard-Kolempas, M.; Topp, E.; Moreno, G.; Huntsinger, L. Dehesas as high nature value farming systems: A social-ecological synthesis of drivers, pressures, state, impacts, and responses. *Ecol. Soc.* **2021**, *26*, 23. [CrossRef]
8. Martín-Delgado, L.-M.; Jiménez-Barrado, V.; Sánchez-Martín, J.-M. Sustainable Hunting as a Tourism Product in Dehesa Areas in Extremadura (Spain). *Sustainability* **2022**, *14*, 10288. [CrossRef]
9. MITECO. Plan de Activación Socioeconómica del Sector Forestal. Available online: https://www.miteco.gob.es/es/biodiversidad/temas/politica-forestal/plan-pasfor/default.aspx (accessed on 12 January 2023).
10. Corcobado, T.; Solla, A.; Madeira, M.A.; Moreno, G. Combined effects of soil properties and *Phytophthora cinnamomi* infections on *Quercus ilex* decline. *Plant Soil* **2013**, *373*, 403–413. [CrossRef]
11. Camilo-Alves, C.S.P.; Clara, M.I.E.; Ribeiro, N.M.C.A. Decline of Mediterranean oak trees and its association with *Phytophthora cinnamomi*: A review. *Eur. J. Forest Res.* **2013**, *132*, 411–432. [CrossRef]
12. Corredoira, E.; Merkle, S.A.; Martínez, M.T.; Toribio, M.; Canhoto, J.M.; Correia, S.I.; Ballester, A.; Vieitez, A.M. Non-Zygotic Embryogenesis in Hardwood Species. *Crit. Rev. Plant Sci.* **2019**, *38*, 29–97. [CrossRef]
13. Corredoira, E.; Toribio, M.; Vieitez, A.M. Clonal propagation via somatic embryogenesis in *Quercus* spp. In *Tree Biotechnology*; Ramawhat, K.G., Mérillon, J.-M., Ahuja, M.R., Eds.; CRC Press: Boca Raton, FL, USA, 2014; pp. 262–302.
14. Hernández, I.; Cuenca, B.; Carneros, E.; Alonso-Blázquez, N.; Ruiz, M.; Celestino, C.; Ocaña, L.; Alegre, J.; Toribio, M. Application of plant regeneration of selected cork oak trees by somatic embryogenesis to implement multivarietal forestry for cork production. *Tree For. Sci. Biotech.* **2011**, *5*, 19–26.
15. Toribio, M.; Fernández, C.; Celestino, C.; Martínez, M.T.; San José, M.C.; Vieitez, A.M. Somatic embryogenesis in mature *Quercus robur* trees. *Plant Cell Tissue Organ Cult.* **2004**, *76*, 283–287. [CrossRef]
16. Valladares, S.; Sánchez, C.; Martínez, M.T.; Ballester, A.; Vieitez, A.M. Plant regeneration through somatic embryogenesis from tissues of mature oak trees: True-to-type conformity of plantlets by RAPD analysis. *Plant Cell Rep.* **2006**, *25*, 879–886. [CrossRef]
17. Corredoira, E.; San-José, M.C.; Vieitez, A.M. Induction of somatic embryogenesis from different explants of shoot cultures derived from young *Quercus alba* trees. *Trees* **2012**, *26*, 881–891. [CrossRef]
18. Mauri, P.; Manzanera, J. Induction, maturation and germination of holm oak (*Quercus ilex* L.) somatic embryos. *Plant Cell Tissue Organ Cult.* **2003**, *74*, 229–235. [CrossRef]
19. Mauri, P.V.; Manzanera, J.A. Protocol of somatic embryogenesis: Holm oak (*Quercus ilex* L.). In *Protocol for Somatic Embryogenesis in Woody Plants*; Jain, S.M., Gupta, P.K., Eds.; Springer: Dordrecht, The Netherlands, 2005; pp. 469–482.

20. Blasco, M.; Barra, A.; Brisa, C.; Corredoira, E.; Segura, J.; Toribio, M.; Arrillaga, I. Somatic embryogenesis in holm oak male catkins. *Plant Growth Regul.* **2013**, *71*, 261–270. [CrossRef]

21. Barra-Jiménez, A.; Blasco, M.; Ruiz-Galea, M.; Celestino, C.; Alegre, J.; Arrillaga, I.; Toribio, M. Cloning mature holm oak trees by somatic embryogenesis. *Trees* **2014**, *28*, 657–667. [CrossRef]

22. Martínez, M.T.; José, M.C.S.; Vieitez, A.M.; Cernadas, M.J.; Ballester, A.; Corredoira, E. Propagation of mature *Quercus ilex* L. (holm oak) trees by somatic embryogenesis. *Plant Cell Tissue Organ Cult.* **2017**, *131*, 321–333. [CrossRef]

23. Martínez, M.T.; Vieitez, F.J.; Solla, A.; Tapias, R.; Ramírez-Martín, N.; Corredoira, E. Vegetative Propagation of *Phytophthora cinnamomi*-Tolerant Holm Oak Genotypes by Axillary Budding and Somatic Embryogenesis. *Forests* **2020**, *11*, 841. [CrossRef]

24. Martínez, M.T.; Arrillaga, I.; Sales, E.; Pérez-Oliver, M.A.; González-Mas, M.d.C.; Corredoira, E. Micropropagation, Characterization, and Conservation of *Phytophthora cinnamomi*-Tolerant Holm Oak Mature Trees. *Forests* **2021**, *12*, 1634. [CrossRef]

25. Martínez, M.T.; San-José, M.C.; Arrillaga, I.; Cano, V.; Morcillo, M.; Cernadas, M.J.; Corredoira, E. Holm oak somatic embryogenesis: Current status and future perspectives. *Front. Plant Sci.* **2019**, *10*, 239. [CrossRef] [PubMed]

26. Gaj, M. Factors influencing somatic embryogenesis induction and plant regeneration with particular reference to *Arabidopsis thaliana* (L.) Heynh. *Plant Growth Regul.* **2004**, *43*, 27–47. [CrossRef]

27. Fehér, A. Somatic embryogenesis—Stress-induced remodeling of plant cell fate. *Biochim. Biophys. Acta Bioenerg.* **2015**, *1849*, 385–402. [CrossRef]

28. Fehér, A. The initiation phase of somatic embryogenesis: What we know and what we don't. *Acta Biol. Szeged.* **2008**, *52*, 53–56.

29. Su, Y.H.; Tang, L.P.; Zhao, X.Y.; Zhang, X.S. Plant cell totipotency: Insights into cellular reprogramming. *J. Integr. Plant Biol.* **2021**, *63*, 228–240. [CrossRef]

30. Guan, Y.; Li, S.-G.; Fan, X.-F.; Su, Z.-H. Application of somatic embryogenesis in woody plants. *Front. Plant Sci.* **2016**, *7*, 938. [CrossRef]

31. Wójcik, A.M.; Wójcikowska, B.; Gaj, M.D. Current Perspectives on the Auxin-Mediated Genetic Network that Controls the Induction of Somatic Embryogenesis in Plants. *Int. J. Mol. Sci.* **2020**, *21*, 1333. [CrossRef] [PubMed]

32. Nic-Can, G.I.; Loyola-Vargas, V.M. The Role of the Auxins During Somatic Embryogenesis. In *Somatic Embryogenesis: Fundamental Aspects and Applications*; Loyola-Vargas, V., Ochoa-Alejo, N., Eds.; Springer: Cham, Switzerland, 2016; pp. 171–182. [CrossRef]

33. Elhiti, M.; Stasolla, C. Transduction of Signals during Somatic Embryogenesis. *Plants* **2022**, *11*, 178. [CrossRef]

34. George, E.F.; Hall, M.A.; Klerk, G.J.D. Plant Growth Regulators I: Introduction; Auxins, their Analogues and Inhibitors. In *Plant Propagation by Tissue Culture*; George, E.F., Hall, M.A., Klerk, G.J.D., Eds.; Springer: Dordrecht, The Netherlands, 2008; pp. 175–204. [CrossRef]

35. Vieitez, A.M.; Corredoira, E.; Martínez, M.T.; José, M.C.S.; Sánchez, C.; Valladares, S.; Vidal, N.; Ballester, A. Application of biotechnological tools to *Quercus* improvement. *Eur. J. For. Res.* **2012**, *131*, 519–539. [CrossRef]

36. Merkle, S.A. Strategies for dealing with limitations of somatic embryogenesis in hardwood trees. *Plant Tissue Cult. Biotech.* **1995**, *1*, 112–121.

37. Ballester, A.; Corredoira, E.; Vieitez, A.M. Limitations of somatic embryogenesis in hardwood trees. In *Vegetative Propagation of Forest Trees*; Park, Y.S., Bonga, J.M., Moon, H.K., Eds.; National Institute of Forest Science (NiFos): Seoul, Republic of Korea, 2016; pp. 56–74.

38. Corredoira, E.; Ballester, A.; Ibarra, M.; Vieitez, A.M. Induction of somatic embryogenesis in leaf and shoot apex explants of shoot cultures derived from adult *Eucalyptus globulus* and *Eucalyptus saligna* × *E. maidenii* trees. *Tree Physiol.* **2015**, *35*, 663–667. [CrossRef]

39. Maillot, P.; Deglène-Benbrahim, L.; Walter, B. Efficient somatic embryogenesis from meristematic explants in grapevine (*Vitis vinifera* L.) cv. Chardonnay: An improved protocol. *Trees* **2016**, *30*, 1377–1387. [CrossRef]

40. Martínez, M.T.; Corredoira, E.; Vieitez, A.M.; Cernadas, M.J.; Montenegro, R.; Ballester, A.; Vieitez, F.J.; José, M.C.S. Micropropagation of mature *Quercus ilex* L. trees by axillary budding. *Plant Cell Tissue Organ Cult.* **2017**, *131*, 499–512. [CrossRef]

41. Lloyd, G.; McCown, B. Commercially-feasible micropropagation of mountain laurel, *Kalmia latifolia*, by use of shoot-tip culture. *Comb. Proc. Int. Plant Prop. Soc.* **1980**, *30*, 421–427.

42. Murashige, T.; Skoog, F. A revised medium for rapid growth and bioassays with tobacco tissue culture. *Physiol. Plant* **1962**, *15*, 473–497. [CrossRef]

43. Bonga, J.M. Vegetative propagation in relation to juvenility, maturity and rejuvenation. In *Tissue Culture in Forestry*; Bonga, J.M., Durzan, D.J., Eds.; Martinus Nijhoff: Dordrecht, The Netherlands; Dr. W. Junk Publishers: Dordrecht, The Netherlands, 1982; pp. 387–412.

44. Corredoira, E.; Martínez, M.T.; Cernadas, M.J.; San José, M.C. Application of Biotechnology in the Conservation of the Genus *Castanea*. *Forests* **2017**, *8*, 394. [CrossRef]

45. Capriotti, L.; Limera, C.; Mezzetti, B.; Ricci, A.; Sabbadini, S. From induction to embryo proliferation: Improved somatic embryogenesis protocol in grapevine for Italian cultivars and hybrid *Vitis* rootstocks. *Plant Cell Tissue Organ Cult.* **2022**, *151*, 221–233. [CrossRef]

46. Martins, J.; Correia, S.; Pinto, G.; Canhoto, J. Cloning adult trees of *Arbutus unedo* L. through somatic embryogenesis. *Plant Cell Tissue Organ Cult.* **2022**, *150*, 611–626. [CrossRef]

47. Wendling, I.; Trueman, S.J.; Xavier, A. Maturation and related aspects in clonal forestry—Part II: Reinvigoration, rejuvenation and juvenility maintenance. *New For.* **2014**, *45*, 473–486. [CrossRef]

48. Martínez, T.; Vidal, N.; Ballester, A.; Vieitez, A.M. Improved organogenic capacity of shoot cultures from mature pedunculate oak trees through somatic embryogenesis as rejuvenation technique. *Trees* **2012**, *26*, 321–330. [CrossRef]

49. Capelo, A.M.; Silva, S.; Brito, G.; Santos, C. Somatic embryogenesis induction in leaves and petioles of a mature wild olive. *Plant Cell Tissue Organ Cult.* **2010**, *103*, 237–242. [CrossRef]

50. Pedroso, M.C.; Pais, M.S. Direct embryo formation in leaves of *Camellia japonica* L. *Plant Cell Rep.* **1993**, *12*, 639–643. [CrossRef] [PubMed]

51. Akula, A.; Dodd, W.A. Direct somatic embryogenesis in a selected tea clone, TRI-2025′(*Camellia sinensis* (L.) O. Kuntze) from nodal explants. *Plant Cell Rep.* **1998**, *17*, 804–809. [CrossRef]

52. Geneye, R.L.; Kester, S.T. The initiation of somatic embryos and adventitious roots from developing zygotic embryo explants of *Cercis canadensis* L. cultured in vitro. *Plant Cell Tissue Organ Cult.* **1990**, *22*, 71–76. [CrossRef]

53. Lad, B.L.; Jayasankar, S.; Pliego-Alfaro, F.; Moon, P.A.; Litz, R.E. Temporal effect of 2,4-D on induction of embryogenic nucellar cultures and somatic embryo development of 'Carabao' Mango. *Cell. Dev. Biol. Plant* **1997**, *33*, 253–257. [CrossRef]

54. Iwase, A.; Mitsuda, N.; Koyama, T.; Hiratsu, K.; Kojima, M.; Arai, T.; Inoue, Y.; Seki, M.; Sakakibara, H.; Sugimoto, K.; et al. The AP2/ERF transcription factor WIND1 controls cell dedifferentiation in *Arabidopsis*. *Curr. Biol.* **2011**, *21*, 508–514. [CrossRef]

55. Grzyb, M.; Kalandyk, A.; Waligórski, P.; Mikula, A. The content of endogenous hormones and sugars in the process of early somatic embryogenesis in the tree fern *Cyathea delgadii* Sternb. *Plant Cell Tissue Organ Cult.* **2017**, *129*, 387–397. [CrossRef]

56. Mallón, R.; Martínez, M.T.; Corredoira, E.; Vieitez, A.M. The positive effect of arabinogalactan on induction of somatic embryogenesis in *Quercus bicolor* followed by embryo maturation and plant regeneration. *Trees* **2013**, *27*, 1285–1296. [CrossRef]

57. San José, M.C.; Corredoira, E.; Martínez, M.T.; Vidal, N.; Valladares, S.; Mallón, R.; Vieitez, A.M. Shoot apex explants for induction of somatic embryogenesis in mature *Quercus robur* L. trees. *Plant Cell Rep.* **2010**, *29*, 661–671. [CrossRef] [PubMed]

58. Zavattieri, M.A.; Frederico, A.M.; Lima, M.; Sabino, R.; Arnholdt-Schmitt, B. Induction of somatic embryogenesis as an example of stress-related plant reactions. *Electron. J. Biotechnol.* **2010**, *13*, 12–13. [CrossRef]

59. Narváez, I.; Martín, C.; Jiménez-Díaz, R.M.; Mercado, J.A.; Pliego-Alfaro, F. Plant Regeneration via Somatic Embryogenesis in Mature Wild Olive Genotypes Resistant to the Defoliating Pathotype of *Verticillium dahliae*. *Front. Plant Sci.* **2019**, *10*, 1471. [CrossRef]

60. Mazri, M.A.; Belkoura, I.; Meziani, R.; Mokhless, B.; Nour, S. Somatic embryogenesis from bud and leaf explants of date palm (*Phoenix dactylifera* L.) cv. Najda. *3 Biotech.* **2017**, *7*, 58. [CrossRef]

61. Rugkhla, A.; Jones, M.G.K. Somatic embryogenesis and plantlet formation in *Santalum album* and *S. spicatum*. *J. Exp. Bot.* **1998**, *49*, 563–571. [CrossRef]

62. Bonga, J.M. Can explant choice help resolve recalcitrance problems in in vitro propagation, a problem still acute especially for adult conifers? *Trees* **2017**, *31*, 781–789. [CrossRef]

63. Hernández, I.; Celestino, C.; Toribio, M. Vegetative propagation of *Quercus suber* L. by somatic embryogenesis: I. Factors affecting the induction in leaves from mature cork oak trees. *Plant Cell Rep.* **2003**, *21*, 759–764. [CrossRef] [PubMed]

64. Loyola-Vargas, V.M.; Avilez-Montalvo, J.R.; Avilés-Montalvo, R.N.; Márquez-López, R.E.; Galaz-Ávalos, R.M.; Mellado-Mojica, E. Somatic Embryogenesis in Coffea spp. In *Somatic Embryogenesis: Fundamental Aspects and Applications*; Loyola-Vargas, V., Ochoa-Alejo, N., Eds.; Springer: Cham, Switzerland, 2016; pp. 241–266. [CrossRef]

Article

Improved Method for Cryopreservation of Embryogenic Callus of *Fraxinus mandshurica* Pupr. by Vitrification

Xueqing Liu [1,†], Yingying Liu [1,†], Xiaoqian Yu [1], Iraida Nikolaevna Tretyakova [2], Alexander Mikhaylovich Nosov [3,4], Hailong Shen [1,5] and Ling Yang [1,5,*]

[1] State Key Laboratory of Tree Genetics and Breeding, School of Forestry, Northeast Forestry University, Harbin 150040, China
[2] Laboratory of Forest Genetics and Breeding, Institute of Forest V.N. Sukachev, Siberian Branch of Russian Academy of Sciences, 660036 Krasnoyarsk, Russia
[3] Laboratory of Cell Biology, Institute of Plant Physiology K.A. Timiryazev, Russian Academy of Sciences, 127276 Moscow, Russia
[4] Department of Plant Physiology, Biological Faculty, Lomonosov Moscow State University, 119991 Moscow, Russia
[5] State Forestry and Grassland Administration Engineering Technology Research Center of Korean Pine, Harbin 150040, China
* Correspondence: yangl-cf@nefu.edu.cn
† These authors contributed equally to this work.

Abstract: In order to simplify the experimental procedure and treatment procedure, we preserved the embryonic callus (EC) of *Fraxinus mandshurica* more efficiently. In this paper, we established a method for cryopreservation of EC of *F. mandshurica* by vitrification. EC was subcultured for 7–10 days (d). Vigorous EC with good growth conditions were selected, and cryopreservation was performed by vitrification. The best pre-culture method was to pre-culture EC on 0.5 mol·L^{-1} sucrose medium for 3 d, load and culture in the liquid woody plant medium (WPM) supplemented with 2 mol·L^{-1} glycerol and 0.4 mol·L^{-1} sucrose for 60 min, then dehydrate in 2 mL of plant vitrification solution 2 (PVS2) (30% glycerol + 15% dimethyl sulfoxide (DMSO) + 15% ethylene glycol + 0.4 mol·L^{-1} sucrose + liquid WPM). EC was rewarmed in a 40 °C water bath for 2 min after cooling in liquid nitrogen. The procedure for cryopreservation of *F. mandshurica* EC by the vitrification method established in this experiment is relatively reliable. The results from the present study provide a technical reference for improving the cryopreservation of *F. mandshurica* EC.

Keywords: *Fraxinus mandshurica*; embryogenic callus; cryopreservation; vitrification; regeneration

Citation: Liu, X.; Liu, Y.; Yu, X.; Tretyakova, I.N.; Nosov, A.M.; Shen, H.; Yang, L. Improved Method for Cryopreservation of Embryogenic Callus of *Fraxinus mandshurica* Pupr. by Vitrification. *Forests* **2023**, *14*, 28. https://doi.org/10.3390/f14010028

Academic Editor: Jorge Manuel P. L. Canhoto

Received: 30 October 2022
Revised: 13 December 2022
Accepted: 19 December 2022
Published: 23 December 2022

1. Introduction

Cryopreservation is the only safe and reliable method for the long-term preservation of germplasm resources [1–5]. Depending on the principle of dehydration, there are two main types of cryopreservation methods: one is the slow-cooling method, the other is the vitrification method. The slow-cooling method has been used for many years and has a complete technical system with relatively stable results. However, it requires expensive cooling devices such as a programmable coolers or continuous coolers for strict temperature control. The effect of slow-cooling on EC of *Fraxinus excelsior* L. was better than other cryopreservation methods [6]. Vitrification is the process whereby the material to be cryopreserved is pre-cultured, loaded and dehydrated in a high concentration of protective agent to form a non-crystalline vitrification as soon as possible during the rapid cooling process and kept safely in this state in liquid nitrogen at −196 °C [7,8]. Before it has time to produce ice crystals, the material to be preserved rapidly goes into a glassy state, thus achieving a long-term preservation result [9,10]. The experimental conditions required for the vitrification method are relatively simple and do not require expensive apparatus. However, the high concentration of cryoprotectants under vitrification treatment can have a toxic effect on the material. Therefore, it is important to choose the right vitrification solution for the vitrification of each species for ultra-low temperature preservation [11,12].

Fraxinus mandshurica Rupr. is distributed in Northeast China, North China, Shaanxi, Gansu, Hubei and other places. It is one of the three most valuable hard broad-leaved species in the northeastern forest region. This species is made of excellent material and is used for high-grade furniture, tools and special construction, but the excellent resources of water willow are limited [13]. In previous studies, we have successfully induced somatic embryos [14] and EC [15] in *F. mandshurica* and have been able to preserve EC in petri dishes for three consecutive years by regular subculture (switching to fresh medium every month). However, in exceptional circumstances (e.g., COVID-19), when regular succession cultures cannot be entered into the laboratory in time, or when contamination is caused by improper manual operation and environmental conditions, callus in vitro is subject to genetic variation, embryogenic loss and cell death. This can result in the loss of a lot of experimental material. This is why we are working on the development of cryoprotectants techniques for *F. mandshurica*. We have recently reported on a new system for the slow-cooling of EC of *F. mandshurica* [16]. However, the lack of a programmable cooler to precisely control the temperature range and rate of cooling has led to some concerns about the stability of EC preservation results. Therefore, we believe it is necessary to continue to develop new, simple and inexpensive methods to provide a reference for efficient and stable conservation of valuable tree species (e.g., *F. mandshurica*) resources through improved techniques for cryopreservation of *F. mandshurica* EC.

To simplify experimental procedures and handling steps and to preserve embryonic guardian tissue effectively, in this paper, we establish a method for cryopreservation of EC of *F. mandshurica* by vitrification and carry out experiments on the differentiation of callus and plant regeneration after cryopreservation. An efficient and stable technique for the cryopreservation of *F. mandshurica* EC by vitrification is established. This technique provides a feasible method for perfecting the preservation of *F. mandshurica* germplasm resources.

2. Materials and Methods

2.1. Plant Materials

The method of obtaining EC of *F. mandshurica* referred to the method of Yu et al. [16]. Specific methods: Immature seeds were collected in early August from free-pollinated parent trees present on the campus of Northeast Forestry University, Harbin, Heilongjiang Province, China (126°37′55″ E, 45°43′16″ N). The cotyledons of sterile immature zygotic embryos were cultured in a woody plant medium (WPM) with 0.1 mg·L^{-1} 6-Benzylaminopurine (6-BA) and 0.15 mg·L^{-1} 2, 4-Dichlorophenoxyacetic Acid (2,4-D) to obtain EC and subcultured every 4 weeks. The EC of Z2 (W2) type of *F. mandshurica* selected by Yu et al. [16] for 7–10 d was selected as the material for cryopreservation by vitrification.

2.2. Experimental Method

2.2.1. Cryopreservation of EC by Vitrification

All the conditions of vitrification cryopreservation were designed by a single factor experiment. The basic culture conditions were the same as the cryopreservation of EC of *Anemarrhena asphodeloides* Bunge by vitrification [17]. The specific methods were as follows:

(1) Sucrose concentration selection: 1.0 g of EC was inoculated on solid WPM with different concentrations of sucrose (0, 0.3, 0.4, 0.5, 0.6, 0.7 and 0.8 mol·L^{-1}). It was cultured in the dark at 25 °C for 1 d. The pre-cultured EC was added to a 1.8 mL cooling tube, along with the loading solution (2 mol·L^{-1} glycerol + 0.4 mol·L^{-1} sucrose + liquid WPM), and treated at room temperature for 40 min. Afterward, the loading solution was removed, and 2 mL of plant vitrification solution 2 (PVS2) was added (30% glycerol + 15% DMSO + 15% ethylene glycol + 0.4 mol·L^{-1} sucrose + liquid WPM). It was dehydrated in the ice water mixture for 40 min, then added to the cooling tube, which was kept in liquid nitrogen immediately. It was then rewarmed for 2 min at 40 °C water bath after 2 h. After rapid removal of PVS2, it was washed 4 times with loading

solution in the horizontal flow clean bench, at intervals of 10–15 min. Finally, the EC was evenly dispersed on the filter paper, and the excess water was absorbed with a pipette and transferred to WPM for dark culture at 25 °C. After 24 h, the relative survival percentage of cells was calculated.

(2) Pre-culture time selection: The sucrose concentration of the pre-culture with the highest relative survival percentage was selected, and different pre-culture times (0, 1, 2, 3, 4 and 5 d) were screened at room temperature, wherein 0 d was the control group. After loading, dehydration, rewarming and washing for 24 h, for different pre-culture times, we compared their effects on the relative survival of cells after vitrification cryopreservation of *F. mandshurica* EC.

(3) Loading time selection: The loading time (30, 40, 50, 60, 70 and 80 min) was determined by the treatments with higher cell relative survival percentage in pre-cultured sucrose concentration and time. The relative survival percentage of cells was detected after dehydration, cooling, rewarming and washing for 24 h.

(4) Dehydration time selection: Vitrification dehydration time was screened based on the highest relative survival percentage in pre-culture and loading time treatment. A total of 2 mL of PVS2 (30% glycerol + 15% DMSO + 15% ethylene glycol + 0.4 mol·L^{-1} sucrose + liquid WPM) was added to the EC mixture, and the dehydration times were 30, 40, 50, 60 and 70 min. We then compared the effects of different dehydration times on the cell relative survival percentage after cryopreservation of *F. mandshurica* EC after vitrification.

(5) Rewarming method selection: The different rewarming methods (25 °C room temperature, 40 °C water bath and running water washing) were determined by the treatment with the highest cell survival percentage with sucrose concentration, pre-culture time, loading time and dehydration time. We then compared the effects of different rewarming methods on the cell relative survival percentage after cryopreservation of *F. mandshurica* EC after vitrification.

(6) Rewarming time selection: Based on the abovementioned experiments, the rewarming time (1, 2, 3, 4 and 5 min) was determined. Then the loading solution was washed for 24 h, and the cell survival percentage was detected.

(7) Recovery culture of EC after resuscitation was based on the Yu et al. [15] recovery culture method for EC of *F. mandshurica*. Specific methods: After cryopreservation and resuscitation by vitrification, the EC of *F. mandshurica* was restored in WPM with 0.1 mg·L^{-1} 6-BA and 0.15 mg·L^{-1} 2,4-D. Subculture multiplication was carried out after 15–20 d. Differentiation culture was performed on $\frac{1}{2}$MS with 1.0 mg·L^{-1} 6-BA medium. Then, it was mature cultured on $\frac{1}{2}$MS with 1.0 mg·L^{-1} ABA medium. Germination and rooting culture was performed on $^{1}/_{3}$MS medium with 0.01 mg·L^{-1} NAA medium, and the images were taken.

2.2.2. Determination of the Cell Relative Survival Percentage and Observation of Recovery Culture

For the fresh weight measurement and 2,3,5-Triphenyltetrazolium chloride (TTC) staining method of callus refer to Yu et al. [16]. The relative survival percentage of cells after cryopreservation is expressed as the ratio of absorbance of treatment and control [17]. The following formula was used:

$$Relative\ cell\ relative\ survival\ percentage\ (\%) = \frac{OD\ value\ of\ cryopreservation\ treatment}{OD\ value\ of\ unprocessed} \times 100$$

Observation and record of recovery culture: According to the method of Liu et al. [15], the proliferation, differentiation and seedling emergence of *F. mandshurica* EC were observed and recorded every 30 d. The following formulae were used:

$$Callus\ proliferation\ coefficient = \frac{weight\ of\ callus\ after\ proliferation}{weight\ of\ calls\ during\ inoculation}$$

$$\text{Callus differentiation percentage } (\%) = \frac{number\ of\ callus\ differentiated\ from\ somatic\ embryos}{number\ of\ callus\ inoculated} \times 100$$

$$\text{Regenerated plant percentage } (\%) = \frac{number\ of\ somatic\ embryos\ germinated\ into\ seedlings}{number\ of\ somatic\ embryos\ inoculated} \times 100$$

2.3. Statistical Analysis

Data were sorted by Microsoft Excel 2007, and one-way ANOVA was conducted by SPSS (2015). Graphs are drawn using sigmapplot software (2011). All data were the mean ± standard deviation of three replicates. Duncan's method was used to compare the significance between the data.

3. Results

3.1. Effects of Sucrose Concentration on Fresh Weight and Cell Survival Percentage of EC

Different sucrose concentrations showed significant effects on the fresh weight of *F. mandshurica* EC after cryopreservation and resuscitation by vitrification ($p < 0.05$) (Table 1). Compared with the fresh weight of EC cultured with sucrose on the 7th and 14th day of culture, it was found that the weight gain of EC was not significant. However, the fresh weight of EC cultured with sucrose increased significantly on the 21st day of culture. The fresh weight of EC cultured with the same sucrose concentration increased gradually with the prolongation of the culture period. On the 60th day, with the increase of concentration, the weight increased significantly at first and then decreased significantly. Therefore, after treatment with 0.5 mol·L^{-1} sucrose solution, the weight of EC increased to 1.82 g, which was the highest. When the concentration of sucrose was 0 mol·L^{-1}, the fresh weight of EC was 1.04 g, and no significant difference was observed compared with that on day 0.

Table 1. Effect of sucrose concentration on EC fresh weight of *F. mandshurica* after cryopreservation by vitrification.

Concentration (mol·L^{-1})	Culture Time (d)					
	0	7	14	21	30	60
0	0.91 ± 0.02 a	0.93 ± 0.01 a	0.94 ± 0.02 a	1.00 ± 0.01 c	1.02 ± 0.02 b	1.04 ± 0.01 c
0.3	0.90 ± 0.01 a	0.93 ± 0.02 a	0.94 ± 0.01 a	1.15 ± 0.02 b	1.24 ± 0.04 b	1.52 ± 0.03 b
0.4	0.93 ± 0.05 a	0.96 ± 0.03 a	0.98 ± 0.02 a	1.16 ± 0.02 b	1.26 ± 0.03 b	1.56 ± 0.04 b
0.5	0.92 ± 0.06 a	0.95 ± 0.05 a	0.97 ± 0.03 a	1.31 ± 0.03 a	1.54 ± 0.04 a	1.82 ± 0.03 a
0.6	0.93 ± 0.01 a	0.95 ± 0.01 a	0.98 ± 0.02 a	1.15 ± 0.02 b	1.25 ± 0.04 b	1.53 ± 0.03 b
0.7	0.93 ± 0.05 a	0.95 ± 0.03 a	0.96 ± 0.01 a	1.13 ± 0.02 b	1.23 ± 0.03 b	1.54 ± 0.04 b

Note: The data in the table represent the fresh weight (g) of EC. The data in the table are expressed as the mean ± SE; different lowercase letters in the same column represent significant differences ($p < 0.05$).

Sucrose treatment significantly affected the cell survival percentage of EC after cryopreservation by vitrification (Figure 1). The cell survival percentage without sucrose pre-culture (0 mol·L^{-1}) was 8.83%, which was the control group. The cell survival percentage of the sucrose pre-culture was higher than that of the control. With the increase of concentration, the cell survival percentage increased significantly at first and then decreased significantly. After pre-culture with 0.3 mol·L^{-1} sucrose, the cell survival percentage was significantly higher than that of the control group. After sucrose culture with 0.5 mol·L^{-1}, the cell survival percentage was the highest (81.86%) ($p < 0.05$). With the increase of sucrose concentration to 0.8 mol·L^{-1}, the cell survival percentage decreased to 20.41%. Therefore, 0.5 mol·L^{-1} sucrose pre-culture is the best choice for vitrification cryopreservation of *F. mandshurica* EC.

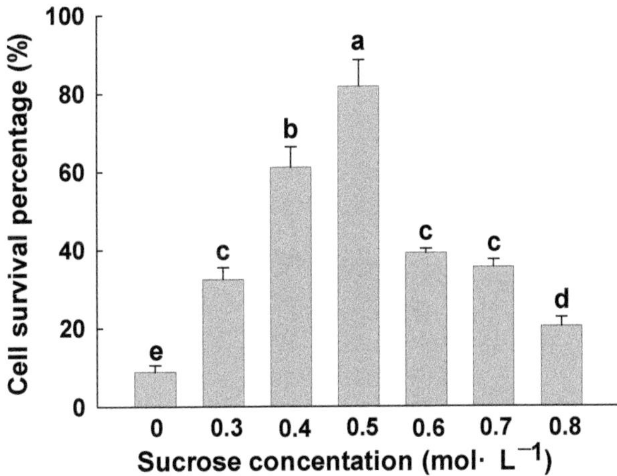

Figure 1. Effects of sucrose concentration on the cell survival of *F. mandshurica* EC after cryopreservation by vitrification. Note: Different letters are significantly different from each other at $p < 0.05$, using Duncan's Multiple Range Test.

3.2. Effect of Pre-Culture Time on Fresh Weight and Cell Survival Percentage of EC

The pre-culture time had a significant effect on the fresh weight of EC after vitrification cryopreservation (Table 2). On the 21st day of culture, compared with the day 0, the fresh weight of EC treated with different sucrose concentrations increased significantly. The fresh weight of each treatment increased significantly. In addition, on the same culture day, with the increase of pre-culture time, the weight increased significantly at first and then decreased significantly. When cultured for the 60th day, the fresh weight of the EC pre-cultured for 3 days was the highest (1.56 g). When the pre-culture time was 0 day, the fresh weight of EC reached the minimum value (1.13 g). This treatment is very different from other treatments ($p < 0.05$).

Table 2. Effect of pre-culture time on fresh weight of *F. mandshurica* EC after cryopreservation by vitrification.

Pre-Culture	Culture Time (d)					
Time (d)	0	7	14	21	30	60
0	0.93 ± 0.02 a	1.00 ± 0.03 a	1.02 ± 0.03 a	1.09 ± 0.02 c	1.11 ± 0.01 c	1.13 ± 0.02 c
1	0.92 ± 0.03 a	0.99 ± 0.05 a	1.01 ± 0.05 a	1.14 ± 0.02 b	1.27 ± 0.03 b	1.35 ± 0.03 b
2	0.93 ± 0.02 a	1.01 ± 0.04 a	1.03 ± 0.03 a	1.18 ± 0.02 a	1.35 ± 0.03 a	1.37 ± 0.03 b
3	0.92 ± 0.01 a	1.00 ± 0.03 a	1.03 ± 0.03 a	1.19 ± 0.04 a	1.36 ± 0.03 a	1.56 ± 0.04 a
4	0.94 ± 0.02 a	1.00 ± 0.04 a	1.02 ± 0.04 a	1.14 ± 0.01 b	1.15 ± 0.02 c	1.35 ± 0.04 b
5	0.93 ± 0.01 a	0.99 ± 0.03 a	1.01 ± 0.02 a	1.14 ± 0.02 b	1.26 ± 0.03 b	1.34 ± 0.05 b

Note: The data in the table represent the fresh weight (g) of EC. The data in the table are expressed as the mean ± SE; different lowercase letters in the same column represent significant differences ($p < 0.05$).

Pre-culture time showed a significant effect on the relative survival percentage of EC during vitrification cryopreservation (Figure 2). When the control group was pre-cultured for day 0, the cell survival percentage was 9.51%. When it was pre-cultured for 3 days, the cell relative survival percentage was the highest (66.64%). The difference between the two groups was significant ($p < 0.05$). With a gradual increase in pre-culture days, the cell relative survival percentage first increased and then decreased. When pre-cultured for 5 days, the cell survival percentage decreased to 28.28%. Thus, the optimal sucrose pre-culture time before cryopreservation of *F. mandshurica* EC was 3 days.

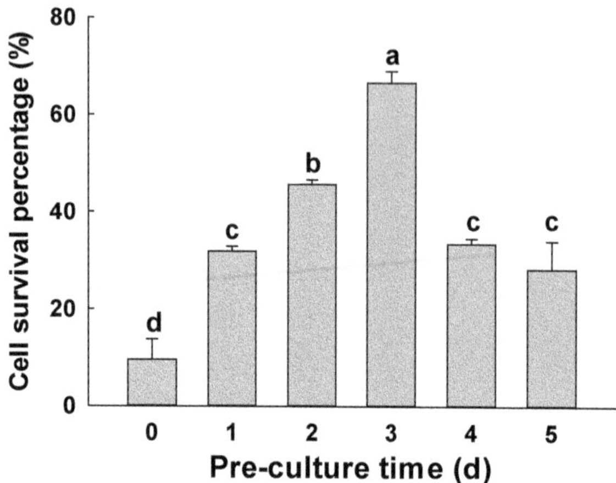

Figure 2. Effects of pre-culture time on cell survival percentage of *F. mandshurica* EC after cryopreservation by vitrification. Note: Different letters are significantly different from each other at $p < 0.05$, using Duncan's Multiple Range Test.

3.3. Effect of Loading Time on Fresh Weight and Cell Survival Percentage of EC

In the process of vitrification cryopreservation, different loading time had a significant effect on the fresh weight of EC (Table 3). On the same culture day, with the increase of loading time, the weight increased significantly at first and then decreased significantly. On the 60th day, the fresh weight of EC treated for 60 min was the highest (1.76 g), which was significantly different from that of other loading time ($p < 0.05$). When the loading time was 30 min, the fresh weight of EC was 1.11 g, which was the smallest.

Table 3. Effect of loading time on fresh weight of *F. mandshurica* EC after vitrification cryopreservation.

Loading Time (min)	Culture Time (d)					
	0	7	14	21	30	60
30	0.96 ± 0.02 a	1.00 ± 0.02 a	1.05 ± 0.01 a	1.09 ± 0.02 c	1.10 ± 0.02 d	1.11 ± 0.01 e
40	0.95 ± 0.03 a	1.00 ± 0.02 a	1.03 ± 0.01 a	1.14 ± 0.03 b	1.27 ± 0.03 bc	1.34 ± 0.07 c
50	0.95 ± 0.02 a	0.99 ± 0.02 a	1.02 ± 0.02 a	1.24 ± 0.02 a	1.31 ± 0.02 a	1.44 ± 0.08 b
60	0.97 ± 0.02 a	1.01 ± 0.02 a	1.03 ± 0.02 a	1.27 ± 0.02 a	1.37 ± 0.04 a	1.76 ± 0.03 a
70	0.96 ± 0.01 a	1.01 ± 0.02 a	1.04 ± 0.03 a	1.12 ± 0.03 b	1.21 ± 0.04 c	1.51 ± 0.03 b
80	0.96 ± 0.01 a	0.98 ± 0.02 a	1.03 ± 0.02 a	1.08 ± 0.04 bc	1.21 ± 0.03 c	1.24 ± 0.03 d

Note: The data in the table represent the fresh weight (g) of EC. The data in the table are expressed as the mean ± SE; different lowercase letters in the same column represent significant differences ($p < 0.05$).

Different loading time had a significant effect on the cell survival percentage of EC during vitrification cryopreservation (Figure 3). With the increase of loading time, the cell survival percentage increased significantly at first and then decreased significantly. When the loading time was 30 min, the cell survival percentage was the smallest (20.24%). When the loading time was 60 min, the cell survival percentage was the highest (61.56%) and significantly different from others ($p < 0.05$). When the loading time was 80 min, the cell survival percentage decreased significantly. Thus, the optimal loading time for cryopreservation of *F. mandshurica* EC by vitrification was 60 min.

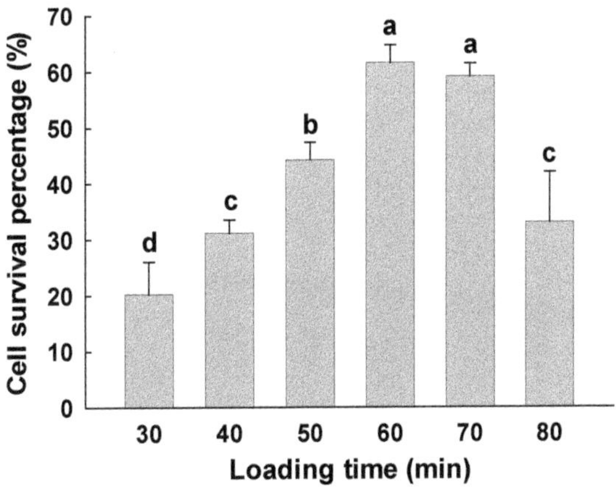

Figure 3. Effects of loading time on cell survival percentage of *F. mandshurica* EC after cryopreservation by vitrification. Note: Different letters are significantly different from each other at *p* < 0.05, using Duncan's Multiple Range Test.

3.4. Effects of Dehydration Time on Fresh Weight and Cell Survival Percentage of EC

In the process of vitrification cryopreservation, dehydration time showed a significant effect on the fresh weight of EC (Table 4). The fresh weight increased with the increase of recovery culture time. On the 60th day of culture, with the extension of dehydration time, the fresh weight of EC showed a trend of first increasing and then decreasing. When treated for 50 min, the maximum fresh weight of EC was 1.75 g, which was 80.41% higher than that of the resuscitation culture on day 0. When treated for 70 min, the lowest fresh weight of 1.29 g was observed, which was 31.63% higher than that of the resuscitation culture on day 0. A significant difference in the fresh weight of the callus was observed between 50 min and 70 min of dehydration time (*p* < 0.05).

Table 4. Effect of dehydration time on fresh weight of *F. mandshurica* EC after cryopreservation by vitrification.

Dehydration Time (min)	Culture Time (d)					
	0	7	14	21	30	60
30	0.98 ± 0.06 a	1.01 ± 0.05 a	1.04 ± 0.03 a	1.12 ± 0.04 b	1.22 ± 0.03 b	1.34 ± 0.05 b
40	0.97 ± 0.02 a	0.99 ± 0.02 a	1.01 ± 0.02 a	1.16 ± 0.03 b	1.32 ± 0.09 b	1.69 ± 0.05 a
50	0.99 ± 0.03 a	1.01 ± 0.04 a	1.02 ± 0.04 a	1.27 ± 0.10 a	1.48 ± 0.12 a	1.75 ± 0.10 a
60	0.97 ± 0.04 a	1.02 ± 0.02 a	1.03 ± 0.02 a	1.11 ± 0.03 b	1.19 ± 0.03 b	1.38 ± 0.02 b
70	0.98 ± 0.04 a	1.00 ± 0.03 a	1.02 ± 0.07 a	1.14 ± 0.02 b	1.21 ± 0.03 b	1.29 ± 0.03 b

Note: The data in the table represent the fresh weight (g) of EC. The data in the table are expressed as the mean ± SE; different lowercase letters in the same column represent significant differences (*p* < 0.05).

With different dehydration time, the cell survival percentage of EC after vitrification cryopreservation is also different (Figure 4). With the increase of dehydration time, the cell survival percentage of EC showed a trend of first increasing and then decreasing. When the dehydration time was 70 min, the cell survival percentage decreased sharply (29.13%). When treated for 50 min, the relative survival percentage was the highest (67.65%), which was significantly different from that of 70 min. Under a 30 min treatment, the cell survival percentage was 35.80%. Thus, the optimal dehydration time for the cryopreservation of *F. mandshurica* EC by vitrification was 50 min, and its relative survival percentage was also the highest.

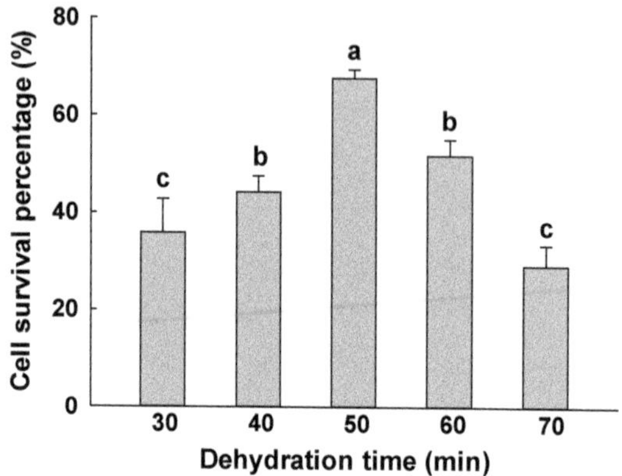

Figure 4. Effects of dehydration time on cell survival percentage of *F. mandshurica* EC after cryopreservation by vitrification. Note: Different letters are significantly different from each other at $p < 0.05$, using Duncan's Multiple Range Test.

3.5. Effects of Rewarming Methods on Fresh Weight and Cell Survival Percentage of EC

The method of rewarming affected the fresh weight of EC after vitrification cryopreservation, but the difference was not significant (Table 5). Under the rewarming treatment at 25 °C, the fresh weight of EC increased gradually with the increase of recovery culture time. On the 60th day of culture, the fresh weight of EC was the lowest (1.05 g), which was 8.25% higher than that on day 0. Under a 40 °C water bath treatment, the fresh weight of EC was the highest (1.33 g), which was 46.15% higher than that on day 0. For *F. mandshurica* EC after vitrification cryopreservation, the cell survival percentage was different with different rewarming methods (Figure 5). Under a 25 °C treatment, the cell survival percentage was 35.80%. Under a 40 °C water bath treatment, the cell survival percentage was the highest (61.84%). When the callus was washed with running water, the cell survival percentage was the lowest (22.71%). Thus, the 40 °C water bath is the most suitable rewarming method for the cryopreservation of *F. mandshurica* EC by vitrification.

Table 5. Effects of rewarming methods on the fresh weight of *F. mandshurica* EC after cryopreservation by vitrification.

Rewarming Method	Culture Time (d)					
	0	7	14	21	30	60
25 °C room temperature	0.97 ± 0.03 a	0.99 ± 0.02 a	0.99 ± 0.01 a	1.01 ± 0.01 c	1.02 ± 0.01 c	1.05 ± 0.02 c
40 °C water bath	0.91 ± 0.02 a	0.94 ± 0.03 a	0.96 ± 0.02 a	1.14 ± 0.01 a	1.19 ± 0.01 a	1.33 ± 0.04 a
Running water	0.92 ± 0.05 a	0.94 ± 0.04 a	0.98 ± 0.04 a	1.05 ± 0.03 b	1.12 ± 0.02 b	1.18 ± 0.03 b

Note: The data in the table represent the fresh weight (g) of EC. The data in the table are expressed as the mean ± SE; different lowercase letters in the same column represent significant differences ($p < 0.05$).

3.6. Effect of Rewarming Time on Fresh Weight and Cell Survival Percentage of EC

Under a 40 °C water bath treatment, the fresh weight of *F. mandshurica* EC was different with different rewarming times. With the extension of rewarming time, the fresh weight of EC showed a trend of first increasing and then decreasing (Table 6). The fresh weight of EC increased gradually with an increase in recovery culture time. When the rewarming time was 2 min, the fresh weight of EC cultured on the 60th day was the highest (1.47 g), which increased by 59.78% compared with the recovery culture on day 0. When treated for 1 min, the minimum fresh weight of EC was 1.17 g, which was 24.47% higher than that on day 0. Different rewarming times showed a significant effect on the relative survival percentage of EC of *F. mandshurica* after cryopreservation by vitrification (Figure 6). The cell survival percentage first increased and then decreased with an increase in rewarming time. When treated for 2 min, the cell survival percentage was the highest (62.13%). However, when the rewarming times were 1 min and 5 min, the cell survival percentages were lower, namely 15.11% and 22.93%. When treated for 4 min, the cell survival percentage decreased sharply and was significantly different from that at 2 min ($p < 0.05$).

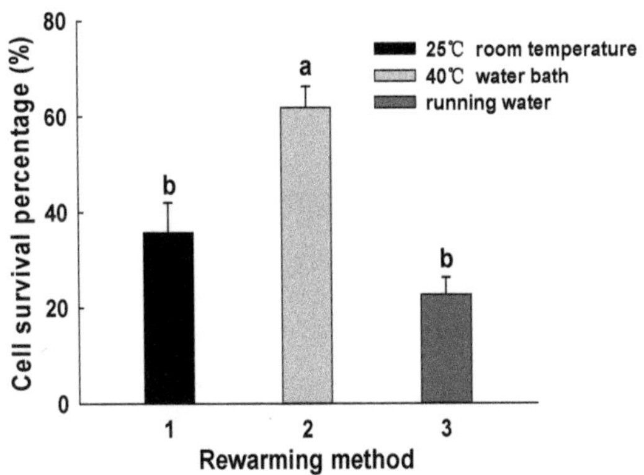

Figure 5. Effects of rewarming method on cell survival percentage of *F. mandshurica* EC after cryopreservation by vitrification. Note: Different letters are significantly different from each other at $p < 0.05$, using Duncan's Multiple Range Test.

Table 6. Effect of rewarming time on fresh weight of *F. mandshurica* EC after cryopreservation by vitrification.

Rewarming Time (min)	Culture Time (d)					
	0	7	14	21	30	60
1	0.94 ± 0.01 a	0.98 ± 0.01 a	0.99 ± 0.02 a	1.09 ± 0.05 ab	1.12 ± 0.06 b	1.17 ± 0.03 b
2	0.92 ± 0.04 a	0.95 ± 0.05 a	0.99 ± 0.04 a	1.14 ± 0.06 a	1.26 ± 0.05 a	1.47 ± 0.12 a
3	0.92 ± 0.02 a	0.94 ± 0.02 a	0.97 ± 0.06 a	1.15 ± 0.09 a	1.21 ± 0.06 a	1.28 ± 0.03 b
4	0.90 ± 0.03 a	0.93 ± 0.03 a	0.98 ± 0.05 a	1.01 ± 0.04 b	1.10 ± 0.02 b	1.22 ± 0.03 b
5	0.94 ± 0.02 a	0.94 ± 0.02 a	0.97 ± 0.04 a	1.03 ± 0.05 b	1.11 ± 0.03 b	1.19 ± 0.03 b

Note: The data in the table represent the fresh weight (g) of EC. The data in the table are expressed as the mean ± SE; different lowercase letters in the same column represent significant differences ($p < 0.05$).

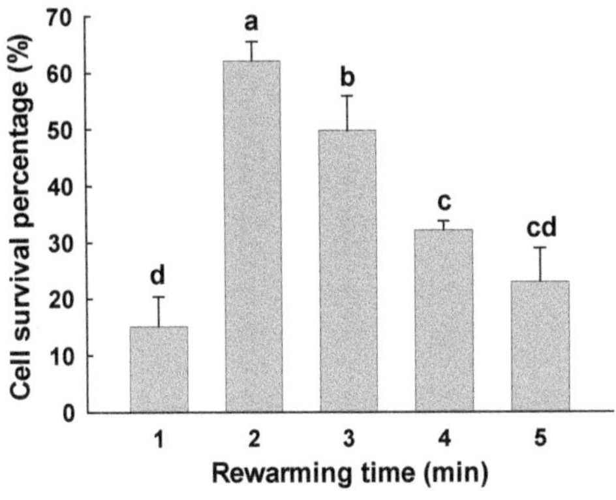

Figure 6. Effects of rewarming time on cell survival percentage of *F. mandshurica* EC after cryopreservation by vitrification. Note: Different letters are significantly different from each other at $p < 0.05$, using Duncan's Multiple Range Test.

3.7. Somatic Embryogenesis and Plant Regeneration after Resuscitation

The proliferation coefficient of the EC of *F. mandshurica* after cryopreservation by vitrification was 2.69, whereas the proliferation coefficient of uncryopreserved EC (control) was 2.86. When the callus was restored on day 60 of culture, the callus was loose and proliferated (Figure 7a). Thereafter, it was transferred to a differentiation medium. Although it could differentiate into somatic embryos normally, the number of somatic embryos was less (Figure 7b). The callus differentiation percentage was 53.87%, and the callus differentiation percentage of the control was 59.44%. After 2 months of germination and rooting culture, the somatic embryos germinated normally into seedlings (Figure 7c). The plant regeneration percentage of somatic embryos after cryopreservation was 20.97%, while that of somatic embryos without cryopreservation was 25.87%. A month later, somatic embryos were cultured in canned bottles for rooting (Figure 7d). To conclude, after cryopreservation of *F. mandshurica* EC by vitrification, the callus proliferation coefficient, callus differentiation percentage, and plant regeneration percentage decreased compared with the control, but the difference was not significant. This showed that the vitrification method is suitable for the preservation of *F. mandshurica* EC.

Figure 7. The process of recovery culture after cryopreservation of *F. mandshurica* EC by the vitrification method: (**a**) the 60th day of recovery of *F. mandshurica* EC; (**b**) callus-differentiated somatic embryo; (**c**) *F. mandshurica* buds generated from somatic embryo; (**d**) Emblings of *F. mandshurica*. bar = 1 cm.

4. Discussion

4.1. Study of Cryopreservation Methods

Cryopreservation ensures the safe and effective conservation of plant genetic resources [18]. Slow-cooling and vitrification are currently common methods of cryopreservation methods. The slow-cooling method involves dehydrating the material, placing it in a gradient cooling box and cooling it down to $-80\ °C$ at $-1\ °C/min$. Finally, the cooled material is quickly plunged into liquid nitrogen for cryopreservation [19]. The temperature of EC of *Fraxinus excelsior* L. was reduced at a rate of $-0.5/-1\ °C/min$ [20]. However, this method is not suitable for laboratories that do not have expensive apparatus to control the temperature. This is because it can lead to a less precise rate of cooling during cryopreservation, which can affect the cooling results. A new cryopreservation technique, cryopreservation by vitrification, was developed for the cryopreservation of *F. mandshurica* EC. This method, in contrast to the published technique for cryopreservation of *F. mandshurica* EC by slow-cooling [16], does not require a temperature-controlled apparatus, does not require a transition procedure to cryogenic cooling and is simple to operate, allowing for cryopreservation of the material in almost any laboratory. Vitrification avoids the formation of intracellular and extracellular ice crystals to a great extent, eliminates the mechanical damage caused by intracellular ice, and facilitates the entry of organs, callus and other parts to a common vitrification state. Moreover, the operation time is short, and the steps are simple. The process is unique in preserving the integrity of organs and tissue structures, and it does not need expensive cooling instruments or equipment [9,21,22]. Compared to the slow-cooling technique [16], the cell survival percentage after cryopreservation by vitrification in this paper was slightly higher than that of the slow-cooling method by 1.04%, and the regeneration percentage was slightly lower than the slow-cooling method by 2.62% (the difference was not significant). Although the percentage of regenerated plants by vitrification was slightly low, we can optimize the technical steps during future studies to improve the percentage of regenerated plants in recovery culture after cryopreservation by vitrification.

4.2. Effect of Pre-Culture on Vitrification Cryopreservation

In the process of cryopreservation by vitrification, steps such as pre-culture, loading, dehydration, rewarming and restoration culture, are important for the survival of materials. The vitrification method uses hyperosmotic pre-culture to reduce the content of free water in plant cells so that the cells can withstand low-temperature stress. In this study, high concentrations of sucrose were not conducive to the survival of *F. mandshurica* EC. Cryopreservation of *F. mandshurica* EC by vitrification was best suited to culture in $0.5 \ mol \cdot L^{-1}$ sucrose pre-culture medium. This finding is similar to that of cryopreservation of *Persea americana* callus [23]. In this paper, the cell survival percentage of *F. mandshurica* EC pre-cultured with $0.5 \ mol \cdot L^{-1}$ sucrose was slightly higher than that of the slow-cooling method, and the fresh weight of EC was slightly lower than that of the slow-cooling method, but the difference between the two was not significant [16]. Loading is an indispensable step in the process of vitrification cryopreservation. Loading time is treated with cryoprotectant solution at room temperature for a certain period during cryopreservation to reduce the water content of cells and the persecution of drastic changes in osmotic pressure. The treatment time of the loading solution also affects the relative survival percentage of materials after cryopreservation. Too short or too long a time causes damage to materials and reduces the survival percentage. During the cryopreservation of *F. mandshurica* EC, the cell survival percentage decreased when the loading time was more than 60 min. This phenomenon is reflected in the callus of *Satureja spiigera* [24], the apical bud of *Smallanthus sonchifolius* [25] and the apical meristem of *Chlorophytum borivilianum* [26]. In future studies, the loading time can be reduced to improve cell survival after loading of *F. mandshurica* EC.

4.3. Effect of Dehydration on Vitrification Cryopreservation

During dehydration of vitrification cryopreservation, the treatment time of plant vitrification solution is the key to cell vitrification cryopreservation [27]. In this study, long dehydration time led to a decrease in the cell survival percentage of *F. mandshurica*, which is consistent with the results of the shoot tip of *Gentiana kurroo* [28] and the shoot tip of *Viola stagnina* [29]. When the dehydration time is too short, the cell dehydration is not less, and it is difficult to reach the vitrification state quickly in the process of cooling treatment. The dehydration time of *Panax ginseng* EC was 90 min [30]. Compared to the slow-cooling method [16], the dehydration time in this study was reduced by 40 min, and the cell survival percentage was increased by 8.81%, indicating that the dehydration time should be strictly controlled during the dehydration process prior to cryopreservation of *F. mandshurica* EC.

4.4. Study of Resuscitation Culture Conditions

The rewarming method and time are important to the cell recovery culture. The optimal rewarming time for the vitrification of *F. mandshurica* EC was 2 min in the water bath at 40 °C. This is similar to the results of preservation of *F. mandshurica* EC by the slow-cooling method [16], the callus of *Satureja spiigera* [24], the apical meristem of *Chlorophytum borivilianum* [26] and the shoot tip of *Viola stagnina* [29]. In addition to water bath rewarming, room temperature rewarming is another alternative. The optimal rewarming method for shoot tips of *Vaccinium myrtillus* and *Allium cepa* after cryopreservation by vitrification should be performed at room temperature for 20 min [31,32]. Future research could further simplify the method of restoring healing tissue culture after cryopreservation by rewarming at room temperature to improve cell survival and regeneration plant percentage.

5. Conclusions

In this study, the relationships between pre-culture sucrose concentration, pre-culture time, loading time, dehydration time, rewarming method, rewarming time, callus fresh weight and cell survival percentage during cryopreservation of *F. mandshurica* EC by vitrification were determined. The technique of vitrification cryopreservation of *F. mandshurica* EC was optimized using various parameters. Specifically, the EC of *F. mandshurica* was pre-cultured in 0.5 mol·L^{-1} sucrose solution at room temperature for 3 days and then cultured in liquid WPM with 2 mol·L^{-1} glycerol and 0.4 mol·L^{-1} sucrose for 60 min. It was then dehydrated for 50 min in PVS2 (30% glycerol + 15% DMSO + 15% glycol + 0.4 mol·L^{-1} sucrose + liquid WPM) and stored in liquid nitrogen. After rewarming for 2 min in the 40 °C water bath, the preserved EC could be recovered and cultured to form intact plants.

Author Contributions: The authors confirm contribution to the paper as follows: L.Y., H.S., I.N.T. and A.M.N. conceived and designed the study. X.L. and Y.L. collected plant materials and prepared samples for analysis. X.L., Y.L. and X.Y. analyzed the results for experiments. X.L. and L.Y. wrote the paper. All authors have read and agreed to the published version of the manuscript.

Funding: The work was supported by the Fundamental Research Funds for the Central Universities (2572018BW02), the Innovation Project of State Key Laboratory of Tree Genetics and Breeding (2021B01), the National Key R&D Program of China (2017YFD0600600) and the National Natural Science Foundation of China (31400535 and 31570596).

Data Availability Statement: Data presented in this study are available from the corresponding author upon request.

Acknowledgments: We thank three anonymous reviewers and the editor for comments that improved an earlier draft of this article.

Conflicts of Interest: The authors declare no conflict of interest.

References

1. Hardstaff, L.K.; Sommerville, K.D.; Funnekotter, B.; Bunn, E.; Offord, C.A.; Mancera, R.L. Myrtaceae in Australia: Use of Cryobiotechnologies for the Conservation of a Significant Plant Family under Threat. *Plants* **2022**, *11*, 1017. [CrossRef] [PubMed]
2. Hasan, M.; Fayter Alice, E.R.; Gibson, M.I. Ice recrystallization inhibiting polymers enable glycerol-free cryopreservation of microorganisms. *Biomacromolecules* **2018**, *19*, 3371–3376. [CrossRef] [PubMed]
3. Kaczmarczyk, A.; Turner, S.R.; Bunn, E.; Mancera, R.L.; Dixon, K.W. Cryopreservation of threatened native Australian species—what have we learned and where to from here? *Vitr. Cell Dev. Biol.-Plant.* **2011**, *47*, 17–25. [CrossRef]
4. Panis, B. Sixty years of plant cryopreservation: From freezing hardy mulberry twigs to establishing reference crop collections for future generations. *Acta Hortic.* **2019**, *1234*, 1–8. [CrossRef]
5. Zhan, L.; Li, M.G.; Hays, T.; Bischof, J. Cryopreservation method for *Drosophila melanogaster* embryos. *Nat. Commun.* **2021**, *12*, 2412–2421. [CrossRef]
6. Ozudogru, E.A.; Capuana, M.; Kaya, E.; Panis, B.; Lambard, M. Cryopreservation of *Fraxinus excelsior* L. embryogenic callus by one-step freezing and slow cooling techniques. *Cryo Lett.* **2010**, *31*, 63–75.
7. Bekheet, S.A.; Sota, V.; EI-shabrawi, H.M.; EI-Minisy, A.M. Cryopreservation of shoot apices and callus cultures of globe artichoke using vitrification method. *J. Genet. Eng. Biotechnol.* **2020**, *18*, 2–9. [CrossRef]
8. Wang, M.R.; Bi, W.L.; Shukla, M.R.; Ren, L.; Hamborg, Z.; Blystad, D.R.; Saxena, P.K.; Wang, Q.C. Epigenetic and genetic integrity, metabolic stability, and field performance of cryopreserved plants. *Plants* **2021**, *10*, 1889. [CrossRef]
9. Fahy, G.M.; Wowk, B. Principles of ice-free cryopreservation by vitrification. *Methods Mol. Biol.* **2021**, *2180*, 27–97.
10. Lin, M.; Cao, H.S.; Meng, Q.H.; Li, J.M.; Jiang, P.X. Insights into the crystallization and vitrification of cryopreserved cells. *Cryobiology* **2022**, *106*, 13–23. [CrossRef]

11. Warner, R.M.; Ampo, E.; Nelson, D.; Benson, J.D.; Eorglu, A.; Higgins, A.Z. Rapid quantification of multi-cryoprotectant toxicity using an automated liquid handling method. *Cryobiology* **2021**, *98*, 219–232. [CrossRef] [PubMed]

12. Zamecnik, J.; Faltus, M.; Bilavcik, A. Vitrification solutions for plant cryopreservation: Modification and properties. *Plants* **2021**, *10*, 2623. [CrossRef]

13. Liang, N.S.; Zhan, Y.G.; Yu, L.; Wang, Z.Q.; Zeng, F.S. Characteristics and expression analysis of *FmTCP15* under abiotic stresses and hormones and interact with DELLA protein in *Fraxinus mandshurica* Rupr. *Forests* **2019**, *10*, 343. [CrossRef]

14. Yang, L.; Wei, C.; Huang, C.; Liu, H.N.; Zhang, D.Y.; Shen, H.L.; Li, Y.H. Role of hydrogen peroxide in stress-induced programmed cell death during somatic embryogenesis in *Fraxinus mandshurica*. *J. For. Res.* **2019**, *30*, 767–777. [CrossRef]

15. Liu, Y.; Wei, C.; Wang, H.; Ma, X.; Shen, H.L.; Yang, L. Indirect somatic embryogenesis and regeneration of *Fraxinus mandshurica* plants via callus tissue. *J. For. Res.* **2021**, *32*, 1613–1625. [CrossRef]

16. Yu, X.Q.; Liu, Y.Y.; Liu, X.Q.; Tretyakova, I.N.; Nosov, A.M.; Shen, H.L.; Yang, L. Cryopreservation of *Fraxinus mandshurica* Rupr. by using the slow cooling method. *Forests* **2022**, *13*, 773. [CrossRef]

17. Hong, S.R.; Yin, M.H. A simple and efficient protocol for cryopreservation of embryogenic calli of the medicinal plant *Anemarrhena asphodeloides* Bunge by vitrification. *Plant Cell Tissue Organ. Cult.* **2012**, *109*, 287–296.

18. Popova, E.; Shukla, M.; Kim, H.H.; Saxena, P.K. Root cryobanking: An important tool in plant cryopreservation. *Plant Cell Tissue Organ. Cult.* **2020**, *144*, 49–66. [CrossRef]

19. Engelmann, F. Use of biotechnologies for the conservation of plant biodiversity. *Vitr. Cell Dev. Biol.-Plant.* **2011**, *47*, 5–16. [CrossRef]

20. Ozudogru, E.A.; Lambardi, M. Cryotechniques for the Long-Term Conservation of Embryogenic Cultures from Woody Plants. *Methods Mol. Biol.* **2016**, *1359*, 537–550.

21. Arav, A. Cryopreservation by directional freezing and vitrification focusing on large tissues and organs. *Cells* **2022**, *11*, 1072. [CrossRef] [PubMed]

22. Zhan, L.; Rao, J.S.; Sethia, N.; Slama, M.Q.; Han, Z.H.; Tobolt, D.; Etheridge, M.; Peterson, Q.P.; Dutcher, C.S.; Bischof, J.C.; et al. Pancreatic islet cryopreservation by vitrification achieves high viability, function, recovery and clinical scalability for transplantation. *Nat. Med.* **2022**, *28*, 798–808. [CrossRef] [PubMed]

23. Guzmán-García, E.; Bradaï, F.; Sánchez-Romero, C. Cryopreservation of avocado embryogenic cultures using the droplet-vitrification method. *Acta Physiol. Plant.* **2013**, *35*, 183–193. [CrossRef]

24. Ghaffarzadeh-Namazi, L.; Joachim Keller, E.R.; Senula, A.; Babaeian, N. Investigations on various methods for cryopreservation of callus of the medicinal plant *Satureja spicigera*. *J. Appl. Res. Med. Aroma. Plants* **2017**, *5*, 10–15. [CrossRef]

25. Hammond, S.D.H.; Viehmannova, I.; Zamecnik, J.; Panis, B.; Faltus, M. Droplet-vitrification methods for apical bud cryo-preservation of yacon [*Smallanthus sonchifolius* (Poepp. and Endl.) H. Rob.]. *Plant Cell Tissue Organ. Cult.* **2021**, *147*, 197–208. [CrossRef]

26. Chauhan, R.; Singh, V.; Keshavkant, S.; Quraishi, A. Vitrification-based cryopreservation of in vitro-grown apical meristems of *Chlorophytum borivilianum* Sant et Fernand: A critically endangered species. *Proc. Natl. Acad. Sci. India Sect. B Biol. Sci.* **2021**, *91*, 471–476. [CrossRef]

27. Kim, H.H.; Lee, S.C. 'Personalisation' of droplet-vitrification protocols for plant cells: A systematic approach to optimising chemical and osmotic effects. *Cryo Lett.* **2012**, *33*, 271–279.

28. Sharma, N.; Gowthami, R.; Devi, S.V.; Malhotra, E.V.; Pandey, R.; Agrawal, A. Cryopreservation of shoot tips of *Gentiana kurroo* Royle—A critically endangered medicinal plant of India. *Plant Cell Tissue Organ. Cult.* **2021**, *144*, 67–72. [CrossRef]

29. Żabicki, P.; Mikuła, A.; Sliwinska, E.; Migdałek, G.; Nobis, A.; Żabicka, J.; Kuta, E. Cryopreservation and post-thaw genetic integrity of *Viola stagnina* Kit., an endangered species of wet habitats—A useful tool in *ex situ* conservation. *Sci. Hortic.* **2021**, *284*, 110056–110066. [CrossRef]

30. Lei, X.J.; Wang, Q.; Yang, H.; Qi, Y.R.; Hao, X.L.; Wang, Y.P. Vitrification and proteomic analysis of embryogenic callus of *Panax ginseng* C. A. Meyer. *Vitr. Cell Dev. Biol.-Plant.* **2021**, *57*, 118–127. [CrossRef]

31. Wang, L.Y.; Li, Y.D.; Sun, H.Y.; Liu, H.G.; Tang, X.D.; Wang, Q.C.; Zhang, Z.D. An efficient droplet-vitrification cryopreservation for valuable blueberry germplasm. *Sci. Hortic.* **2017**, *219*, 60–69. [CrossRef]

32. Wang, M.R.; Zhang, Z.B.; Zámečník, J.; Bilavčík, A.; Blystad, D.R.; Haugslien, S.; Wang, Q.C. Droplet-vitrification for shoot tip cryopreservation of shallot (*Allium cepa* var. *aggregatum*): Effects of PVS3 and PVS2 on shoot regrowth. *Plant Cell Tissue Organ. Cult.* **2020**, *140*, 185–195.

Article

Adult Trees *Cryptomeria japonica* (Thunb. ex L.f.) D. Don Micropropagation: Factors Involved in the Success of the Process

Alejandra Rojas-Vargas [1,2], Itziar A. Montalbán [2,†] and Paloma Moncaleán [2,*,†]

1   Instituto de Investigación y Servicios Forestales, Universidad Nacional, Heredia 86-3000, Costa Rica
2   NEIKER-BRTA, Department of Forestry Sciences, 01192 Arkaute, Spain
*   Correspondence: pmoncalean@neiker.eus
†   These authors contributed equally to this work.

Abstract: *Cryptomeria japonica* (Thunb. ex L.f.) D. Don is a commercial tree native to Japan and is one of the most important forest species in that country and the Azores (Portugal). Because of the quality of *C. japonica* timber, several genetic improvement programs have been performed. Recently, some studies focusing on *C. japonica* somatic embryogenesis have been carried out. However, in this species, this process uses immature seeds as initial explants, and for this reason, it is not possible to achieve the maximum genetic gain (100% genetic of the donor plant). Although some studies have been made applying organogenesis to this species, the success of the process in adult trees is low. For this reason, our main goal was to optimize the micropropagation method by using trees older than 30 years as a source of plant material. In this sense, in a first experiment, we studied the effect of different types of initial explants and three basal culture media on shoot induction; then, two sucrose concentrations and two light treatments (LEDs versus fluorescent lights) were tested for the improvement of rooting. In a second experiment, the effects of different plant growth regulators (6-benzylaminopurine, meta-topolin, and thidiazuron) on shoot induction and the subsequent phases of the organogenesis process were analyzed. The cultures produced the highest number of shoots when QL medium (Quoirin and Lepoivre, 1977) and long basal explants (>1.5 cm) were used; the shoots obtained produced a higher number of roots when they were grown under red LED lights. Moreover, root induction was significantly higher in shoots previously induced with meta-topolin.

Keywords: carbohydrates; conifers; cytokinins; LEDs; plant growth regulators; rooting; shoot induction

Citation: Rojas-Vargas, A.; Montalbán, I.A.; Moncaleán, P. Adult Trees *Cryptomeria japonica* (Thunb. ex L.f.) D. Don Micropropagation: Factors Involved in the Success of the Process. *Forests* **2023**, *14*, 743. https://doi.org/10.3390/f14040743

Academic Editor: Dušan Gömöry

Received: 14 March 2023
Revised: 29 March 2023
Accepted: 1 April 2023
Published: 5 April 2023

1. Introduction

The Japanese cedar, *Cryptomeria japonica* (Thunb. ex L.f.) D. Don, subfamily Taxodioideae, family Cupressaceae, is a monoecious conifer distributed across East Asia [1,2]. *C. japonica* covers approximately 4.5 million ha, representing 44% of the total reforested area [3]. This conifer is one of the most important timber tree species in Japan, and it is traditionally used for construction wood and for obtaining biomass [4].

Tsubomura and Taniguchi (2008) [5] mentioned that Japanese cedar is clonally propagated by cuttings, but this type of propagation requires many hours of manual labor, therefore it is difficult to establish a short-term propagation protocol. For the abovementioned reasons, biotechnological approaches for *C. japonica* clonal propagation, such as in vitro methods including somatic embryogenesis (SE) or shoot organogenesis, are valuable tools for the propagation of this conifer [6]. In this sense, SE is a recognized technique for the large-scale propagation of conifers [7]. However, SE is a complex, multistage process initiated from immature seeds, so it is not possible to reproduce the genotype of the donor plant. In contrast, micropropagation by nodal tissue culture is faster, and higher multiplication rates are possible [5,8]. Micropropagation starting from nodal segments uses smaller explants than conventional techniques, and selection of juvenile explants is possible [8]. With regard to explants coming from mature trees, they are less likely to dedifferentiate and reprogram [9], but the use of explants from mature trees for conifer micropropagation has also been reported [10,11]. However, because these trees are selected after reaching maturity, there has been limited success reported in the vegetative propagation of mature conifers, and procedures should be improved.

Although the benefits of tissue culture for the propagation of forest trees have been recognized, the success of such methods is still highly dependent on the species, the explant quality, the age of the donor plant, the culture medium, plant growth regulators, and/or the interaction among all these different factors. As a result, morphogenesis determines the growth and development of plant tissues, and it is influenced by several physico-chemical factors [12]. In relation to culture media, DCR [13], MS [14], and QL [15] are commonly used basal media for in vitro regeneration of conifers [16,17]. Furthermore, the cytokinins added to the culture medium have a direct effect on the endogenous phytohormone balance, provoking a response to the induction of axillary shoot buds and affecting the organogenesis of the culture [6,18].

Carbohydrates in plants are basic elements; they constitute substrates for respiration and are essential for many other processes related to plant development or gene expression, and in many species they favor rooting, acting mainly as a source of energy [19–21].

Plant growth and development are also influenced by different physical factors, with light being one of the most important [22]. The traditional light source used in in vitro culture in the growth chambers is fluorescent tubes (FL), with irradiances between 25 and 150 mmol m^{-2}s^{-1} for a 16 h photoperiod [23]. FL emit a broad light spectrum, and their physiological effects on plants are not specific [24]. Furthermore, the power consumption of FL is high as the heat emitted needs to be removed from growth chambers using air conditioners, making the process expensive [25]. Light-emitting diodes (LEDs) are available today as an alternative to conventional light sources for in vitro plant growth [26], and they present advantages over FL such as small size, longer lifespan, less power consumption, high energy conversion efficiency, and adjustable light spectra [27].

The analysis of the combined effects morphological (type of explants), chemical (culture media, sucrose concentrations), and physical factors (LED lights) at different micropropagation stages of adult trees has not been carried out in *Cryptomeria japonica*. Moreover, the use of ventilated culture containers is not widely used for conifer species. For these reasons, with the main objective of improving the micropropagation protocol for adult Japanese cedar, we focused on optimizing (1) the shoot induction stage using different types of explants, cytokinins, and culture media as well as (2) the rooting stage using different sucrose concentrations, Ecobox containers$^{®}$, and light treatments (fluorescent versus LEDs).

2. Materials and Methods

2.1. Plant Material

Actively growing *C. japonica* twigs were collected in October 2019 from two healthy >30-year-old adult trees located in Arkaute (Spain; 42°51′9.35″ N, 2°37′30.55″ W) to carry out Experiment 1 (Figure 1a). In October 2020, actively growing twigs were collected from four healthy >30-year-old adult trees located in Urnieta (Spain; 43°13′13.03″ N, 1°57′40.91″ W), and one adult tree (Spain; 43°13′50.32″ N, 1°59′25.37″ W) was chosen to carry out Experiment 2.

2.2. Sterilization

The plant material was first washed with commercial detergent, then rinsed under running water for 5 min immersed in 70% ethanol for 2 min, and then washed two times with sterile distilled water in the laminar flow unit. Finally, the actively growing twigs were disinfected in commercial bleach (30% *v/v*) (active chlorine 37 gL^{-1} sodium hypochlorite) for 20 min and rinsed three times in sterile distilled water for 5 min. each.

Figure 1. Plant material at different stages of *Cryptomeria japonica* micropropagation process: (**a**) actively growing twigs used as starting explants; arrows show the explants used for in vitro culture establishment, bar = 30 mm; (**b**) explant type; apical explants > 1.5 cm (AT), apical explants < 1.0 cm (AS), basal explants > 1.5 cm (BT) and basal explants < 1.0 cm (BS), bar = 10 mm; (**c**) basal explants of 2.0 cm length after 4 weeks cultured in QL medium [15] supplemented with 8.8 μM BA, bar = 30 mm; (**d**) elongated shoots after 6 weeks in hormone-free QL medium supplemented with 2 gL^{-1} activated charcoal, bar = 30 mm; (**e**) rooted shoots after six weeks in hormone-free QL medium supplemented with 2 gL^{-1} activated charcoal and under red LEDs, bar = 30 mm; (**f**) plantlets after four weeks in ex vitro conditions in the greenhouse, bar = 30 mm.

After the sterilization protocol, four types of explants were tested: apical explants > 1.5 cm (AT), apical explants < 1.0 cm (AS), basal explants > 1.5 cm (BT), and basal explants < 0 cm (BS) (Figure 1b).

2.3. Micropropagation Process

2.3.1. Experiment 1

After sterilization, the four types of explants were cultured vertically on 25×145 mm test tubes with polypropylene caps (Lab Associates, Oudenbosch, The Netherlands) containing 15 mL of bud induction medium (IM) (Supplementary Table S1). Three basal media were assayed: DCR [13], MS [14], and QL [15]. All media were supplemented with 3% (*w/v*) sucrose, 6-benzyladenine (BA, 8.8 µM, Duchefa Biochemie, Haarlem, The Netherlands), and 8 gL^{-1} Difco Agar$^{®}$ (Becton and Dickinson, Madrid, España) granulated. The pH of all media was adjusted to 5.8, and then they were autoclaved at 121 °C for 20 min. All cultures were placed in the growth chamber at a photoperiod of 16 h with 120 µmol m^{-2} s^{-1} light intensity provided by cool white fluorescent tubes (TLD 58 W/33; Philips, Suresnes, France) and a temperature of 21 ± 1 °C.

As soon as shoot induction was observed (after four weeks) (Figure 1c), four to five explants were transferred to baby food glass jars with Magenta^{TM} b-cap lids filled with 25 mL of elongation medium (EM) (Supplementary Table S1). EM was composed of hormone-free DCR, MS, or QL supplemented with 2 gL^{-1} activated charcoal, 3% (*w/v*) sucrose, and solidified with 8.5 gL^{-1} Difco Agar$^{®}$ granulated; pH and autoclaving conditions were those mentioned for IM.

The shoots were transferred to fresh EM every six weeks. Shoots were cultivated individually in a fresh EM when they reached 10–15 mm (Figure 1d). The conditions in the growth chamber were the same as those described above.

2.3.2. Root Induction and Acclimatization of Rooted Plants

Elongated shoots of at least 20–25 mm long were used for root induction. Based on the results of Experiment 1, QL basal medium was selected. The explants were transferred to Ecoboxes (Eco2box/green filter, consisting of a polypropylene oval vessel with a "breathing" hermetic cover, Duchefa$^{®}$) with 100 mL of root induction medium (RIM) (Supplementary Table S1), which was composed of half-strength macronutrient QL medium with 50 µM 1-naphthaleneacetic acid (NAA, Duchefa Biochemie, Haarlem, The Netherlands), 8 gL^{-1} Difco Agar$^{®}$, and 3% sucrose or 1.5% (*w/v*) sucrose. The pH and autoclaving conditions were those previously described. The shoots were placed under dim light for eight days. Then, two different light treatments were tested for four weeks: (A) white fluorescent light (FL) (color temperature 4000 K), 120 µmol m^{-2}s^{-1} light intensity provided by cool white fluorescent tubes (TLD 58 W/33; Philips, Suresnes, France); and (B) red light (peak wavelength 630 nm), 60 µmol m^{-2}s^{-1} light intensity provided by adjustable LEDs (RB4K Grow Light LEDs). The photoperiod and the temperature of the growth chamber were the same as previously described.

After five weeks of culture in RIM, shoots were cultured for six weeks in Ecoboxes with 100 mL of root expression medium (REM) (Supplementary Table S1); this medium consisted of half-strength macronutrient QL medium supplemented with 2 gL^{-1} activated charcoal, 3% sucrose or 1.5% (*w/v*) sucrose, and 8.5 gL^{-1} Difco Agar$^{®}$. Then, the rooted plants (Figure 1e) were planted ex vitro, transferring them to moist peat moss (Pindstrup, Aarhus, Denmark) with vermiculite at a proportion of 8:2 (*v/v*); acclimatization was carried out in a greenhouse under controlled conditions at a temperature of 21 ± 1 °C and progressively decreasing the humidity during a month from 95 to 80% (Figure 1f).

2.3.3. Experiment 2

Based on the results of Experiment 1, basal explants of >1.5 cm length and QL medium were selected to perform this experiment (Supplementary Table S2). The medium was supplemented with one of these three types of cytokinin (CK): BA, meta-topolin (m-T), or thidiazuron (TDZ, Duchefa Biochemie, Haarlem, The Netherlands), all of them at 8.8 μM. The explants were placed in the growth chamber at the same conditions described above (Section 2.3.1. Experiment 1).

As soon as shoot induction was observed (after four weeks), four to five shoots were transferred into baby food glass jars with Magenta^{TM} b-cap lids and 25 mL of EM (Supplementary Table S2), and they were subcultured every six weeks. Shoots that reached 10–15 mm were separated and individually cultivated in fresh EM. The growth chamber temperature and photoperiod were the same as those previously described (Section 2.3.1. Experiment 1).

2.3.4. Root Induction and Acclimatization of Rooted Plants

Shoots at least 20–25 mm long from the EM were employed for root induction. Based on the results of Experiment 1, the shoots were cultivated in Ecoboxes filled with half-strength macronutrient QL basal medium supplemented with 50 μM NAA, 1.5% (*w/v*) sucrose, and 8 gL^{-1} Difco Agar$^{®}$ (Supplementary Table S2). In addition, based on the results from Experiment 1, red light was selected for the rooting stage; the shoots were placed under dim light for eight days, followed by four weeks under a 16 h photoperiod with red light (peak wavelength 630 nm) and 60 μmol m^{-2}s^{-1} light intensity provided by adjustable LEDs (RB4K Grow Light LEDs). After these five weeks, shoots were cultured for six weeks in Ecoboxes with 100 mL of REM (Supplementary Table S2). The photoperiod and the temperature of the growth chamber were the same as those described in previous sections. Then, rooted plants were acclimatized as described above (Section 2.3.2).

2.4. Data collection and Statistical Analysis

2.4.1. Experiment 1

Twenty-four to forty-eight test tubes and one explant per test tube (AT, AS, BT, or BS) per each tree (two trees) were cultured in each culture medium. After two months of culture, the contamination, survival, and shoot induction percentages for each condition tested were measured. In the case of the shoot induction (%) and the mean number of shoots per explant (NS/E), these were calculated with respect to the non-contaminated explants. The effect of the explant type and culture medium on survival and shoot induction (%) was analyzed using a logistic regression model, and when necessary, differences were assessed by Tukey's post hoc test ($\alpha = 0.05$).

Data for the NS/E were analyzed by analysis of variance (ANOVA). When necessary, multiple comparisons were made using Tukey's post hoc test ($\alpha = 0.05$).

After the root expression stage, data for the root induction percentage, the mean number of roots per explant (NR/E), and the length of the longest root (LLR) (cm) were recorded. A completely randomized design using seven to twenty-four plantlets per sucrose concentration and light treatment was performed.

The effect of the sucrose concentration and light treatment on root induction (%) was analyzed with a logistic regression model. Data for NR/E and LLR were analyzed by ANOVA, and when necessary, differences were assessed by Tukey's post hoc test ($\alpha = 0.05$). To evaluate the effect of the sucrose concentration and light treatment on the acclimatization percentage, a logistic regression model was applied to plantlets after four weeks of growth in the greenhouse. Data processing was done using R Core Team software$^{®}$ (version 4.2.1, Vienna, Austria).

2.4.2. Experiment 2

Forty test tubes and one explant per test tube (BT) were cultured in each culture medium per tree (five trees). After two months of culture, the contamination, survival, and shoot induction percentages were recorded for each condition tested.

The shoot induction percentage (%) and the mean number of shoots per explant (NS/E) were calculated with respect to the non-contaminated explants after the elongation stage. The effect of the cytokinin type on survival and shoot induction (%) was analyzed using a logistic regression model; when necessary, differences were assessed by Tukey's post hoc test ($\alpha = 0.05$).

Data for the NS/E were analyzed by analysis of variance (ANOVA), and when necessary, differences were assessed by Tukey's post hoc test ($\alpha = 0.05$).

After the root expression stage, data for the root induction percentage, the mean number of roots per explant (NR/E), and the length of the longest root (LLR) (cm) were recorded. A completely randomized design using forty-one to fifty plantlets per cytokinin type applied during the shoot induction stage was performed.

The effect of the cytokinin type applied during the shoot induction stage on the root induction was analyzed using a logistic regression model. Data for NR/E and LLR were analyzed by ANOVA, and when necessary, differences were assessed by Tukey's post hoc test ($\alpha = 0.05$). To evaluate the effect of the cytokinin type applied during the shoot induction stage on the acclimatization percentage, a logistic regression was used. As mentioned above, all data were processed using R Core Team software[®].

3. Results

3.1. Micropropagation Process

3.1.1. Experiment 1

The rates of contamination were registered eight weeks after sterilization, showing general values of 39%. Explants' survival was significantly affected by the explant type used; AS explants showed significantly higher survival rates (90%) than AT (62%) and BT (32%; Figure 2, Supplementary Table S3).

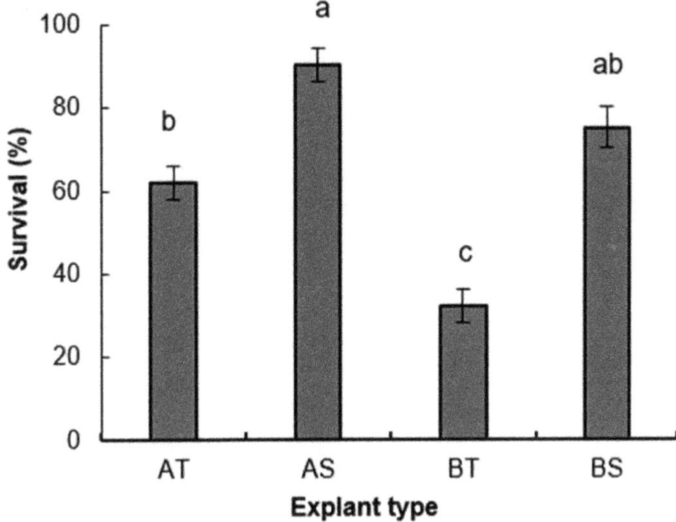

Figure 2. Survival (%) in different explant types of *Cryptomeria japonica* cultured in DCR medium [13], MS medium [14], and QL medium [15]. Apical explants > 1.5 cm (AT), apical explants < 1.0 cm (AS), basal explants > 1.5 cm (BT), and basal explants < 1.0 cm (BS). Data are presented as mean values ± S.E. Significant differences are indicated by different letters according to Tukey's post hoc test ($p < 0.05$).

The basal medium and the interaction between explant type and basal medium did not show significant differences for survival (%) (Supplementary Table S3). The survival percentage ranged from 57% in explants grown in MS and QL media to 63% in explants cultured in DCR medium.

When the variables explant type and basal medium and the interaction between them were analyzed after the induction stage, statistically significant differences were only found for the percentage of shoot induction depending on the explant type (Figure 3 and Supplementary Table S4). A significantly higher shoot induction percentage was obtained in AS (69%) and AT (56%), compared to the rest of the explants tested (Figure 3).

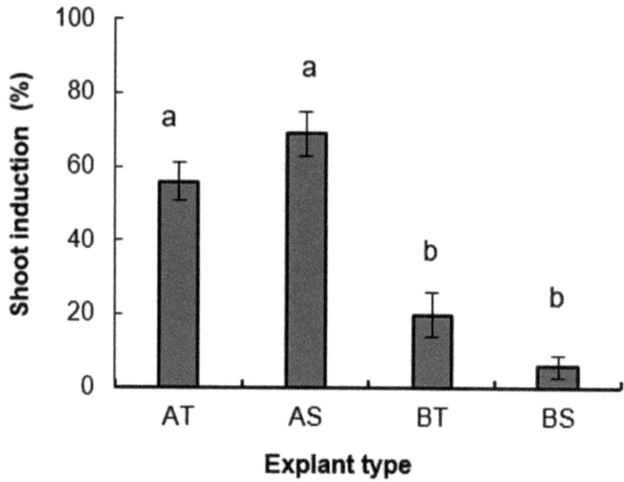

Figure 3. Shoot induction (%) in different explant types of *Cryptomeria japonica* cultured in DCR medium [13], MS medium [14], and QL medium [15]. Apical explants > 1.5 cm (AT), apical explants < 1.0 cm (AS), basal explants > 1.5 cm (BT), and basal explants < 1.0 cm (BS). Data are presented as mean values ± S.E. Significant differences are indicated by different letters according to Tukey's post hoc test ($p < 0.05$).

The shoot induction percentage ranged from 38% in explants grown in DCR medium to 46% in explants cultured in QL medium. Explants developed in MS medium showed an intermediate value of shoot induction (43%).

Regarding the NS/E, significant differences were found for the variables explant type and basal medium and the interaction between them (Figure 4, Supplementary Table S4). BT explants cultured in QL medium produced a significantly higher response than the other explant and medium combinations tested (Figure 4). Based on the results of Experiment 1, basal explants of >1.5 cm length and QL medium were selected to carry out the micropropagation process in Experiment 2.

3.1.2. Root Induction and Acclimatization of Rooted Plants

When the effect of the sucrose concentration, the light treatment, or the interaction between them on the root induction (%) was analyzed, no statistically significant differences were observed. (Supplementary Table S5). The root induction percentage ranged from 36% in shoots cultured in 3% sucrose concentration under FL to 54% in shoots, independently of the sucrose concentration in the culture medium and exposure to red LEDs.

The root number was significantly affected by the light treatment applied in the rooting phase (Figure 5 and Supplementary Table S5). In this sense, explants exposed to red LEDs showed significantly higher NR/E (6.5 ± 0.5) than those under fluorescent light (2.7 ± 0.4) (Figure 5). The sucrose concentration and the interaction between sucrose concentration and light treatment did not show statistically significant differences (Supplementary Table S5). The NR/E ranged from 4.5 ± 0.6 for shoots grown with 1.5% sucrose to 5.4 ± 0.7 for shoots cultured in QL medium supplemented with 3% sucrose.

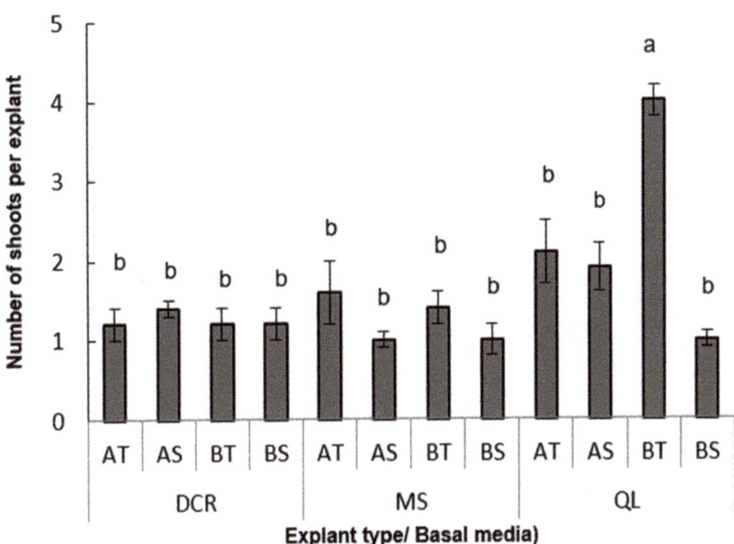

Figure 4. Number of shoots per explant in different explant types of *Cryptomeria japonica* cultured in DCR medium [13], MS medium [14], and QL medium [15] supplemented with 6-benzyladenine (BA, 8.8 µM). Apical explants > 1.5 cm (AT), apical explants < 1.0 cm (AS), basal explants > 1.5 cm (BT), and basal explants < 1.0 cm (BS). Data are presented as mean values ± S.E. Significant differences are indicated by different letters according to Tukey's post hoc test ($p < 0.05$).

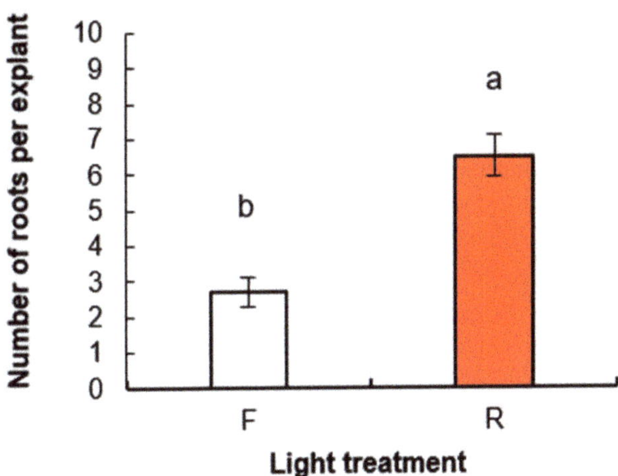

Figure 5. Number of roots per explant in shoots of *Cryptomeria japonica* cultured in QL medium [15], supplemented with 3% (*w/v*) sucrose or 1.5% (*w/v*) sucrose under light treatments (fluorescent light (F) and red LEDs (R)). Data are presented as mean values ± S.E. Significant differences are indicated by different letters according to Tukey's post hoc test ($p < 0.05$).

The different sucrose concentrations (1.5% or 3.0%) and light treatments tested for shoot induction showed a statistically significant effect on LLR, whereas the interaction between them did not have a significant effect (Figure 6 and Supplementary Table S5). A significantly higher LLR was observed in shoots cultured in the presence of 3% (*w/v*) sucrose compared with shoots grown at the lowest sucrose concentration (Figure 6a). In the same way, shoots exposed to fluorescent light showed significantly higher LLR than those under red LEDs (Figure 6b).

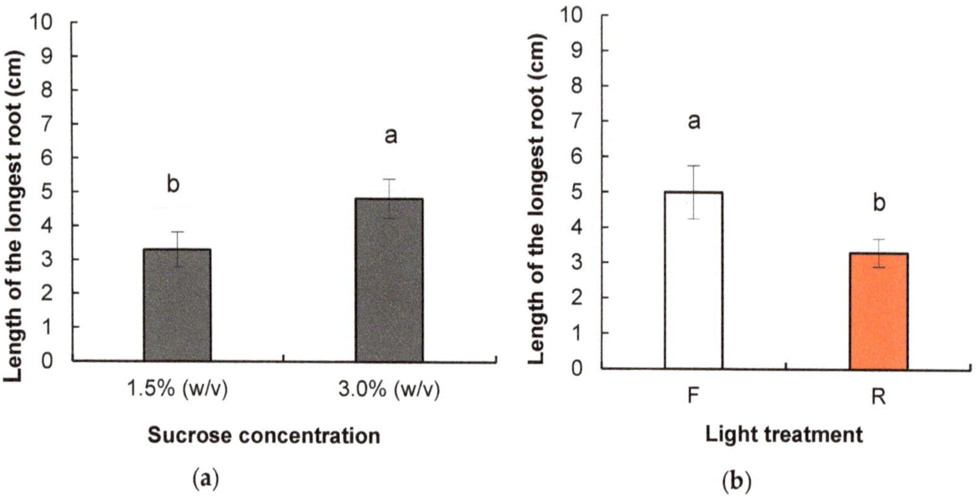

Figure 6. Length of the longest root in shoots of *Cryptomeria japonica* cultured in QL medium [15], according to sucrose concentration (3% and 1.5% (*w/v*) (**a**) and light treatments (fluorescent light (F) and red LEDs (R)) (**b**). Data are presented as mean values ± S.E. Significant differences are indicated by different letters according to Tukey's post hoc test ($p < 0.05$).

When the effect of sucrose concentration and light treatment applied during the rooting phase on the acclimatization percentage of rooted shoots was analyzed, no statistically significant differences were observed (Supplementary Table S6). Statistically significant differences were found in the acclimatization percentage for the interaction between sucrose concentration and light treatment (Supplementary Table S6). Nevertheless, as the p-value was bordering on significance, Tukey's post hoc test could not detect them. The acclimatization percentage ranged from 30% in shoots cultured in medium supplemented with 3% (*w/v*) and exposed to FL to 80% in those growing in the same sucrose concentration supplemented the culture medium under red LEDs.

3.2. Experiment 2

Four weeks after sterilization, the contamination rates were at 28%. No statistically significant differences were observed for the explant survival percentage considering the CK type tested (Supplementary Table S7). Explant survival percentages ranged from 66% in explants cultured in medium with m-T to 76% in explants grown with BA treatment.

The CK type showed a significant effect on the shoot induction (%) (Figure 7, Supplementary Table S8). A significantly higher shoot induction percentage was observed in explants induced with BA and m-T treatments (50% and 48%, respectively) than in explants grown on QL medium supplemented with TDZ.

No statistically significant differences were observed when the effect of CK type on the NS/E was analyzed (Supplementary Table S8). It was not possible to obtain shoots from explants induced with TDZ treatment due to tissue necrosis. Explants induced with m-T and BA treatments produced 2.3 ± 0.1 and 2.4 ± 0.1 NS/E, respectively.

Regarding the root induction percentage, significant differences were observed depending on the CK used for shoot induction (Figure 8 and Supplementary Table S9). A significantly higher root induction percentage was recorded in shoots induced with m-T (54%) than in shoots grown with BA (33%) during the shoot induction stage (Figure 8).

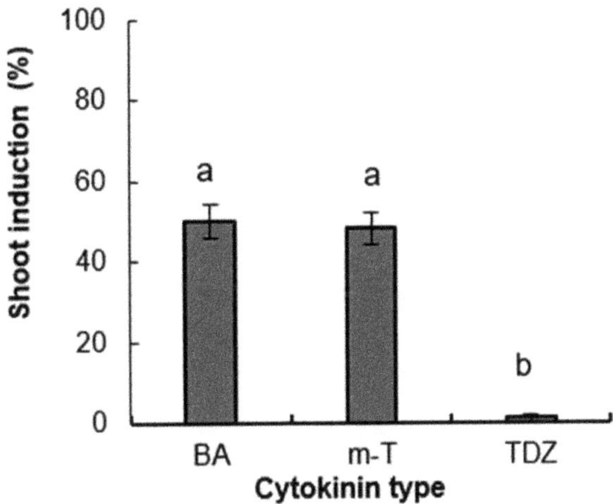

Figure 7. Shoot induction (%) in explants of *Cryptomeria japonica* cultured on QL medium [15], supplemented with 6-benzyladenine (BA), meta-topolin (m-T), and thidiazuron (TDZ) (8.8 µM). Data are presented as mean values ± S.E. Significant differences are indicated by different letters according to Tukey's post hoc test ($p < 0.05$).

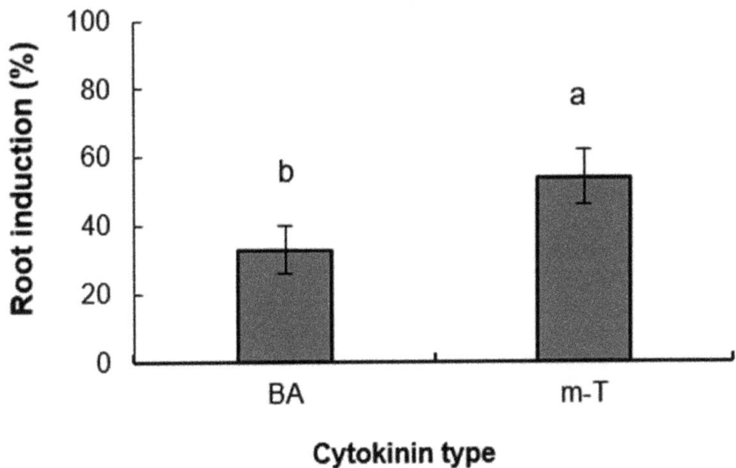

Figure 8. Root induction (%) in shoots of *Cryptomeria japonica* cultured in QL medium [15], supplemented with 50 µM 1-naphthaleneacetic acid (NAA), according to cytokinin type (6-benzyladenine (BA) and meta-topolin (m-T) (8.8 µM) use during shoot induction stage. Data are presented as mean values ± S.E. Significant differences are indicated by different letters according to Tukey's post hoc test ($p < 0.05$).

When the effect of the CK type utilized for shoot induction was evaluated for the NR/E and LLR parameters, no statistically significant differences were observed (Supplementary Table S9). The NR/E was 3.1 ± 0.5 for shoots previously induced with BA and 3.9 ± 0.6 for those from m-T treatment.

Explants cultured with BA and m-T treatment during the induction stage showed a LLR of 3.3 cm ± 0.5 and 3.7 cm ± 0.3, respectively.

No statistically significant differences were observed for shoots coming from different CK treatments when acclimatization percentage was analyzed (Supplementary Table S10). The ex vitro survival rate was 92% for shoots induced with BA and 94% for those induced with m-T treatment.

4. Discussion

As was reviewed in [28], contamination is considered a crucial obstacle that prohibits the successful establishment of an aseptic in vitro culture. In this study, contamination rates ranging from 28% to 39% were obtained using sodium hypochlorite. In this sense, our work obtained lower contamination rates than those recorded by [29] for *C. japonica*, who obtained 46% using calcium hypochlorite for surface sterilization of apical explants from adult trees. The sterilization protocol applied in Experiment 1 and Experiment 2 resulted in an optimal in vitro establishment of *C. japonica*. Therefore, our results suggest that it can be used for the establishment of cultures from other mature conifer explants [17].

Factors such as the type, size, or age of the explant, the physiological state of the donor plant, and the type of disinfectant and its concentration can influence the effectiveness of the sterilization protocol [30]. In our case, the survival percentage was significantly higher in AS explants in Experiment 1. A similar tendency was observed in *Cedar libani*, where shoot apices (<1.0 cm) from adult trees showed the best survival response [31]. Similarly, in *Taxus mairei*, the highest survival percentage was obtained in small stems (<1.0 cm) from cutting [32]. In our study, the smaller size and lower exposure of the explants' surface could have favored the culture establishment [33], as supported by [34], who mentioned that smaller explants may be easier plant material to sterilize from microorganisms.

In our experiments, the efficiency of the organogenic response was assessed by considering the shoot induction percentage and the NS/E. In Experiment 1, the results showed statistically significant differences for AS and AT explants for shoot induction percentage. In this sense, our study confirmed that the explant size is an important factor affecting axillary bud proliferation [31,34], and probably, the larger explants have more mineral nutrient reserves and endogenous hormones to support the culture [35]. Additionally, George et al. [34] explained that bigger explants from extensive parts of the shoot apex or stem segments with lateral buds could have advantages over smaller explants. Furthermore, it is probable that the morphogenetic gradient, where variations in the levels of endogenous plant growth regulators such as auxins (from the basal to the apical explant region) [34] are responsible for the different morphogenesis responses of the explants tested.

Cytokinins play several recognized roles in plant development, through the suppression of apical dominance and promotion of the development of axillary buds, the promotion of cell division, or the stimulation of plant protein synthesis [16,36]. BA is the most commonly used plant growth regulator; it is applied alone or in combination with other CK to promote in vitro shoot induction due to its effectiveness and affordability [37–39]. During Experiment 2, explants cultured with BA or m-T showed an efficient organogenic response regarding shoot induction percentage, and although not significant, a slightly higher response in BT explants induced with BA was observed. Analogous results were obtained in several organogenesis protocols from adult trees of *Pinus pinea*, *P. radiata*, *Sequoia sempervirens*, and *P. halepensis*, where, to obtain an in vitro shoot response, BA at concentrations ranging from 4.4 to 50 μM was utilized [10,40–43].

The highest NS/E was obtained in BT explants cultured in QL medium (Experiment 1). In accordance with this result, in previous experiments in our lab, the highest number of NS/E was obtained in *S. sempervirens* when explants bigger than 1.5 cm in length were used [10]. Similarly, Hine et al. [29] and Rafi and Salehi [44] developed an in vitro protocol for *C. japonica* and *Cedrus deodara*, using explants of mature trees from 1.0 to 2.5 cm in length, respectively. On the other hand, NS/E was higher on basal QL medium than on basal MS and DCR medium. According to Maruyama et al., 2021 [45], this may be attributable to a higher level of inorganic nitrogen present in those media when compared to QL medium. This hypothesis is supported by Tuskan et al. [46] who explained that a nitrate excess could have a negative impact on the organogenic response. Similar to our experiment on *C. japonica*, the lower nitrogen content of QL medium promoted organogenesis in *Cedrus deodara* and *P. ponderosa* [44,47]. Recently, for improving the micropropagation protocol of plants, artificial neural networks algorithms have been used to build models to determine the effect of mineral nutrients, vitamins, and plant growth regulators on several growth and quality parameters of micropropagated plants [48].

Several studies attributed the improvement in multiplication rates or rooting percentages and the alleviation of physiological disorders to the use of topolins in plant tissue culture [49]. However, in our study, no significant differences for the effect of CK on NS/E were found, but a slightly higher response in explants cultured with BA was observed (Experiment 2). Analogous results were obtained in *P. radiata*, *S. sempervirens*, and *P. ponderosa*, where the induction of axillary buds was not improved when applying m-T instead of BA [10,41,47,50].

In recent years, TDZ has received more attention due to its ability to aid in vitro regeneration of woody plants [17,40,51], and it has been recommended in explants coming from adult trees to induce regeneration via axillary shoot proliferation and adventitious shoot organogenesis [52]. However, in our work, it was not possible to obtain shoots coming from explants induced with TDZ due to tissue necrosis. The reason for this result may be due to an inadequacies concentration of TDZ, as supported by [53], who mentioned that the TDZ concentration and exposure time at their optimum depends on the species. Nevertheless, TDZ at low concentrations, in pulse treatments, or during short exposure periods can be effective in circumventing TDZ-induced abnormalities such as tissue necrosis [53].

Adventitious rooting can be influenced by physical and chemical factors, among them plant growth regulators, light quality, temperature, medium composition, and carbohydrates [51]. When root induction percentage was analyzed, no significant effect of sucrose concentration (1.5 or 3.0%) and light treatment was found. A similar result was observed in the organogenesis protocols of *C. japonica*, where the culture medium was supplemented with sucrose (from 1.5 to 3.0%) to obtain rooting [5,29,54].

As mentioned above, in the last few years, m-T has been proven as an alternative to conventional CK for in vitro propagation of plants [55]. In this sense, the advantages of m-T to promote rhizogenesis have been described [49]. In Experiment 2, when we focused on the study of the effect of the cytokinin type on root induction, we found that shoots induced with m-T promoted the highest rooting percentage. Similar to our results, Naaz et al. [56], studying *Syzygium cumini*, reported that shoots induced with m-T increased rhizogenic competency compared with shoots coming from kinetin, 2-isopentyl adenine, or BA. Similarly, in *Caralluma umbellata*, shoots derived from a culture medium supplemented with m-T and NAA showed the highest in vitro rooting activity [57]. In contrast to our work, *in S. sempervirens* and *Juniperus drupacea*, growing shoots with m-T did not improve the rooting response [10,58]. Summarizing, the effect of m-T to promote rhizogenesis at different concentrations may be species-specific and depends on the starting material used to establish the in vitro culture.

Plant growth and development are strongly influenced by the quality of the light in their environment [59]. In the last few years, LEDs have shown a favorable response in in vitro culture when compared with the results obtained using fluorescent tubes [60,61]. Additionality, Ragonezi et al. [62] mentioned that the light type and wavelength specificity influence adventitious rooting. In this sense, statistically significant differences in NR/E were found when shoots were cultured under red LEDs (Experiment 1). Similarly, in *C. japonica* and *Populus sieboldii* × *Populus grandidentata*, red LEDs showed a higher response to in vitro rooting [5,63]. In *Pinus pseudostrobus* shoots exposed to red LEDs, the best rooting rates were observed at 30 days of evaluation [26]. Contrary to our results, *P. radiata* and *P. ponderosa* shoots grown under white fluorescent light and white LEDs, respectively, displayed the highest rooting responses [43,47]. Regarding LLR, shoots under fluorescent light showed longer primary roots. Analogous results were observed by Ishii et al. [64] in *C.*

japonica and Rojas-Vargas et al. [47] in *P. ponderosa,* where shoots growing under fluorescent light showed the longest root length. Contrary to our result, the longest roots were recorded in plants of *P. radiata* and *P. pseudostrobus* exposed to red LEDs, [26,43]. Summarizing, different plant species respond differently, even when using the same light treatment, which may be due to the genotypic characteristics of the plants or the physiological state of the explants [30,65].

Regarding the acclimatization stage, no significant differences were observed for the effect of CK on acclimatization percentage, but a slightly higher response in explants cultured with m-T was observed. These results agreed with those observed in *Aloe polyphylla, S. cumini,* and *C. umbellate,* where the plantlets induced with m-T were successfully acclimatized [39,56,57]. The reason for this may be due to the rapid uptake and transport of mT into the plant system and the production of reversibly sequestered metabolites [57].

5. Conclusions

The regeneration of *C. japonica* through micropropagation of adult trees using basal explants of >1.5 cm length was achieved and depended on physico-chemical factors. The optimal result in terms of shoot induction was obtained when basal explants were cultured in QL medium supplemented with BA treatment.

Our results suggest that the use of m-T and a 1.5% sucrose concentration favored root induction. In this sense, the use of red LEDs was better for the number of roots per explant. Finally, in the greenhouse, the shoots, independently of the cytokinins used in the shooting stage, showed high acclimatization success. In order to optimize the micropropagation efficiency in *C. japonica,* our results suggest the use of fluorescent light for the shoot induction stage and red LEDs for the rooting stage.

Supplementary Materials: The following supporting information can be downloaded at: https://www.mdpi.com/article/10.3390/f14040743/s1, Table S1: Variations of basal DCR [13], MS [14] and QL [15] media tested at different stages of *Cryptomeria japonica* micropropagation process, Experiment 1; Table S2: Variations of basal QL [15] medium tested at different stages of *Cryptomeria japonica* micropropagation process, Experiment 2; Table S3: Statistical analysis for the survival (%) showed in *Cryptomeria japonica* per explant type (apical explants > 1.5 cm, apical explants < 1.0 cm, basal explants > 1.5 cm, and basal explants < 1.0 cm) and basal media (DCR [13], MS [14] and QL [15]); Table S4: Statistical analysis for shoot induction (%) and number of shoots per explant showed in *Cryptomeria japonica* per explant type (apical explants > 1.5 cm, apical explants < 1.0 cm, basal explants > 1.5 cm, and basal explants < 1.0 cm) and basal media (DCR [13], MS [14] and QL [15]); Table S5: Statistical analysis for root induction (%), number roots per explant and length of the longest root of *Cryptomeria japonica* shoots cultured in QL medium [15], supplemented with 3% (*w/v*) sucrose or 1.5% (*w/v*) sucrose, according to light treatment; Table S6: Statistical analysis for the survival (%) of rooted shoots propagated in vitro coming from *Cryptomeria japonica* adult trees, after four weeks under ex vitro conditions; Table S7: Statistical analysis for the survival (%) showed in *Cryptomeria japonica* explants (basal explants of >1.5 cm length) cultured in QL medium [15], supplemented with 6-benzyladenine (BA), meta-topolin (m-T) or thidiazuron (TDZ) at 8.8 µM; Table S8: Statistical analysis for shoot induction (%) and number of shoots per explant showed in *Cryptomeria japonica* explants (basal explants of >1.5 cm length) cultured in QL medium [15], supplemented with 6-benzyladenine (BA), meta-topolin (m-T) or thidiazuron (TDZ) at 8.8 µM; Table S9: Statistical analysis for root induction (%), number root per explant and length of longest root showed in *Cryptomeria japonica* explants (basal explants of >1.5 cm length) cultured in QL medium [15], supplemented with 50 µM 1-naphthaleneacetic acid (NAA), according to cytokinin type [6-benzyladenine (BA), meta-topolin (m-T) or thidiazuron (TDZ) at 8.8 µM]; Table S10: Statistical analysis for the survival (%) in *Cryptomeria japonica* explants (basal explants > 1.5 cm) cultured in QL medium [15], supplemented with 50 µM 1-naphthaleneacetic acid (NAA), according to cytokinin type [6-benzyladenine (BA) or meta-topolin (m-T) at 8.8 µM].

Author Contributions: Conceptualization, P.M. and I.A.M.; methodology, P.M. and I.A.M.; formal analysis, A.R.-V.; investigation, A.R.-V.; data curation, A.R.-V.; writing—original draft preparation, A.R.-V.; writing—review and editing, A.R.-V., P.M. and I.A.M. All authors have read and agreed to the published version of the manuscript.

Funding: This research has been funded by Universidad Nacional de Costa Rica and Instituto de Investigación y Servicios Forestales (INISEFOR), DECO (Basque government), MINECO (AGL2016-76143-C4-3R), MICINN (PID2020-112627RB-C32), CYTED (P117RT0522), and MULTIFOREVER project, under the umbrella of ERA-NET Cofund ForestValue by ANR (FR), FNR (DE), MINCyT (AR), MINECO-AEI (ES), MMM (FI), VINNOVA (SE), ForestValue has received funding from the European Union's Horizon 2020 Research and Innovation Programme under grant agreement no. 773324.

Acknowledgments: The authors sincerely thank Ander Isasmendi (NEIKER-BRTA) for his technical assistance.

Conflicts of Interest: The authors declare no conflict of interest.

Abbreviations

ANOVA—analysis of variance; AS—apical explants < 1.0 cm; AT—apical explants > 1.5 cm; BA—6-benzyladenine; BS—basal explants < 1.0 cm; BT—basal explants > 1.5 cm; CK—cytokinin; EM—elongation medium; FL—fluorescent light; IM—induction medium; LEDs—light-emitting diodes; LLR—length of the longest root; DCR—DCR medium (Gupta and Durzan, 1985); m-T—meta-topolin; MS—MS medium (Murashige and Skoog, 1962); NAA—1-naphthalene acetic acid; NR/E—number of roots per explant; NS/E—number of shoots per explant; QL—QL medium (Quoirin and Lepoivre, 1977); REM—root expression medium; RIM—root induction medium; TDZ—thidiazuron.

References

1. Kusumi, J.; Tsumura, Y.; Yoshimaru, H.; Tachida, H. Phylogenetic relationships in Taxodiaceae and Cupressaceae sensu stricto based on matK gene, chlL gene, trnLtrnF IGS region, and trnL intron sequences. *Am. J. Bot.* **2000**, *87*, 1480–1488. [CrossRef] [PubMed]
2. Onuma, Y.; Uchiyama, K.; Kimura, M.; Tsumura, Y. Genetic analysis distinguished new natural population and old plantations of *Cryptomeria japonica*. *Trees For. People* **2023**, *11*, 100365. [CrossRef]
3. Forestry Agency. *Statistical Handbook of Forest and Forestry*; Forestry Agency, Ministry of Agriculture, Forestry and Fisheries: Tokyo, Japan, 2020; pp. 8–9. (In Japanese)
4. Taniguchi, T.; Konagaya, K.; Nanasato, Y. Somatic embryogenesis in artificially pollinated seed families of 2nd generation plus trees and cryopreservation of embryogenic tissue in *Cryptomeria japonica* D. Don (sugi). *Plant Biotechnol.* **2020**, *37*, 239–245. [CrossRef]
5. Tsubomura, M.; Taniguchi, T. Micropropagation of the male sterile 'Soushun' Japanese cedar. *Comb. Proc. Int. Plant. Prop. Soc.* **2008**, *58*, 421–427.
6. Phillips, G.C. In vitro morphogenesis in plants-recent advances. *In Vitro Cell. Dev. Biol. Plant.* **2004**, *40*, 342–345. [CrossRef]
7. Park, J.S.; Barret, J.D.; Bonga, J.M. Application of somatic embryogenesis in high-value clonal forestry: Deployment, genetic control, and stability of cryopreserves clones. *Cell. Dev. Biol. Plant.* **1998**, *34*, 231–239. [CrossRef]
8. Sahu, J.; Sahu, R.K. A review on low cost methods for in vitro micropropagation of plant through tissue culture technique. *UK J. Pharm. Biosci.* **2013**, *1*, 38–41. [CrossRef]
9. Narváez, I.; Martín, C.; Jiménez-Díaz, R.M.; Mercado, J.A.; Pliego-Alfaro, F. Plant regeneration via somatic embryogenesis in mature wild olive genotypes resistant to the defoliating pathotype of *Verticillium dahliae*. *Front. Plant Sci.* **2019**, *10*, 1471. [CrossRef]
10. Rojas-Vargas, A.; Castander-Olarieta, A.; Moncaleán, P.; Montalbán, I.A. Optimization of the micropropagation of elite adult trees of *Sequoia sempervirens*: Forest species of interest in the Basque Country, Spain. *Rev. Bionatura.* **2021**, *6*, 1511–1519. [CrossRef]
11. Juan-Vicedo, J.; Serrano-Martinez, F.; Cano-Castillo, M.; Casas, J.L. In vitro propagation, genetic assessment, and, medium-term conservation of coastal endangered species *Tetraclinis articulata* (Vahl) Masters (Cupressaceae) from adult trees. *Plants* **2022**, *11*, 187. [CrossRef]
12. Rai, M.K.; Asthana, P.; Jaiswal, V.S.; Jaiswa, U. Biotechnological advances in guava (*Psidium guajava* L.): Recent developments and prospects for further research. *Trees* **2010**, *24*, 1–12. [CrossRef]
13. Gupta, P.K.; Durzan, D.J. Shoot multiplication from mature trees of Douglas-fir (*Pseudotsuga menziesii*) and sugar pine (*Pinus lambertiana*). *Plant Cell Rep.* **1985**, *4*, 177–179. [CrossRef]
14. Murashige, T.; Skoog, F. A revised medium for rapid growth and bio assays with *Tobacco* tissue cultures. *Physiol. Plant.* **1962**, *15*, 474–497. [CrossRef]

15. Quoirin, M.; Lepoivre, P. Études des milieux adaptés aux cultures in vitro de *Prunus*. *Acta Hortic.* **1977**, *78*, 437–442. [CrossRef]
16. Arab, M.M.; Yadollahi, A.; Shojaeiyan, A.; Shokri, S.; Ghojah, S.M. Effects of nutrient media, different cytokinin types and their concentrations on in vitro multiplication of G × N15 (hybrid of almond × peach) vegetative rootstock. *J. Genet. Eng. Biotechnol.* **2014**, *12*, 81–87. [CrossRef]
17. De Diego, N.; Montalbán, I.A.; Moncaleán, P. In vitro regeneration of adult *Pinus sylvestris* L. trees. *S. Afr. J. Bot.* **2010**, *76*, 158–162. [CrossRef]
18. De Almeida, M.; de Almeida, C.V.; Graner, E.M.; Brondani, G.E.; de Abreu-Tarazi, M.F. Pre-procambial cells are niches for pluripotent and totipotent stem-like cells for organogenesis and somatic embryogenesis in the peach palm: A histological study. *Plant Cell Rep.* **2012**, *31*, 1495–1515. [CrossRef] [PubMed]
19. Gibson, S.I. Plant sugar-response pathways. part of a complex regulatory web. *Plant Physiol.* **2000**, *124*, 1532–1539. [CrossRef]
20. Calamar, A.; De Klerk, G.J. Effect of sucrose on adventitious root regeneration in apple. *Plant Cell Tissue Organ Cult.* **2002**, *70*, 207–212. [CrossRef]
21. Dewir, Y.H.; Murthy, H.N.; Ammar, M.H.; Alghamdi, S.S.; Al-Suhaibani, N.A.; Alsadon, A.A.; Paek, K.Y. In vitro rooting of leguminous plants: Difficulties, alternatives, and strategies for improvement. *Hortic. Environ. Biotechnol.* **2016**, *57*, 311–322. [CrossRef]
22. Alallaq, S.; Ranjan, A.; Brunoni, F.; Novák, O.; Lakehal, A.; Bellini, C. Red light controls adventitious root regeneration by modulating hormone homeostasis in *Picea abies* seedlings. *Front. Plant Sci.* **2020**, *11*, 586140. [CrossRef]
23. Larraburu, E.E.; Correa, G.S.; Llorente, B.E. In vitro development of yellow lapacho (Bignoniaceae) using high-power lightemitting diode. *Rev. Arvore.* **2018**, *42*, e420508. [CrossRef]
24. Rocha, P.S.G.; Oliveira, R.P.; Scivittaro, W.B.; Santos, U.L. Diodos emissores de luz e concentrações de BAP na multiplicaçãoin vitro de morangueiro. *Ciênc. Rural* **2010**, *40*, 1922–1928. [CrossRef]
25. Kulus, D.; Woźny, A. Influence of light conditions on the morphogenetic and biochemical response of selected ornamental plant species under in vitro conditions: A mini-review. *BioTechnologia* **2020**, *101*, 75–83. [CrossRef]
26. Marín-Martínez, L.A.; Iglesias-Andreu, L.G. Effect of LED lights on the in vitro growth of *Pinus pseudostrobus* Lindl., plants. *J. For. Sci.* **2022**, *68*, 311–317. [CrossRef]
27. Río-Álvarez, I.; Rodríguez-Herva, J.J.; Martínez, P.M.; González-Melendi, P.; García-Casado, G.; Rodríguez-Palenzuela, P.; López-Solanilla, E. Light regulates motility, attachment, and virulence in the plant pathogen *Pseudomonas syringae* pv tomato DC3000. *Environ. Microbiol.* **2014**, *16*, 2072–2085. [CrossRef] [PubMed]
28. Payghamzadeh, K.; Kazemitabar, S.K. In vitro propagation of walnut—A review. *Afr. J. Biotechnol.* **2011**, *10*, 290–311.
29. Hine-Gómez, A.; Valverde-Cerdas, L. Establecimiento in vitro de *Cryptomeria japónica* (Taxocidaceae). *Rev. Biol. Trop.* **2003**, *51*, 683–690, (In Espanish with English Abstract).
30. Da Silva, J.A.T.; Winarto, B.; Dobránszki, J.; Cardoso, J.C.; Zeng, S. Tissue disinfection for preparation of Dendrobium in vitro culture. *Folia Horticult.* **2016**, *28*, 57–75. [CrossRef]
31. Renau-Morata, B.; Ollero, J.; Arrillaga, I.; Segura, J. Factors influencing axillary shoot proliferation and adventitious budding in cedar. *Tree Physiol.* **2005**, *25*, 477–486. [CrossRef]
32. Chang, S.H.; Ho, C.K.; Chen, Z.Z.; Tsay, J.Y. Micropropagation of *Taxus mairei* from mature trees. *Plant Cell Rep.* **2001**, *20*, 496–502. [CrossRef]
33. Sathyagowri, S.; Seran, T.H. In vitro plant regeneration of ginger (*Zingiber officinale* Rosc.) with emphasis on initial culture establishment. *J. Med. Aromat. Plants* **2011**, *1*, 195–202.
34. George, E.F.; Debergh, P.C. Micropropagation: Uses and Methods. In *Plant Propagation by Tissue Culture*; George, E.F., Hall, M.A., De Klerk, G.J., Eds.; Springer: Dordrecht, The Netherlands, 2008; pp. 29–64. [CrossRef]
35. Desai, C.; Inghalihalli, R.; Krishnamurthy, R. Micropropagation of *Anthurium andraeanum* an important tool in floriculture. *J. Pharmacogn. Phytochem.* **2015**, *4*, 112–117.
36. Rodríguez, S.M.; Ordás, R.J.; Alvarez, J.M. Conifer Biotechnology: An Overview. *Forests* **2022**, *13*, 1061. [CrossRef]
37. Phillips, G.C.; Garda, M. Plant tissue culture media and practices: An overview. *In Vitro Cell. Dev. Biol. Plant.* **2019**, *55*, 242–257. [CrossRef]
38. De Diego, N.; Montalbán, I.A.; Fernandez de Larrinoa, E.; Moncaleán, P. In vitro regeneration of *Pinus pinaster* adult trees. *Can. J. For. Res.* **2008**, *38*, 2607–2615. [CrossRef]
39. Bairu, M.W.; Stirk, W.A.; Dolezal, K.; Van Staden, J. Optimizing the micropropagation protocol for the endangered *Aloe polyphylla*: Can meta-topolin and its derivatives serve as replacement for benzyladenine and zeatin? *Plant Cell Tissue Organ Cult.* **2007**, *90*, 15–23. [CrossRef]
40. Cortizo, M.; de Diego, N.; Moncaleán, P.; Ordás, R.J. Micropropagation of adult stone pine (*Pinus pinea* L.). *Trees Struct. Funct.* **2009**, *23*, 835–842. [CrossRef]
41. Montalbán, I.A.; Novák, O.; Rolčik, J.; Strnad, M.; Moncaleán, P. Endogenous cytokinin and auxin profiles during in vitro organogenesis from vegetative buds of *Pinus radiata* Adult Trees. *Physiol. Plant.* **2013**, *148*, 214–231. [CrossRef] [PubMed]
42. Pereira, C.; Montalbán, I.A.; Pedrosa, A.; Tavares, J.; Pestryakov, A.; Bogdanchikova, N.; Canhoto, J.; Moncaleán, P. Regeneration of *Pinus halepensis* (Mill.) through organogenesis from apical shoot buds. *Forests* **2021**, *12*, 363. [CrossRef]
43. Rojas-Vargas, A.; Castander-Olarieta, A.; Montalbán, I.A.; Moncaleán, P. Influence of physico-chemical factors on the efficiency and metabolite profile of adult *Pinus radiata* D. Don bud organogenesis. *Forests* **2022**, *13*, 1455. [CrossRef]

44. Rafi, Z.N.; Salehi, H. Factors affecting in-vitro propagation of some genotypes of Himalayan cedar [*Cedrus deodara* (Roxb. ex Lamb) G. Don.]. *Adv Hortic Sci.* **2018**, *32*, 479–486.

45. Maruyama, T.E.; Ueno, S.; Mori, H.; Kaneeda, T.; Moriguchi, Y. Factors influencing somatic embryo maturation in sugi (Japanese Cedar, *Cryptomeria japonica* (Thunb. ex L.f.) D. Don). *Plants* **2021**, *10*, 874. [CrossRef]

46. Tuskan, G.A.; Sargent, W.A.; Rensema, T.; Walla, J.A. Influence of plant growth regulators, basal media and carbohydrate levels on the in vitro development of *Pinus ponderosa* (Dougl. Ex Law.) cotyledon explants. *Plant Cell Tissue Organ Cult.* **1990**, *20*, 47–52. [CrossRef]

47. Rojas-Vargas, A.; Castander-Olarieta, A.; do Nascimento, A.M.M.; Vélez, M.L.; Pereira, C.; Martins, J.; Zuzarte, M.; Canhoto, J.; Montalbán, I.A.; Moncaleán, P. Testing Explant Sources, Culture Media, and Light Conditions for the Improvement of Organogenesis in *Pinus ponderosa* (P. Lawson and C. Lawson). *Plants* **2023**, *12*, 850. [CrossRef]

48. Arteta, T.A.; Hameg, R.; Landin, M.; Gallego, P.P.; Barreal, M.E. Artificial Neural Networks Elucidated the Essential Role of Mineral Nutrients versus Vitamins and Plant Growth Regulators in Achieving Healthy Micropropagated Plants. *Plants* **2022**, *11*, 1284. [CrossRef] [PubMed]

49. Aremu, A.O.; Bairu, M.W.; Doležal, K.; Finnie, J.F.; Van Staden, J. Topolins: A panacea to plant tissue culture challenges? *Plant Cell Tissue Organ Cult.* **2012**, *108*, 1–16. [CrossRef]

50. Montalbán, I.A.; De Diego, N.; Moncaleán, P. Testing novel cytokinins for improved in vitro adventitious shoots formation and subsequent ex vitro performance in *Pinus radiata*. *Forestry* **2011**, *84*, 363–373. [CrossRef]

51. Zarei, M.; Salehi, H.; Jowkar, A. Controlling the barriers of cloning mature *Picea abies* (L.) H. Karst. via tissue culture and co-cultivation with *Agrobacterium rhizogenes*. *Trees* **2020**, *34*, 637–647. [CrossRef]

52. Vinoth, A.; Ravindhran, R. In vitro morphogenesis of woody plants using thidiazuron. In *Thidiazuron: From Urea Derivative to Plant Growth Regulator*; Ahmad, N., Faisal, M., Eds.; Springer: Singapore, 2018; pp. 211–230. [CrossRef]

53. Dewir, Y.H.; Nurmansyah; Naidoo, Y.; da Silva, J.A.T. Thidiazuron-induced abnormalities in plant tissue cultures. *Plant Cell Rep.* **2018**, *37*, 1451–1470. [CrossRef]

54. Koguta, K.V.; Flores, P.C.; Alcantara, G.B.; Higa, A.R. Serial micropropagation and cuttings as rejuvenation methods for *Cryptomeria japonica* (L.F) D.Don. *Rev. Espac.* **2017**, *38*, 7.

55. Gantait, S.; Mitra, M. Role of Meta-topolin on in vitro shoot regeneration: An insight. In *Me-Tatopolin: A Growth Regulator for Plant Biotechnology and Agriculture*; Ahmad, N., Strnad, M., Eds.; Springer: Singapore, 2021. [CrossRef]

56. Naaz, A.; Hussain, S.A.; Anis, M.; Alatar, A.A. Meta-topolin improved micropropagation in *Syzygium cumini* and acclimatization to ex vitro conditions. *Biol. Plant.* **2019**, *63*, 174–182. [CrossRef]

57. Jayaprakash, K.; Manokari, M.; Cokulraj, M.; Dey, A.; Faisal, M.; Alatar, A.A.; Joshee, N.; Shekhawat, M.S. Improved organogenesis and micro-structural traits in micropropagated plantlets of *Caralluma umbellata* Haw. in response to Meta-Topolin. *Plant Cell Tissue Organ Cult.* **2023**, *in press*. [CrossRef]

58. Ioannidis, K.; Tomprou, I.; Panayiotopoulou, D.; Boutsios, S.; Daskalakou, E.N. Potential and Constraints on In Vitro Micropropagation of *Juniperus drupacea* Labill. *Forests* **2023**, *14*, 142. [CrossRef]

59. Sarropoulou, V.; Sperdouli, I.; Adamakis, I.D.; Grigoriadou, K. The use of different LEDs wavelength and light intensities for in vitro proliferation of cherry rootstock: Influence on photosynthesis and photomorphogenesis. *Plant Cell Tissue Organ Cult.* **2023**, *152*, 317–330. [CrossRef]

60. Chen, L.; Zhang, K.; Gong, X.; Wang, H.; Gao, Y.; Wang, X.; Zeng, Z.; Hu, Y. Effects of different LEDs light spectrum on the growth, leaf anatomy, and chloroplast ultrastructure of potato plantlets in vitro and minituber production after transplanting in the greenhouse. *J. Integr. Agric.* **2020**, *19*, 108–119. [CrossRef]

61. Gupta, S.; Jatothu, B. Fundamentals and applications of light-emitting diodes (LEDs) in in vitro plant growth and morphogenesis. *Plant Biotechnol. Rep.* **2013**, *7*, 211–220. [CrossRef]

62. Ragonezi, C.; Klimaszewska, K.; Castro, M.R.; Lima, M.; de Oliveira, P.; Zavattieri, M.A. Adventitious rooting of conifers: Influence of physical and chemical factors. *Trees Struct. Funct.* **2010**, *24*, 975–992. [CrossRef]

63. Kobayashi, H.; Hashimoto, K.; Ohba, E.; Kurata, Y. Light-emitting Diode–Induced Root Photomorphogenesis and Root Retention in *Populus*. *HortScience* **2022**, *57*, 872–876. Available online: https://journals.ashs.org/hortsci/view/journals/hortsci/57/8/article-p872.xml (accessed on 28 February 2023). [CrossRef]

64. Ishii, K.; Hosoi, Y.; Taniguchi, T.; Tsubomura, M.; Kondo, T.; Yamada, H.; Saito, M.; Suda, T.; Fukisawa, T.; Tanaka, K. In vitro culture of various genotypes of male sterile Japanese cedar (*Cryptomeria japonica* D. Don). *Plant Biotechnol.* **2011**, *28*, 103–106. [CrossRef]

65. Lai, C.S.; Kho, Y.H.; Chew, B.L.; Raja, P.B.; Subramaniam, S. Organogenesis of *Cucumis metuliferus* plantlets under the effects of LEDs and silver nanoparticles. *S. Afr. J. Bot.* **2022**, *148*, 78–87. [CrossRef]

 *forests*

Article

Potential and Constraints on In Vitro Micropropagation of *Juniperus drupacea* Labill.

Kostas Ioannidis [1,*], Ioanna Tomprou [1], Danae Panayiotopoulou [2], Stefanos Boutsios [3] and Evangelia N. Daskalakou [3]

[1] Laboratory of Sylviculture, Forest Genetics and Biotechnology, Institute of Mediterranean and Forest Ecosystems, Hellenic Agricultural Organization "DIMITRA", Ilissia, 11528 Athens, Greece
[2] Library and Documentation Center, Institute of Mediterranean & Forest Ecosystems, Hellenic Agricultural Organization "DIMITRA", Ilissia, 11528 Athens, Greece
[3] Laboratory Forest Management & Forest Economics, Institute of Mediterranean & Forest Ecosystems, Hellenic Agricultural Organization "DIMITRA", Ilissia, 11528 Athens, Greece
* Correspondence: ioko@fria.gr; Tel.: +30-210-7783-750

Abstract: *Juniperus drupacea* Labill. (Cupressaceae) is a species with ecological and medicinal value. In Europe, it is native only in southern Greece, and is listed as endangered. Due to its uniqueness, this study attempted, for the first time, an in vitro propagation effort of Syrian juniper. Explants of the lateral shoot tips were surface-sterilized and cultured on Murashige and Skoog (MS) medium. The cultures were subcultured on MS, woody plant medium (WPM), and Driver and Kuniyaki Walnut (DKW) supplemented with different concentrations of 6-benzylaminopurine (BA), thidiazuron (TDZ), or meta-topolin [6-(3-hy-droxybenzylamino)purine] for shoot induction. Explants derived from female trees exhibited 54.17% bud proliferation on DKW medium with 4 μM meta-topolin or 4 μM TDZ and on WPM with 4 μM meta-topolin or 4 μM BA. A total of 62.50% of the male tree derived explants produced multiple shoots on DKW with 4 μM BA. The maximum average number of shoots per explant were 1.17 per explant in both cases. The length of the shoot derived from explants of female origin was 2.94 mm compared to 2.69 mm of the in vitro shoots from the explants of male trees. Overall, the best medium and plant growth regulator combination for the explants derived from both female and male trees, for the traits under study, was proven to be DKW + 4 μM TDZ. Our experiments show that *Juniperus drupacea*, under in vitro conditions, shows recalcitrance in rooting, as the applications of IBA, NAA, and IAA concentrations were proven to be ineffective treatments. Although the results show low values, this avant-garde study provides a foundation for further research on the in vitro regeneration of *Juniperus drupacea*.

Keywords: Syrian juniper; in vitro culture; shoot induction; recalcitrant rooting species

Citation: Ioannidis, K.; Tomprou, I.; Panayiotopoulou, D.; Boutsios, S.; Daskalakou, E.N. Potential and Constraints on In Vitro Micropropagation of *Juniperus drupacea* Labill. *Forests* **2023**, *14*, 142. https://doi.org/10.3390/f14010142

Academic Editors: Ricardo Javier Ordás and José Manuel Álvarez Díaz

Received: 12 December 2022
Revised: 30 December 2022
Accepted: 4 January 2023
Published: 12 January 2023

1. Introduction

Juniperus drupacea Labill., commonly known as Syrian juniper, belongs to the Cupressaceae family. It is a dioecious tree, 10–20 m in height, forming a conical crown. The needles are acicular, up to 25 mm long and 4 mm wide, with two white bands on the top, arranged in alternate whorls. The cones of Syrian juniper, being the largest among juniper species, are ovoid to globose, 20–25 mm in diameter, brownish, glaucosus, and pruniose in maturity, and have three seeds in a characteristic drupe-like strobile [1–3]. It is considered as a relict species with a disjunct geographical range. The location of its divergence and evolution remains unknown [3]. Rare fossil data only include remnants known from Miocene and Pliocene deposits in Europe [3–7] and concern the *Juniperus* spp. in general.

Currently, the distribution range of the species extends mainly to SE Turkey, western Syria, Israel, and Lebanon . In Europe, its natural populations are restricted only to Greece, in the SE part of the Peloponnese Peninsula [8–12]. Specifically, more than 95% of *J. drupacea* populations are found on Mt. Parnon [9] and a few have been recorded in a limited area of Mt Taygetos in small patches [10–12]. Syrian juniper can be used to enhance the biodiversity in oak and cedar forest restoration [13] and protect the soil from erosion [14]. It is considered as a very interesting ornamental tree because of its columnar shape, good growth rate, and resistance to frost [15].

In Greece, *J. drupacea*, due to its decay-resistant timber, used to be exploited for carpentry as well as fuel. However, its endangered status has prevented any extensive use today as they are included in natural habitat types of community interest whose conservation requires the designation of special areas of conservation [16]. According to the International Union for the Conservation of Nature (IUCN) [17], *J. drupacea* is considered worldwide as a species of Least Concern (LC). However, in Europe, it is listed as Endangered (EN) [18] under the criteria B1ab(iii) + 2ab(iii) [19]. In Greece, its ecological value has been acknowledged since 1980 when *Juniperus drupacea* forests has were declared as a "Natural Monument under Preservation", according to FEK 121D/1980 [20]. In 1992, it was included in Annex I of Directive 92/43/EEC as a priority habitat type and the Mt Parnon summit and Malevi Monastery were included in Natura 2000 as Special Protected Areas (code: GR 2520006) [16].

Natural reproduction occurs through seeds and it has been proven to be a very slow procedure, especially due to its seed's deep dormancy [21]. Juniper species have sexual reproductive capacity but their seed number varies. They mainly not only have low seed production, but also present low germination percentage, physiological dormancy, and lessened embryos viability [13,22,23]. In particular, Syrian juniper seeds present germination morphophysiological barriers and can delay natural germination for 4–5 years [24]. On the other hand, the cone flesh alone can postpone it for 1 to 2 years [25].

Several attempts have been made to in vitro regenerate *Juniperus* as the species shows a general recalcitrance in natural regeneration. Micropropagation by axillary shoots, among others, is considered as an effective method for the accomplishment of one of the aims of vegetative propagation, that is, the mass production of plants. Although it has shown positive results in many forest species, it presents more difficulties in conifers, particularly in the genus Juniperus, being challenging and demanding as well as arduous [14,26–31]. There are physical and chemical factors that stimulate the different conifer species to develop shoots and adventitious roots with the progress not always being triumphant. According to Ragonezi et al. [32], these factors include plant growth regulators, carbohydrates, light quality, temperature, and the rooting medium.

Gomez and Segura [33] first reported the successful application of this method for *Juniperus oxycedrus* L., which resulted in the induction of shoots but with very limited rooting frequency. Rooting of the in vitro regenerated microshoots is a very laborious, crawling and inefficient process in conifers [14,28,32,34]. The efficiency of shoot induction is varied among Juniperus species and depends on the types of explants [28,35]. The type of medium as well as the type of cytokinin and auxin and their concentrations were revealed to play determined roles in blastogenesis and rhizogenesis in *Juniperus oxycedrus* [33,36], *Juniperus phoenicea* [27,37], *Juniperus navicularis* [35], *Juniperus excelsa* M. Bieb, *Juniperus horizontalis* Moench and *Juniperus chinensis* L. [38], *Juniperus excelsa* [39], *Juniperus polycarpos* L. [40], and *Juniperus thulifera* L. [28]. Some studies have shown that the proliferation response of some juniper species increased during subsequent subcultures [27,35,38] due to overcoming the first shock after the first establishment in in vitro conditions [27].

Rizhogenesis is influenced by factors such as donor plant age and health, shoot vigor and juvenility genotype, and type of explant, auxin treatment, and environmental conditions on rooting, decreasing the mineral, sucrose, and agar concentrations in the medium [41–45]. In junipers, a very high rooting rate has only been reported in *Juniperus oxycedrus* L. and *Juniperus cedrus* Webb & Berthel. [26,34]. Varying rooting rates were observed in *Juniperus excels* M.Bieb., *Juniperus horizontalis* Moench and *Juniperus chinensis* L. [38], *Juniperus navicularis* Gand. [35], *Juniperus thulifera* L. [28], *Juniperus oxycedrus* L. [36], while the rooting of *Juniperus polycarpos* K. Koch [40] were not satisfactory. In *Juniperus phoenicea* L., the results were contradictory as some studies reported small to moderate rooting [27,37] and some failed in rhizogenesis [46].

In this context, the aim of the study was to investigate, for the first time, whether in vitro micropropagation would overcome the in vitro regeneration recalcitrance of *Juniperus drupacea*. The scientific team investigated the in vitro culture establishment, shoot proliferation as well as the potential for rooting. We incorporated three types of media and three types of plant growth regulators in several concentrations in our experiments in order to achieve blastogenesis and rhizogenesis. Although there were promising results on shooting, they lacked rooting, which, like other juniper species, was not feasible.

2. Materials and Methods

2.1. Plant Material—Explants Sterilization—Culture Establishment

Healthy mature *Juniperus drupacea* male and female individuals growing on Mt. Parnon were selected as explant source trees. Eight trees per sex from three different areas were selected as explant donors. The age of the selected trees ranged from 30 to 50 years. The lateral shoots of the actively growing stems of the source trees were the explant donors. These were collected during April to May, placed in damp cotton cloth, stored at 4 °C, and transferred to the laboratory until subsequent manipulations. The following day, the explants (i.e., nodal segments and apical shoot tips of 1.5–2.5 cm long) were excised from the explant donors collected during the vegetative growth stage. The explants were distinguished as female and male in relation to their tree gender.

The explant surface was successfully disinfected by successive immersions in two different aqueous solutions: the first was a solution of 70% ethanol with continuous stirring for 1 min, and the second was sodium hypochlorite (10% NaOCl, Fluka, Germany) at a concentration of 1.0% (v/v), complemented with 0.05% (v/v) Tween-20 (Fisher Bioreagents, USA) with continuous stirring for 15 min. After immersion, the explants were rinsed three times with sterile deionized water for three minutes each.

Each explant was placed in a 25 mm × 150 mm culture tube containing 20 mL of the nutrient medium. Three media were used to establish the in vitro culture: the MS of Murashige and Skoog [47] (Duchefa Biochemie, Haarlem, The Netherlands), the wood plant medium (WPM) of Lloyd and McCown [48] (Duchefa Biochemie, Haarlem, The Netherlands), and the Driver and Kuniyaki Walnut (DKW) of Driver and Kuniyuki [49] (Duchefa Biochemie, Haarlem, The Netherlands). Each medium contained 3% (w/v) sucrose (Duchefa Biochemie, Haarlem, The Netherlands) solidified with 6 g L^{-1} agar (Duchefa Biochemie, Haarlem, The Netherlands) and their pH was adjusted to 5.8 before agar addition and autoclaving at 121 °C and 122 kPa for 20 min. All cultures were incubated in a growth chamber at 23 ± 1 °C with a 16 h light/8 h dark photoperiod at a 50 μmol m^{-2} s^{-1} photosynthetic photon flux density (culture level) provided by cool-white fluorescent lamps.

2.2. Shoot Regeneration, Multiplication, and Elongation

After 10 days, the healthy non-contaminated explants were subcultured in full-strength mediums (i.e., MS, WPM and DKW) containing 6-benzylaminopurine (BA) (Sigma Chemicals, St. Louis, MO, USA) or thidiazuron (TDZ) (Cayman Chemicals, Ann Arbor, MI, USA) or meta-topolin [6-(3-hydroxybenzylamino)purine] (Duchefa Biochemie, Haarlem, The Netherlands) at various concentrations (1.0, 2.0, 4.0 and 8.0 μM) for multiple shoot induction. Each medium contained 3% (w/v) sucrose and was solidified with 6 g L^{-1} agar and their pH adjusted at 5.8 before agar addition and autoclaving at 121 °C and 122 kPa for 20 min. The treatments used in the shoot regeneration experiments are shown in Table 1. The plantlets established in in vitro conditions were transferred every 2–3 weeks to new nutrient media of the same composition. After an 8-week period (three subcultures), the effect of the various concentrations that the plant growth regulators had on the average shoot formation percentage (%), average shoot number, and length per explant were evaluated. Every treatment (i.e., medium—plant growth regulator combination) included three replicates. Each replication constituted eight tubes with one explant per tube. In total,

1872 explants were incorporated in our shoot regeneration, multiplication, and elongation experiments, not counting all the explants used in the establishment of cultures. The cultures of each experiment was arranged in a completely randomized design in a growth chamber at 23 ± 1 °C with a 16 h light/8 h dark photoperiod at a 50 μmol m^{-2} s^{-1} photosynthetic photon flux density (culture level) provided by cool-white fluorescent lamps.

Table 1. Treatments used in the shoot regeneration experiments for both the female and male explants.

	Treatment Concentration		
PGR	**MS Medium**	**WPM Medium**	**DKW Medium**
BA or TDZ or m-T	Control	Control	Control
	1.0 μM	1.0 μM	1.0 μM
	2.0 μM	2.0 μM	2.0 μM
	4.0 μM	4.0 μM	4.0 μM
	8.0 μM	8.0 μM	8.0 μM

MS: Murashige and Skoog medium, WPM: wood plant medium, DKW: Driver and Kuniyaki Walnut medium, BA: 6-benzylaminopurine, TDZ: thidiazuron, mT: meta-topolin.

2.3. In Vitro Rooting of Shoots

Shoots of 2.0–2.5 cm long, derived from the shoot regeneration step, were transplanted on culture tubes containing full-strength of the same media as in the previous stage, supplemented with several auxins for rooting. Explants from each treatment of the shoot regeneration, multiplication, and elongation stage were transplanted to each rooting treatment. In order to satisfy the required number of eight explants per replicate and per treatment, where necessary, we used explants from the establishment cultures stage. Plant growth regulators were IBA (indole-3-butyric acid) (Sigma Chemicals, Saint Louis, MO, USA) at concentrations of 1.0, 2.0, 4.0, 8.0, 16.0, and 32.0 μM; NAA (α-naphthalene acetic acid) (Sigma Chemicals, Saint Louis, MO, USA) at concentrations of 1.0, 2.0, 4.0, and 8.0 μM; and IAA (3-indoleacetic acid) (Sigma Chemicals, Saint Louis, MO, USA) at concentrations of 0.5, 1.0, 2.0, and 4.0 μM. All treatments used in the rooting experiments are shown in Table 2. The nutrient media were solidified with 7 g L^{-1} agar, and supplemented with 3% (*w/v*) sucrose. The conditions of the cultures were the same as above-mentioned. After a 4-week period, the effect of the concentrations of the plant growth regulators on the rooting percentage (%), root number, and length per shoot were evaluated. The experimental design was the same as that above-mentioned.

Table 2. Treatments used in the rooting experiments for both the female and male explants.

	Treatment Concentration				
PGR	**MS WPM DKW Media**	**PGR**	**MS WPM DKW Media**	**PGR**	**MS WPM DKW Media**
IBA	Control	NAA	Control	IAA	Control
	1.0 μM				
	2.0 μM		1.0 μM		0.5 μM
	4.0 μM		2.0 μM		1.0 μM
	8.0 μM		4.0 μM		2.0 μM
	16.0 μM		8.0 μM		4.0 μM
	32.0 μM				

MS: Murashige and Skoog medium, WPM: wood plant medium, DKW: Driver and Kuniyaki Walnut, IBA: indole-3-butyric acid, NAA: α-naphthalene acetic acid, IAA: 3-indoleacetic acid.

All cultures were incubated in a growth chamber at 23 ± 1 °C with a 16 h light/8 h dark photoperiod at a 50 μmol m^{-2} s^{-1} photosynthetic photon flux density (culture level) provided by cool-white fluorescent lamps.

2.4. Statistical Analysis

Analysis was based on individual values of the average shooting percentage, the mean number, the mean length of shoots per explant, the proportion of rooted microcuttings, the number, and the mean length of roots per explant. The following linear model was used in the analysis to specify the impact of the gender, the plant growth regulator treatment, and the interaction between the gender and the plant growth regulator treatment:

$$y_{ijkl} = \mu + g_j + m_i + t_k + g_j{}^*m_i + g_j{}^*t_k\, e_{ijkl}$$

where y_{ijkm} is the measurement for a trait of the *l*th explant, the *k*th plant growth regulator treatment, the *i*th nutrient medium and the *j*th explant gender, as dependent variables; μ is the fixed population mean of all explants; g_j is the fixed effect of the *j*th gender; m_i is the random effect of the *i*th nutrient medium; t_k is the random effect of the *k*th plant growth regulator treatment; $g_j{}^*t_k$ is the interaction of the *j*th gender with the *k*th plant growth regulator treatment; $g_j{}^*m_i$ is the interaction of the *j*th gender with the *i*th nutrient medium; and e_{ijkl} is the random residual error of the *l*th explant, the *k*th plant growth regulator treatment, the *i*th nutrient medium, and the *j*th gender. The restricted maximum likelihood (REML) method was used to estimate the variance components. Moreover, a binomial logistic regression was performed to predict the probability of the explant to shoot, which was considered as the dichotomous dependent variable by using the type of the medium (categorical), type of the plant growth regulator (categorical) and its concentration (ordinal), and gender (categorical) as explanatory variables (covariates). Descriptive statistics, analysis of variance (ANOVA) as well as the Duncan's multiple range test (MRT) based on the 0.05 level of significance were performed on the number and average shoot length per explant and the shooting proportion per treatment. Data in percentages were subjected to appropriate transformation in order to statistically analyze and were transformed back to percentages for presentation in the tables and graphs. All statistical analyses were performed using SPSS v.20 software for Windows (IBM SPSS Statistics 2011, IBM Corp., Armonk, NY, USA).

3. Results

3.1. Shoot Regeneration, Multiplication, and Elongation

Shoot formation was affected by the kind and concentration of plant growth regulators and nutrient medium but not the tree gender. Throughout the experiment, no significant interactions between gender and nutrient medium and between gender and plant growth regulator treatment was observed regarding the average shoot formation percentage, the average shoot number per explant, and the average shoot length per explant. The impact of different media and plant growth regulators and their concentration in the treatments on the average number of shoots per explant, the average shoot length, and the frequency of shoot formation of female and male *Juniperus drupacea* explants are presented in Table 3, Supplementary Material Tables S1–S3, and Figures 1–3.

Table 3. Effect of the medium on the average percentage of blastogenesis (%), average number of shoots, and average shoot length of *Juniperus drupacea* explants in relation to their gender (means followed by the same letter did not differ statistically at $p \leq 0.05$ according to the Duncan test).

		Average Percentage of Blastogenesis (%)			Average Number of Shoots per Explant			Average Shoot Length per Explant (mm)		
Explant Gender		**Female**	**Male**	**Overall**	**Female**	**Male**	**Overall**	**Female**	**Male**	**Overall**
N		312	312	624	312	312	624	312	312	624
Nutrient Medium	DKW	41.35 [a]	44.23 [a]	42.79 [a]	0.78 [a]	0.76 [a]	0.77 [a]	1.08 [a]	1.04 [a]	1.06 [a]
	WPM	39.74 [ab]	39.74 [ab]	39.74 [ab]	0.72 [ab]	0.66 [ab]	0.68 [ab]	0.95 [ab]	0.95 [ab]	0.95 [ab]
	MS	35.90 [b]	36.86 [b]	36.38 [b]	0.66 [b]	0.64 [b]	0.66 [b]	0.80 [b]	0.86 [b]	0.83 [b]

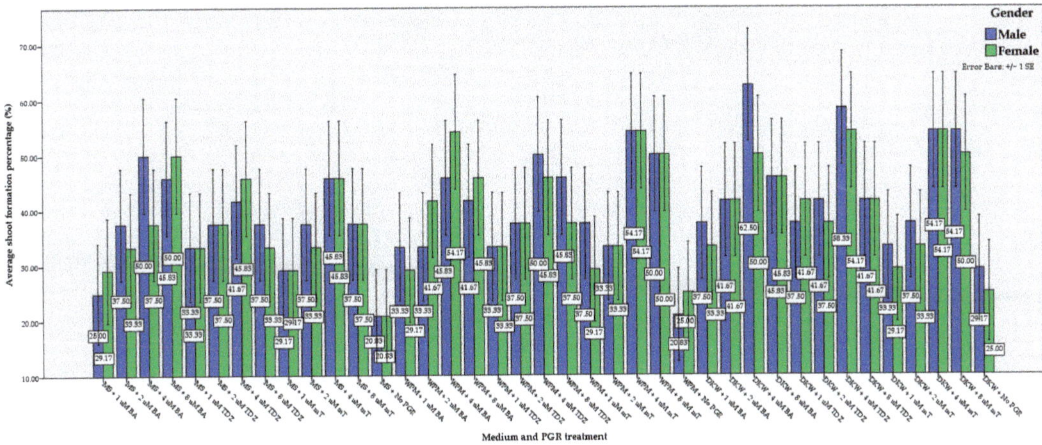

Figure 1. Effect of the medium and plant growth regulator types and their concentrations on the average percentage of blastogenesis (%) of *Juniperus drupacea* explants in relation to their gender.

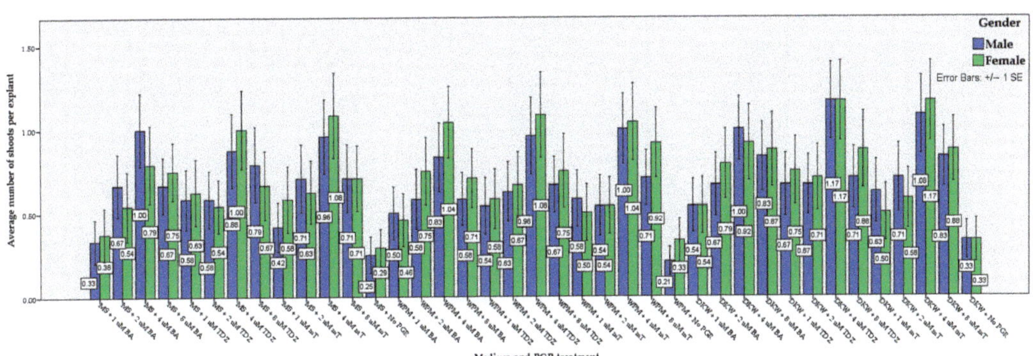

Figure 2. Effect of the medium and plant growth regulator types and their concentrations on the average shoot number per *Juniperus drupacea* explant in relation to their gender.

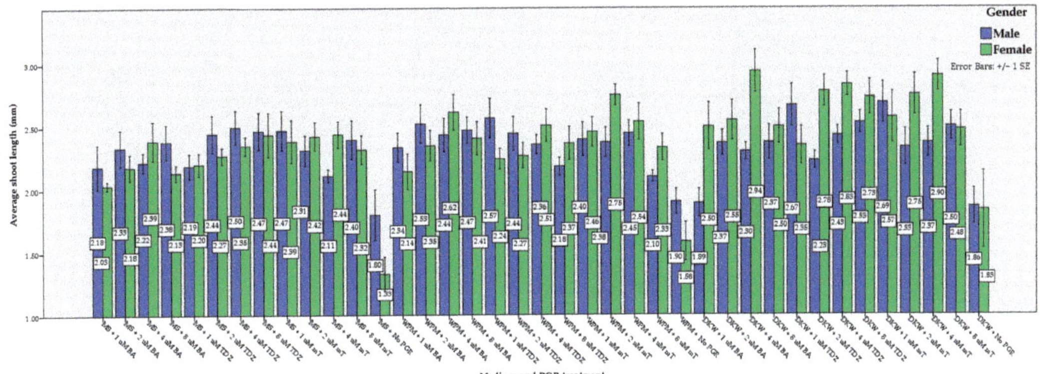

Figure 3. Effect of the medium and plant growth regulator types and their concentrations on the average shoot length per *Juniperus drupacea* explant in relation to their gender.

Shoot induction was achieved after three weeks of culture, depending on the treatment applied. The type of nutrient medium significantly affected the average percentage of blastogenesis, the average number of shoots, and the average shoot length of the *Juniperus drupacea* explants (Table 3). Explants growing in DKW medium presented the best values compared to the relevant ones in WPM and MS, respectively. In contrast, MS presented the lowest values that differed statistically compared to DKW and WPM, which did not show statistically significant differences ($p \leq 0.05$). Within each gender and among the different treatments, statistically significant differences ($p \leq 0.05$) were observed in the mean percentage of blastogenesis (Supplementary Material Table S1), the mean number of shoots per explant (Supplementary Material Table S2), and the mean length of shoots (Supplementary Material Table S3). The binomial logistic regression model adequately fit the data as the logistic regression model was statistically significant ($\chi^2(4) = 30.005$, $p \leq 0.000$) and the Hosmer and Lemeshow test resulted in $\chi^2 = 9.211$ ($p \leq 0.238$). The model correctly classified 59.9% of cases. The explained variation in the dependent variable based on our model ranged from 16.6 to 22.2%, depending on whether we referenced the Cox and Snell R^2 or Nagelkerke R^2 methods, respectively. From the analysis, we could see that the medium ($p \leq 0.030$) and plant growth regulator ($p \leq 0.000$) significantly enhanced the model/prediction. In contrast, gender did not contribute statistically significantly to the model.

Regarding the female explants, the maximum percentage of shooting (54.17%) was achieved in the DKW and WPM media when supplemented with 4 µM mT or 4 µM TDZ in the former and with 4 µM mT or 4 µM BA in the latter. The media with no growth regulators showed the lowest values of the percentage of blastogenesis (20.83%). Regarding the male explants, the maximum percentage of blastogenesis (62.50%) was attained in the DKW medium when it was supplemented with 4 µM BA. The media without the addition of growth regulators showed the lowest values in shooting percentage (20.83%).

The maximum average shoot number in female explants was achieved by supplementing the DKW medium with 4 µM mT or 4 µM TDZ (1.17). Similarly, regarding male explants, the maximum average shoot number was achieved using the DKW nutrient medium combined with 4 µM TDZ (1.17). The DKW, WPM, and MS media with no growth regulators presented the lowest values in the mean number of shoots (i.e., 0.33, 0.33, and 0.29, respectively) regarding the female explants. Likewise, in relation to the male explants, DKW, WPM, and MS media containing no growth regulators exhibited the lowest values in the mean number of shoots (i.e., 0.33, 0.25, and 0.21, respectively). Moreover, the treatment with MS medium supplemented with 1 µM BA also showed a very low value (0.33).

Maximum average shoot length per explant in the female explants was achieved when DKW medium was supplemented with 4 µM BA (2.94 mm) or 4 µM mT (2.90 mm). The difference between these treatments was not significant. The results were similar in the male explants, where the maximum values in average shoot length were obtained when the DKW nutrient medium was supplemented with 1 µM mT (2.69) or 1 µM TDZ (2.67). The difference between these treatments was also not significant. The DKW, WPM, and MS media with no growth regulators presented the lowest values in the average shoot length per female explant (i.e., 1.83, 1.58, and 1.33, respectively). Additionally, in relation to the male explants, all types of DKW, WPM, and MS media containing no growth regulators exhibited the lowest values in the average shoot length per explant (i.e., 1.90, 1.86, and 1.80, respectively). Moreover, the treatment of the DKW medium supplemented with 1 µM BA also showed a very low value (1.89).

In every treatment, many explants browned and showed necrosis, which eventually led to the death of many explants (Figure 4).

(a)   (b)

Figure 4. Culture establishment and shoot formation: (**a**) explants of *Juniperus drupacea* on MS medium containing 4 µM BA after 10 days of culture; (**b**) shoot formation on DKW medium containing 4 µM BA after 4 weeks of culture. Explant discoloration (browning) and necrotic zones of the explants are common among *Juniperus* species. Test tube diameter = 25 mm.

3.2. In Vitro Explant Rooting

Root initiation of the in vitro cultured explants was inspected immediately after they were subcultured on the rooting treatment media supplemented with auxins. At this rooting stage, all of the plant growth regulators were tested in terms of their concentrations. More specifically, IBA was tested at concentrations of 1.0, 2.0, 4.0, 8.0, 16.0, and 32.0 µM, NAA at 1.0, 2.0, 4.0, and 8.0 µM, and finally, IAA at 0.5, 1.0, 2.0, and 4.0 µM, respectively. However, all treatments were insufficient to induce rooting. The effectiveness of the nutrient media alone, containing no auxin, was also inadequate to stimulate root formation.

4. Discussion

The shooting percentage of all cultures was moderate. A small number of new shoots was developed showing insignificant elongation. In each medium, many explants browned and showed apical and lateral necrosis, which eventually led to death. Explant discoloration and necrosis is common among *Juniperus* species. Salih et al. [50], after a three-week culturing period of *Juniperus procera* Hoechst. Ex Endl. in several media, observed the yellowing of explants, which led to their necrosis two weeks later. Khater and Benbouza [28], during in vitro culture of *Juniperus thurifera* L., reported that the explants' color changed from green to yellow, and finally to brown, and necrosis reached approximately 90%. The authors attributed the necrosis to an inappropriate medium and PGR combinations. The same phenomenon of explant discoloration as well as high percentages of necrosis was observed by Momeni et al. [40] and Castro et al. [35] in the in vitro culture of *Juniperus polycarpos* L. and *Juniperus navicularis* Gand., respectively. Al-Ramamneh et al. [37] stated that micro-cuttings of *Juniperus phoenicea* L. failed to show any morphogenic response, browned, and showed progressively necrotic areas by the end of the culture. Loureiro et al. [27], in the in vitro culture of *Juniperus phoenicea* L., attributed the browning and the necrotic zones to the inappropriate selection of the nutrient medium they used.

In vitro regeneration depends on the composition and concentration of basal salts, growth regulators, and organic components [51]. In particular, the nitrogen content in the medium seems to influence the shoot formation of the explants [40]. Our best results concerning blastogenesis were achieved in the DKW medium, which had an intermediate concentration of nitrogen compared to the other two media used. Compared to MS, WPM also presented better results, as it was the medium with the lowest nitrogen availability. Our findings were consistent with those of Loureiro et al. [27], who reported that *Juniperus phoenicea* L. explants growing in DKW presented significantly better results than those growing in WPM or MS. Several studies in *Juniperus thurifera* L. [28], *Juniperus procera* Hoechst. Ex Endl. [50], *Juniperus polycarpos* K.Koch [52], and *Juniperus phoenicea* L. [37,53] have shown that the best results concerning blastogenesis were produced in media with a lower nitrogen content. The type of nutrient medium also significantly influenced the mean number of shoots formed in *Juniperus oxycedrus* L. [33,36].

In contrast, Bertsouklis et al. [46] faced difficulties using the poor in nitrogen MS medium in the in vitro propagation of *Juniperus phoenicea*. The percentage of shoot formation in MS was moderate, while the lowest values were achieved using the poorer WPM. Our results concerning shooting percentage in MS were slightly lower, while those for WPM appeared approximately twice as high in comparison. On the other hand, Bertsouklis et al. [46] reported that the use of DKW increased shoot formation, findings similar to ours. The results of Castro et al. [35] showed that WPM was less appropriate for culture. According to our study, similar results were obtained using the MS medium.

The addition of several plant growth regulators significantly affected the *Juniperus drupacea* explant shooting response. In our study, the addition of BA, mT, and TDZ increased the average shooting percentage, shoot number per explant, and the average shoot length. Explants of *Juniperus phoenicea* L. also showed the best response while using the medium supplemented with TDZ [37]. The promoting effect of TDZ was also stressed in our study and our results were in accordance with part of Al-Ramamneh et al.'s findings [53] regarding *Juniperus phoenicea* L. According to Salih et al. [50], the highest average shoot number and the longest average shoot length were obtained in WPM containing IAA, BAP, or IBA. Bertsouklis et al. [46] reported that the use of 2iP induced higher shooting responses in *J. phoenicea* while the addition of ZEA, NAA, or NAA in combination with BA presented moderate results. Khater and Benbouza [28] documented the stimulating effects of 2,4-D and BAP, alone or in combination with IBA or IAA, in the shoot development of *Juniperus thurifera* L. explants. In the in vitro culture of *Juniperus polycarpos* L., media supplementation with KIN and BA also had a significant effect [40]. Several studies on *Juniperus excelsa* M.Bieb., *Juniperus horizontalis* Moench. and *Juniperus chinensis* Roxb. [27], *Juniperus navicularis* Gand. [35], and *Juniperus oxycedrus* L. [33,36] have reported the shoot inducing effect of BA, NAA, and KIN.

Rooting is the primary bottleneck in the in vitro culture of most conifer species [32,35,45,54–56], thus restraining the establishment of commercial protocols [28]. Additionally, previous studies have documented the difficulty the *Juniperus* species has toward inducing adventitious roots under in vitro conditions [27,28,33,40,46,50,56,57]. In coniferous species, auxins such as NAA, IBA, and IAA are the most commonly used to achieve successful rooting under in vitro conditions. Our inability to achieve rooting is consistent with the results of several studies. Al-Ramamneh et al. [37] also did not report successful rooting in the in vitro culture. Loureiro et al. [27] reported that many treatments with IBA failed to stimulate rooting in *Juniperus phoenicea* L. explants. IBA, alone or in combination with NAA, was insufficient to induce rooting in *Juniperus thurifera* L. [28], *Juniperus polycarpos* K.Koch [40], and *Juniperus phoenicea* L. [46]. Negussie [56] found that treatment with IBA and NAA failed to stimulate shoot rooting, results that were in accordance with ours. According to Castro et al. [35], in vitro rooting experiments using NAA were also unsuccessful. Rooting of *Juniperus polycarpos* K.Koch was also not satisfactory [40]. Salih et al. [50] did not even try to root the in vitro cultured shoots of *Juniperus procera* Hoechst. Ex Endl., reporting that more research should be conducted on this species' root regeneration.

Various explanations have been expressed concerning the recalcitrance of in vitro culture, and particularly, the rooting difficulty of Juniperus species. Micropropagation of *Juniperus* species depends on the specific species/ecotype [27], the age of the trees used as explant donors [27], the genotype [33,57], or even too low levels of internal auxin [58].

Further research is needed regarding blastogenesis and rooting ability of the in vitro cultures of *Juniperus drupacea* L. Different nutrient media, at several strengths as well as other plant growth regulators or infection with *Agrobacterium rhizogenes*, must be applied in order to achieve rooting and better shoot induction and elongation. The decrease in macronutrients in culture media has been found to stimulate rooting in many plants [55,59–62]; indeed, decreased concentrations of nutrients in the medium, in particular, lessening the nitrogen, seem to promote adventitious rooting [27,32,63,64]. Complementary or substitute regeneration methods could be in vitro techniques of somatic embryogenesis, which have been implemented with greater or less success in *Pinus* spp. [65], *Juniperus communis* [66], *Picea abies* [67,68], and *Pinus nigra* and Abies hybrids [69]. Moreover, ex vitro rooting systems, an alternative rooting methodology, can be used to induce rooting. Such systems are commonly used in conifers as they present positive effects in developing roots [34] compared to in vitro systems [26].

5. Conclusions

This was the first attempt to study the in vitro propagation of *Juniperus drupacea* L., found in Greece, where shoot induction has been achieved. The results showed that DKW was the most suitable medium regardless of any plant growth regulators used. The addition of BA, mT, and TDZ promoted the average shooting percentage, the number of shoots per explant, and the average length of shoots. However, the results of this pioneering study did not provide a protocol for the root development of in vitro cultured *Juniperus drupacea* L. explants, even though different auxins (i.e., IAA, IBA, and NAA) in different concentrations were used. Nevertheless, it provides a basis for further research where all alternatives of in vitro or ex vitro rooting should be examined and thoroughly analyzed, in order to successfully induce adventitious roots of in vitro cultured shoots of *Juniperus drupacea* L.

Supplementary Materials: The following supporting information can be downloaded at: https://www.mdpi.com/article/10.3390/f14010142/s1, Table S1: Effect of the medium and plant growth regulator types and their concentrations on the average percentage of blastogenesis (%) of *Juniperus drupacea* explants in relation to their gender (means followed by the same letter did not differ statistically at $p \leq 0.05$, according to the Duncan test); Table S2: The effect of the medium and plant growth regulator types and their concentrations on the average number of shoots per *Juniperus drupacea* explants in relation to their gender (means followed by the same letter did not differ statistically at $p \leq 0.05$, according to the Duncan test); Table S3: Effect of the medium and plant growth regulator types and their concentrations on the average shoot length per *Juniperus drupacea* explants in relation to their gender (means followed by the same letter did not differ statistically at $p \leq 0.05$, according to the Duncan test).

Author Contributions: Conceptualization, K.I. and E.N.D.; Methodology, K.I.; Validation, K.I.; Formal analysis, K.I..; Investigation, K.I., D.P. and S.B.; Resources, K.I. and E.N.D.; Data curation, K.I. and I.T. Writing—original draft preparation, K.I. and D.P.; Writing—review and editing, K.I. and D.P.; Visualization, K.I. and E.N.D.; Supervision, K.I.; Project administration E.N.D.; Funding acquisition, E.N.D. All authors have read and agreed to the published version of the manuscript.

Funding: This research was co-financed by the European Regional Development Fund of the European Union and Greek National Funds (2020-2023) through the Operational Program Competitiveness, Entrepreneurship and Innovation, under the call RESEARCH—CREATE—INNOVATE (project code: T2EDK-03378).

Data Availability Statement: Not applicable.

Conflicts of Interest: The authors declare no conflict of interest.

References

1. Vidakovic, M. *Conifers. Morphology and Variation.*; Grafičko Zavod Hrvatske: Zagreb, Yugoslavia, 1991; ISBN 0-85198-807-5.
2. Adams, R.P. *Junipers of the World: The Genus Juniperus*; Trafford Publishing: Bloomington, IN, USA, 2014; ISBN 1-4907-2325-0.
3. Sobierajska, K.; Boratyńska, K.; Jasińska, A.; Dering, M.; Ok, T.; Douaihy, B.; Bou Dagher-Kharrat, M.; Romo, Á.; Boratyński, A. Effect of the Aegean Sea Barrier between Europe and Asia on Differentiation in *Juniperus drupacea* (Cupressaceae). *Bot. J. Linn. Soc.* **2016**, *180*, 365–385. [CrossRef]
4. Coode, M.; Cullen, J. *Juniperus L. Flora of Turkey and the East Aegean Islands*; Edinburgh University Press: Edinburgh, UK, 1965; Volume 1, pp. 78–84.
5. Axelrod, D.I. Evolution and Biogeography of Madrean-Tethyan Sclerophyll Vegetation. *Ann. Mo. Bot. Gard.* **1975**, *62*, 280–334. [CrossRef]
6. Kvaček, Z. A New Juniper from the Palaeogene of Central Europe. *Feddes Repert. Z. Bot. Taxon. Geobot.* **2002**, *113*, 492–502. [CrossRef]
7. Palamarev, E. Paleobotanical Evidences of the Tertiary History and Origin of the Mediterranean Sclerophyll Dendroflora. *Plant Syst. Evol.* **1989**, *162*, 93–107. [CrossRef]
8. Boratyński, A.; Browicz, K.; Zieliński, J. *Chorology of Trees and Shrubs in Greece*; Institute of Dendrology, Polish Academy of Sciences: Kornik, Poland, 1992; ISBN 83-85599-04-5.
9. Constantinidis, T.A.; Kalpoutzakis, E. *Plant Guide to Mount Parnon and Moustos Wetland Protected Area: Endemic, Rare and Threatened Species*; Management Body Mount Parnon: Moustos Wetland, Greece, 2015; ISBN 960-99424-4-X.
10. Tan, K.; Iatrou, G.; Johsen, B. *Endemic Plants of Greece: The Peloponnese*; Gads Forlag: Copenhagen, Denmark, 2001; Volume 1.
11. Bergmeier, E. Plant Communities and Habitat Differentiation in the Mediterranean Coniferous Woodlands of Mt. Parnon (Greece). *Folia Geobot.* **2002**, *37*, 309–331. [CrossRef]
12. Dimopoulos, P.; Raus, T.; Bergmeier, E.; Constantinidis, T.; Iatrou, G.; Kokkini, S.; Strid, A.; Tzanoudakis, D. *Vascular Plants of Greece: An Annotated Checklist*; Botanic Garden and Botanical Museum Berlin-Dahlem: Berlin, Germany, 2013.
13. Douaihy, B.; Tarraf, P.; Stephan, J. *Juniperus drupacea* Labill. Stands in Jabal Moussa Biosphere Reserve, a Pilot Study for Management Guidelines. *Plant Sociol.* **2017**, *54*, 39–45.
14. Hazubska-Przybył, T. Propagation of Juniper Species by Plant Tissue Culture: A Mini-Review. *Forests* **2019**, *10*, 1028. [CrossRef]
15. Farjon, A. *A Handbook of the World's Conifers*; Brill: Leiden, The Netherlands, 2010; Volume 1.
16. Directive, H. Council Directive 92/43/EEC of 21 May 1992 on the Conservation of Natural Habitats and of Wild Fauna and Flora. *Off. J. Eur. Union* **1992**, *206*, 7–50.
17. Gardner, M. *Juniperus drupacea.* The IUCN Red List of Threatened Species 2013: E. T30311A2792553. 2013. Available online: https://doi.org/10.2305/IUCN.UK.2013-1.RLTS.T30311A2792553.en (accessed on 28 October 2021).
18. Rivers, M.; Beech, E.; Bazos, I.; Bogunić, F.; Buira, A.; Caković, D.; Carapeto, A.; Carta, A.; Cornier, B.; Fenu, G. *European Red List of Trees*; International Union for Conservation of Nature and Natural Resources (IUCN): Fontainebleau, France, 2019; ISBN 2-8317-1986-0.
19. Gardner, M. *Juniperus drupacea.* The IUCN Red List of Threatened Species 2017: E.T30311A83932989. 2017. Available online: https://doi.org/10.2305/IUCN.UK.2013-1.RLTS.T30311A2792553.en (accessed on 1 November 2022).
20. Greek Biotope & Wetland Centre. Available online: https://www.ekby.gr/ekby/en/EKBY_PA_en_.html (accessed on 1 August 2022).
21. Yucedag, C.; Cetin, M.; Ozel, H.B.; Abo Aisha, A.E.S.; Alrabiti, O.B.M.; AL JAMA, A.M.O. The Impacts of Altitude and Seed Pretreatments on Seedling Emergence of Syrian Juniper (*Juniperus drupacea* (Labill.) Ant. et Kotschy). *Ecol. Process.* **2021**, *10*, 1–6. [CrossRef]
22. Gaussen, H. *Les Gymnospermes Actuelles et Fossiles, Fascicle 10, Les Cuppressacées*; Centre National de la Recherche Scientifique, Faculté des Sciences de Toulouse: Toulouse, France, 1968.
23. Maerki, D.; Frankis, M. *Juniperus drupacea* in the Peloponnese (Greece). Trip Report and Range Map, with Notes on Phenology, Phylogeny, Palaeontology, History, Types and Use. *Bull. CCP* **2015**, *3*, 3–31.
24. Christensen, K.; Juniperus, L. *Flora Hellenica*; Strid, A., Tan, K., Eds.; Koeltz Scientific Books: Königstein, Germany, 1997; Volume 1, pp. 10–14.
25. Gültekin, H.C.; Gültekin, Ü.G.; Divrik, A. Andız (*Arceuthos drupacea* (Labill.) Ant. et. Kotschy.) Tohumlarının Çimlenmesi, Diğer Tohum ve Fidan Özelliklerine İlişkin Bazı Tespit ve Öneriler. *J. Artvin Coruh Univ. For. Fac.* **2004**, *5*, 48–54.
26. Gómez, M.P.; Segura, J. Morphogenesis in Leaf and Single-Cell Cultures of Mature *Juniperus oxycedrus*. *Tree Physiol.* **1996**, *16*, 681–686. [CrossRef] [PubMed]
27. Loureiro, J.; Capelo, A.; Brito, G.; Rodriguez, E.; Silva, S.; Pinto, G.; Santos, C. Micropropagation of *Juniperus phoenicea* from Adult Plant Explants and Analysis of Ploidy Stability Using Flow Cytometry. *Biol. Plant.* **2007**, *51*, 7–14. [CrossRef]
28. Khater, N.; Benbouza, H. Preservation of *Juniperus thurifera* L.: A Rare Endangered Species in Algeria through in Vitro Regeneration. *J. For. Res.* **2019**, *30*, 77–86. [CrossRef]
29. Kentelky, E. The Analysis of Rooting and Growth Peculiarities of *Juniperus* Species Propagated by Cuttings. *Bull. UASVM Hortic* **2011**, *68*, 380–385.
30. Hartmann, H.T.; Kester, D.E.; Davies, F.T.; Geneve, R.L. *Plant Propagation: Principles and Practices.*; Prentice-Hall Inc.: Hoboken, NJ, USA, 1997; ISBN 0-13-261488-X.

31. Berhe, D.; Negash, L. Asexual Propagation of Juniperus Procera from Ethiopia: A Contribution to the Conservation of African Pencil Cedar. *For. Ecol. Manag.* **1998**, *112*, 179–190. [CrossRef]

32. Ragonezi, C.; Klimaszewska, K.; Castro, M.R.; Lima, M.; de Oliveira, P.; Zavattieri, M.A. Adventitious Rooting of Conifers: Influence of Physical and Chemical Factors. *Trees* **2010**, *24*, 975–992. [CrossRef]

33. Gómez, M.P.; Segura, J. Axillary Shoot Proliferation in Cultures of Explants from Mature *Juniperus Oxycedrus* Trees. *Tree Physiol.* **1995**, *15*, 625–628. [CrossRef]

34. Harry, I.S.; Pulido, C.M.; Thorpe, T.A. Plantlet Regeneration from Mature Embryos of *Juniperus cedrus*. *Plant Cell Tissue Organ Cult.* **1995**, *41*, 75–78. [CrossRef]

35. Castro, M.R.; Belo, A.F.; Afonso, A.; Zavattieri, M.A. Micropropagation of *Juniperus navicularis*, an Endemic and Rare Species from Portugal SW Coast. *Plant Growth Regul.* **2011**, *65*, 223–230. [CrossRef]

36. Gómez, M.P.; Segura, J. Factors Controlling Adventitious Bud Induction and Plant Regeneration in Mature *Juniperus oxycedrus* Leaves Cultured In Vitro. *Plant* **1994**, *30*, 210–218. [CrossRef]

37. Al-Ramamneh, E.A.; Dura, S.; Daradkeh, N. Propagation Physiology of *Juniperus phoenicea* L. from Jordan Using Seeds and in Vitro Culture Techniques: Baseline Information for a Conservation Perspective. *Afr. J. Biotechnol.* **2012**, *11*, 7684–7692. [CrossRef]

38. Zaidi, M.A.; Khan, S.; Jahan, N.; Yousafzai, A.; Mansoor, A. Micropropagation and Conservation of Three *Juniperus* Species (Cupressaceae). *Pak. J. Bot.* **2012**, *44*, 301–304.

39. Kashani, S.A.; Asrar, M.; Leghari, S.K. In Vitro Callus Induction and Shoot Formation of *Juniperus excelsa* of Ziarat, Balochistan, Pakistan. *Fuuast J. Biol.* **2018**, *8*, 203–208.

40. Momeni, M.; Ganji-Moghadam, E.; Kazemzadeh-Beneh, H.; Asgharzadeh, A. Direct Organogenesis from Shoot Tip Explants of *Juniperus Polycarpos* L.: Optimizing Basal Media and Plant Growth Regulators on Proliferation and Root Formation. *Plant Cell Biotechnol. Mol. Biol* **2018**, *19*, 40–50.

41. Fett-Neto, A.G.; Fett, J.P.; Goulart, L.W.V.; Pasquali, G.; Termignoni, R.R.; Ferreira, A.G. Distinct Effects of Auxin and Light on Adventitious Root Development in Eucalyptus Saligna and Eucalyptus Globulus. *Tree Physiol.* **2001**, *21*, 457–464. [CrossRef] [PubMed]

42. Greenwood, M.S.; Cui, X.; Xu, F. Response to Auxin Changes during Maturation-related Loss of Adventitious Rooting Competence in Loblolly Pine (*Pinus taeda*) Stem Cuttings. *Physiol. Plant.* **2001**, *111*, 373–380. [CrossRef]

43. Bielenin, M. Rooting and Gas Exchange of Conifer Cuttings Treated with Indolebutyric Acid. *J. Fruit Ornam. Plant Res.* **2003**, *11*, 99–106.

44. Henrique, A.; Campinhos, E.N.; Ono, E.O.; de Pinho, S.Z. Effect of Plant Growth Regulators in the Rooting of Pinus Cuttings. *Braz. Arch. Biol. Technol.* **2006**, *49*, 189–196. [CrossRef]

45. Harry, I.S.; Thorpe, T.A. In Vitro Culture of Forest Trees. In *Plant Cell and Tissue Culture*; Vasil, I.K., Thorpe, T.A., Eds.; Springer Netherlands: Dordrecht, The Netherlands, 1994; pp. 539–560. ISBN 978-94-017-2681-8.

46. Bertsouklis, K.; Paraskevopoulou, A.; Zarkadoula, N. In Vitro Propagation of *Juniperus phoenicea* L. *Acta Hortic.* **2019**, *1298*, 331–334. [CrossRef]

47. Murashige, T.; Skoog, F. A Revised Medium for Rapid Growth and Bio Assays with Tobacco Tissue Cultures. *Physiol. Plant.* **1962**, *15*, 473–497. [CrossRef]

48. Lloyd, G.; McCown, B. Commercially-Feasible Micropropagation of Mountain Laurel, Kalmia Latifolia, by Use of Shoot-Tip Culture. *Comb. Proc.-Int. Plant Propagator's Soc.* **1980**, *30*, 421–427.

49. Driver, J.A.; Kuniyuki, A.H. In Vitro Propagation of Paradox Walnut Rootstock. *HortScience* **1984**, *19*, 507–509. [CrossRef]

50. Salih, A.M.; Al-Qurainy, F.; Khan, S.; Tarroum, M.; Nadeem, M.; Shaikhaldein, H.O.; Alabdallah, N.M.; Alansi, S.; Alshameri, A. Mass Propagation of *Juniperus procera* Hoechst. Ex Endl. From Seedling and Screening of Bioactive Compounds in Shoot and Callus Extract. *BMC Plant Biol.* **2021**, *21*, 192. [CrossRef]

51. Ge, X.; Chu, Z.; Lin, Y.; Wang, S. A Tissue Culture System for Different Germplasms of Indica Rice. *Plant Cell Rep.* **2006**, *25*, 392–402. [CrossRef]

52. Thorpe, T.A.; Harry, I.S.; Kumar, P.P. Application of Micropropagation to Forestry. In *Micropropagation: Technology and Application*; Debergh, P.C., Zimmerman, R.H., Eds.; Springer Netherlands: Dordrecht, The Netherlands, 1991; pp. 311–336. ISBN 978-94-009-2075-0.

53. Al-Ramamneh, E.A.-D.; Daradkeh, N.; Rababah, T.; Pacurar, D.; Al-Qudah, M. Effects of Explant, Media and Growth Regulators on in Vitro regeneration and Antioxidant Activity of *Juniperus phoenicea*. *Aust. J. Crop Sci.* **2017**, *11*, 828–837. [CrossRef]

54. Oliveira, P.; Barriga, J.; Cavaleiro, C.; Peixe, A.; Potes, A.Z. Sustained in Vitro Root Development Obtained in *Pinus pinea* L. Inoculated with Ectomycorrhizal Fungi. *For. Int. J. For. Res.* **2003**, *76*, 579–587. [CrossRef]

55. Renau-Morata, B.; Ollero, J.; Arrillaga, I.; Segura, J. Factors Influencing Axillary Shoot Proliferation and Adventitious Budding in Cedar. *Tree Physiol.* **2005**, *25*, 477–486. [CrossRef]

56. Negussie, A. In Vitro Induction of Multiple Buds in Tissue Culture of *Juniperus excelsa*. *For. Ecol. Manag.* **1997**, *98*, 115–123. [CrossRef]

57. Brito, G.M.O. Micropropagaçao de Duas Espécies Autóctones Da Ilha de Porto Santo (*Olea europaea* L. ssp. Maderensis Lowe e *Juniperus phoenica* L.) e Estudo Da Resposta de Rebentos in Vitro a Stress Osmótico. Master's Thesis, Universidade de Aveiro, Aveiro, Portugal, 2000.

58. Smith, N.G.; Wareing, P.F. The Rooting of Actively Growing and Dormant Leafy Cuttings in Relation to Endogenous Hormone Levels and Photoperiod. *New Phytol.* **1972**, *71*, 483–500. [CrossRef]

59. Hyndman, S.F.; Hasegawa, P.M.; Bressan, R.A. Of Mineral Salts. *HortScience* **1982**, *17*, 82–83. [CrossRef]
60. Sha Valli Khan, P.S.; Hausman, J.F.; Rao, K.R. Effect of Agar, MS Medium Strength, Sucrose and Polyamines on in Vitro Rooting of *Syzygium Alternifolium*. *Biol. Plant.* **1999**, *42*, 333–340. [CrossRef]
61. Fadel, D. Effect of Different Strength of Medium on Organogenesis, Phenolic Accumulation and Antioxidant Activity of Spearmint (*Mentha spicata* L.). *TOHORTJ* **2010**, *3*, 31–35. [CrossRef]
62. Ioannidis, K.; Dadiotis, E.; Mitsis, V.; Melliou, E.; Magiatis, P. Biotechnological Approaches on Two High CBD and CBG *Cannabis sativa* L. (Cannabaceae) Varieties: In Vitro Regeneration and Phytochemical Consistency Evaluation of Micropropagated Plants Using Quantitative 1H-NMR. *Molecules* **2020**, *25*, 5928. [CrossRef]
63. Ordas, R.; Rodríguez, A.; Rodríguez, R.; Sanchez Tames, R. Desarrollo de Tecnicas de Cultivo «in Vitro» Para La Micropropagation de Variedades de Manzana Sidrera. In *Anales de edafologia y agrobiologia*; Consejo Superior de Investigaciones Científicas. Instituto Nacional de Edafología y Agrobiología: Madrid, Spain, 1984; Volume 43, pp. 905–917.
64. Trewavas, A.J.; Cleland, R.E. Is Plant Development Regulated by Changes in the Concentration of Growth Substances or by Changes in the Sensitivity to Growth Substances? *Trends Biochem. Sci.* **1983**, *8*, 354–357. [CrossRef]
65. Klimaszewska, K.; Trontin, J.-F.; Becwar, M.R.; Devillard, C.; Park, Y.-S.; Lelu-Walter, M.-A. Recent Progress in Somatic Embryogenesis of Four *Pinus* Spp. *Tree For. Sci. Biotechnol.* **2007**, *1*, 11–25.
66. Helmersson, A.; von Arnold, S. Embryogenic Cell Lines of *Juniperus communis*; Easy Establishment and Embryo Maturation, Limited Germination. *Plant Cell Tissue Organ Cult. PCTOC* **2009**, *96*, 211–217. [CrossRef]
67. Von Arnold, S.; Egertsdotter, U.; Ekberg, I.; Gupta, P.; Mo, H.; Nörgaard, J. Somatic Embryogenesis in Norway Spruce (*Picea abies*). *Somat. Embryog. Woody Plants* **1995**, *3*, 17–36.
68. Von Arnold, S.; Clapham, D. Spruce Embryogenesis. In *Plant Embryogenesis*; Springer: Berlin/Heidelberg, Germany, 2008; pp. 31–47.
69. Salaj, T.; Klubicová, K.; Matusova, R.; Salaj, J. Somatic Embryogenesis in Selected Conifer Trees *Pinus nigra* Arn. and *Abies* Hybrids. *Front. Plant Sci.* **2019**, *10*, 13. [CrossRef]

Article

Improvement of Rooting Performance in Stem Cuttings of Savin Juniper (*Juniperus sabina* L.) as a Function of IBA Pretreatment, Substrate, and Season

Maliheh Abshahi [1], Francisco Antonio García-Morote [2,*], Hossein Zarei [3], Bahman Zahedi [1] and Abdolhossein Rezaei Nejad [1]

[1] Department of Horticultural Sciences, Faculty of Agriculture, Lorestan University, P.O. Box 465, Khorramabad 68151-44316, Iran

[2] Department of Science and Agroforestry Technology and Genetics, Higher Technical School of Agricultural and Forest Engineering, University of Castilla-La Mancha, 02071 Albacete, Spain

[3] Department of Horticulture, College of Plant Production, Gorgan University of Agricultural Sciences and Natural Resources, Gorgan 49138-15739, Iran

* Correspondence: fcoantonio.garcia@uclm.es

Abstract: *Juniperus sabina* is an interesting species for forest restoration and ornamental purposes. The seeds of this plant have several types of dormancies; therefore, seed propagation is difficult and time consuming. The production of cuttings can be an alternative way to produce plants more quickly. The main objective of this experiment was to determine the best propagation conditions (indole butyric acid dose, substrate, and season) for this species using stem cuttings. Rooting performance of the cuttings was evaluated based on the rooting percentage (%), root biomass, and specific root length (SRL). In addition, we examined the internal composition (auxin and peroxidase content) in treated stem cuttings. Cuttings were pretreated with five doses of indole butyric acid (IBA; 0 (control), 1000, 2000, 4000, and 8000 ppm) and were rooted in four substrates (perlite, perlite-cocopeat, pumice, and mixed substrate) during the four seasons (winter, spring, summer, and autumn). The best treatments, with more than 60% rooting, were applied in spring, and IBA at 1000 ppm in perlite–cocopeat substrate obtained 62% rooting. The highest rooting percentage correlated with the highest root biomass production and the lowest SRL. IBA pretreatment decreased the concentration of peroxidase in spring (coinciding with maximum rooting), representing an indicator of rooting performance. Based on these results, we recommend a new protocol for *Juniperus sabina* production: (i) prepare cuttings in spring, (ii) treat cutting bases with 1000 ppm IBA, and (iii) plant cuttings in a substrate of perlite–cocopeat (1:1).

Keywords: auxin concentration; juniper production; peroxidase concentration; root development; vegetative propagation

Citation: Abshahi, M.; García-Morote, F.A.; Zarei, H.; Zahedi, B.; Nejad, A.R. Improvement of Rooting Performance in Stem Cuttings of Savin Juniper (*Juniperus sabina* L.) as a Function of IBA Pretreatment, Substrate, and Season. *Forests* **2022**, *13*, 1705. https://doi.org/10.3390/f13101705

Academic Editors: Carol Loopstra, Ricardo Javier Ordás and José Manuel Álvarez Díaz

Received: 30 July 2022
Accepted: 13 October 2022
Published: 16 October 2022

1. Introduction

The genus *Juniperus* comprises about 50 species of coniferous trees and shrubs widely distributed throughout the temperate and subtropical regions of the Northern Hemisphere. Savin juniper, or Savin (*Juniperus sabina* L.; Cupressaceae), is native to the mountains of Central and Southern Europe and Western and Central Asia, from Spain to Eastern Siberia, typically growing at altitudes of 700–3300 m [1,2]. The species has its origin in the remote times of the Tertiary, when it grew in colder climates. It is well adapted to continental climates with very cold winters and hot, dry summers, which are typical of semi-arid climates [3].

This species can be utilized for forest restoration on poor sites with low potential productivity, such as arid and semi-arid areas. This plant is native to mainly steep slopes, and its increased strength in these conditions is due to its deep roots) which can penetrate up to one meter), making the plant last on slopes over 30° [4]. In addition, Savin is one of the most beautiful juniper species and is suitable for ornamental use [5]. Thus, information about *Juniperus sabina* plant production could be useful for forest managers and plant producers in some areas of Europe and Asia.

Seed propagation of *Juniperus* is difficult and time consuming [2]. The sexual regeneration of *Juniperus sabina* is low due to poor seed quality [6], and reproductive propagation is difficult due to the prolonged period between pollen production and fertility (fertilization), seed loss, and the long duration of seed dormancy [7]. The success of asexual propagation using rooting cuttings of juniper branches is often reported as less than 30% [8]. To enhance vegetative propagation in juniper, using a tissue culture system based on micropropagation and in vitro culture can be an option to produce plants [8]. However, in conifers, the rapid loss of the callus multiplication capacity under in vitro culture is a serious problem [9]. In this sense, the production of cuttings can be an alternative way to produce plants of this conifer more quickly in a greenhouse. However, problems with the asexual propagation of *Juniperus* species by rooting vegetative cuttings has not been well studied, and the rooting efficiency of juniper branch cuttings is often less than 30% [9]. For this reason, we think that it is necessary to investigate new techniques or protocols for the vegetative reproduction of this species.

To propagate plants via cuttings, the growth regulator indole butyric acid (IBA) has been widely used as pretreatment [10]. In general, treated cuttings have shown higher rooting percentage and number of roots and longer root length than untreated cuttings, which indicates a strong influence of IBA on rooting capacity [11]. Thus, to produce junipers by stem cuttings, IBA was used in previous studies on *Juniperus* genus [12,13]. A certain amount of indole acetic acid (IAA) may also be necessary for root formation [11]. IAA is the most important auxin, and seems to be an important internal factor for the regulation of adventitious root formation, and it is a stimulant hormone [14]. During the propagation of plants, the amount of IAA (as an internal hormone) can vary based on the time the cuttings are prepared for rooting [15,16]. Measuring the amount of internal IAA in treated cuttings can show the effect of the treatment. The peroxidase enzyme level is another indicator because a high level of peroxidase decreases rooting percentage [17]. In fact, peroxidases are often introduced as biochemical markers [18]. Moreover, in plants with hydric stress, increasing the peroxidase, will decrease the percentage of rooting [17].

Previous studies also reported differences in rooting of cuttings affected by substrate [19]. Analyzing the rooting of *Juniperus horizontalis*, Hong-wei et al. [20] found that the best substrate was 1:3 (v/v) vermicelli and 2:3 (v/v) perlite, with 36% rooting. In the study of Sabina et al. [21] on *Juniperus sabina*, the results showed that a substrate containing 60% foliar soil (fragmented leaves of the trees) and 40% fragmented oats provided a suitable growth rate. The size and age of cuttings can also be important factors to maximize rooting and cutting growth. Maintaining stock plants at 15 cm height sometimes reduced the production of stem cuttings, but often increased the percentage of cuttings that formed roots by 30–53% [22]. Choosing the right time to achieve the best rooting is also important. For example, the best rooting of *Juniperus virginiana* was from cuttings collected in winter [13].

Finally, to evaluate the level of rooting of cuttings, it is necessary to measure root morphology and development. Roots take up minerals and water from the soil, and larger amounts of adventitious roots could improve the root system symmetry, stability, survival, and growth rate [23]. Thus, root biomass should be a good indicator of the growth strategy of root development and the capacity to endure water stress in juniper trees [24]. Another indicator is the specific root length (SRL; calculated as root length per unit of root biomass), which depends on root diameter and tissue density [25]. Roots with high SRL have a high surface-to-volume ratio for the same C investment. This maximizes the root–soil interface, and a high SRL can be achieved by having roots with a small diameter to maximize water uptake [26]. For this, both root production and structure can be considered fundamental parameters to assess the level of rooting in plant material.

Therefore, the present study was intended to investigate an efficient method (dose of IBA and type of substrate) for vegetative propagation of Savin juniper using stem cuttings, and the effects on morphological (root development) and phytochemical (auxin and peroxidase) characteristics. We also hypothesized that several parameters of roots (root biomass and specific root length) and the levels of auxins and peroxidase could be indicators of rooting performance in cuttings of *Juniperus sabina*. Thus, the main objective of this research was to analyze the effects of five concentrations of IBA as pretreatment and four substrate types (perlite, perlite–cocopeat, pumice, and mixed substrate) on the rooting performance and levels of auxin and peroxidase in cuttings. The experiment was conducted in the four seasons of the year to determine the effects of harvesting time on the rooting capacity of cuttings.

2. Materials and Methods

2.1. Cutting Preparation, Pretreatment with Indole Butyric Acid (IBA), and Substrate Composition

Cuttings of *Juniperus sabina* were sampled from its natural habitat in the Chaharbagh mountains of Gorgan (northern Iran; Figures 1 and 2), one of the main Mediterranean populations at higher altitude (2700 m.a.s.l.).

Figure 1. Worldwide distribution of different populations of *Juniperus sabina*, in gray [27]; sampling area of collected stem cuttings is in red circle (Northern Iran).

The climate in the study area is cold and semi-arid, type BSk [28]. Based on 30-year averages, the mean annual temperature at the site is 9.2 °C and the mean annual precipitation is 429 mm. Temperature extremes (in summer and winter) range from 23 °C to −5 °C (data from Gorgan climatic station, 46°06′ N, 28°00′ W, 2600 m.a.s.l.). Soils are sandy loam. In this area, the mean age of Savin juniper plants is 50 years. The size of crowns is approximately 2×2 m (length × width; Figure 2). The ring diameter of shrubs is on average 20.0 cm, and the height is 1.5 m (these are a type of old and horizontal shrub; Figure 2). Generally, 20 male shrubs were used for this experiment, growing in the same area with a similar environment (climate and soil). The experiment was conducted at Gorgan University of Agricultural Sciences and Natural Resources in winter 2016 and spring, summer, and autumn 2017. Stem cuttings were only collected from the upper crown in male trees.

The harvesting of cuttings took place in the morning. After harvesting, stem cuttings 15 cm in length and 0.5–0.7 cm in diameter [29] were prepared for treatment and cultivation in a greenhouse (Figure 2). Substrates were prepared, and cuttings were placed in the greenhouse equipped with an automatic system to control humidity (micro-irrigation) and temperature at the root level. The average daily temperature during the experiment was 22 °C, and average relative humidity was 77%. The photoperiod was based on nature in each season.

Figure 2. (**a**) Example of Savin juniper plants for collection of stem cuttings in study area; (**b**) greenhouse used to culture stem cuttings under different treatments; (**c**) cuttings growing on substrate; (**d**) example of rooted and unrooted stem cuttings of Savin juniper.

For the pretreatment of stem cuttings, 5 IBA concentrations (Merk, Darmstadt, Germany) were used: 0 (control), 1000, 2000, 4000, and 8000 parts per million (ppm). The base of each cutting was placed in IBA for 5 s and then inserted into the substrate. The 4 substrates were (i) perlite; (ii) mixed rooting substrate, a combination of sand (20%), perlite (20%), cocopeat (20%), vermicompost (20%), and potash (20%); (iii) perlite–cocopeat (1:1); and (iv) mineral pumice (all substrates were developed at the University of Gorgan, Golestan, Iran). For each treatment (combination of pretreatment and substrate), 3 replicates were prepared, with 9 cuttings per replicate. Thus, a total of 540 cuttings were planted in the greenhouse each season (Figure 2).

2.2. Rooting Performance: Rooting (%), Root Biomass, and Specific Root Length (SRL)

To determine the rooting percentage, roots were counted in all rooted cuttings in each treatment (Figure 2). The root length (in mm) was obtained by separating the roots from the stem cuttings and then measuring them with a digital caliper (Mitutoyo, Kanagawa, Japan). To measure the root biomass (dry weight in grams), the roots of each cutting after separation were placed in an oven (Unico, E. Dayton, NJ, USA) at 70 °C, and after 24 h, they were weighed separately. To measure average root biomass, the dry weights of all roots of cuttings in each treatment were combined, and the total was divided by the number of cuttings. To calculate the specific root length (SRL; cm g^{-1}), the length of roots (cm) was divided by the root biomass (g) [30].

2.3. Internal Chemical Compounds of Stem Cuttings: Auxin and Peroxidase Enzyme Concentration

Internal auxin (μg g^{-1}) was measured following the methodology of Sridhar et al. [31], with slight modification. First, 5 g of plant sample (bark or periderm of the end part of stem cuttings for each treatment) was ground into powder and filtered with 10 mL methanol (Merk, Darmstadt, Germany). After adding methanol several times, it was evaporated in a rotary evaporator (Unico, E. Dayton, NJ, USA) at 30 °C.

In the next step, the followed materials were added (all made by Tetrachem Company, Cardiff, UK): 10 mL of potassium phosphate, 10 mL of petroleum spirit (3 times), 10 mL of petroleum spirit, and 3 mL of phosphoric acid. Then 10 mL of this solution was extracted, and these materials were added: 10 mL of potassium phosphate solution, phosphoric acid (0.28 M), and 10 mL of ethyl ether. The ether was evaporated in the rotary evaporator. The residue was dissolved in 5 mL of methanol at low temperature, and 0.2 mL of tri-fluoro acetic acid and 3 mL water were added to each sample.

The amount of internal auxin was measured with a spectrophotometer (Unico, E. Dayton, NJ, USA) at 440 and 490 nm wavelengths. Samples of unrooted and rooted cuttings totaled 152 and 88, respectively.

To measure the level of peroxidase enzyme in stem cuttings (mg g^{-1}), 1 g of leaves from cuttings of each treatment was weighted. The following combination was prepared (all chemical materials made by Tetrachem Company, Cardiff, UK) [31]: 1.2 g of Tris, 2 g of ascorbic acid, 3.8 g of sodium borate, 2 g of ethylene diamine tetra acetate (EDTANa), and 50 g of polyethylene glycol 2000, mixed with distilled water to achieve a volume of 100 mg L^{-1}.

Each plant sample was mixed with 4 mL of the above solution and placed on a shaker, then kept in a refrigerator (LG, Seoul, South Korea) at 4 °C. In the next step, the samples were centrifuged (Unico, E. Dayton, NJ, USA), then filtered. In the final step, 0.1 mg g^{-1} of each plant sample was combined with 2 mg of buffer acetate 0.3 M (pH = 5.0), 0.4 mL of 3% hydrogen peroxide, and 0.2 mL of benzidine dissolved in alcohol at 50 °C (0.01 M). Using the spectrophotometer at a wavelength of 530 nm, they were measured against the control. The total numbers of samples for unrooted and rooted cuttings were 122 and 67, respectively.

Auxin and peroxidase were measured in rooted and unrooted stem cuttings to detect differences in the internal compounds between cuttings that have the capacity for rooting and those that do not. To evaluate the treatments and compare the chemical compounds in cuttings at the beginning of sampling and the amount of increase or decrease between the time of planting and rooting (between the beginning and end of each season), samples were taken from freshly harvested cuttings in each season (the first of each season) and compared with the results at the end of the season.

2.4. Statistical Analysis

A factorial arrangement of treatments [32] was applied to analyze the effects of the 3 main factors on the 5 dependent variables. The first factor was pretreatment or a concentration of IBA (5 levels: 0, 1000, 2000, 4000, and 8000 ppm), the second was substrate (4 types: perlite, perlite–cocopeat, pumice, and mixed substrate), and the third was season (winter, spring, summer, and autumn). This represents a 5 × 4 × 4 factorial design. The dependent variables were the indicators of rooting performance (% of rooting, root biomass, and SRL) and internal compounds of unrooted and rooted cuttings (auxin and peroxidase enzyme).

In addition, for the dependent variables concerning internal chemical compounds, another level was added as pretreatment, fresh samples (stem cuttings not planted and prepared at the beginning of each season), in order to compare the effects of treatments between untreated and treated cuttings at the end of each season.

SAS® statistical software (SAS Institute Inc., Cary, NC, USA) was used to detect significant factors and to compare mean values between factors and treatments. Means were compared using the PROC GLM procedure. We utilized multifactor analysis of variance (3-way ANOVA) at a probability level of 5% ($p < 0.05$). The analysis within seasons was performed by 2-way ANOVA (excluding season as a main factor in the complete model). We performed independent ANOVAs (not mixed-design or repeated-measures) because the measurements were independent (different stem cuttings were used for each treatment and season). Three-level interactions were not performed because the degrees of freedom were cero in all cases.

Fisher's least significant difference (LSD) test ($p < 0.05$) was used to determine significant differences between treatments [33]. To apply this statistical method, it is desirable for the data to be normally distributed. This is not the case for proportions, which have values that range between 0 and 1. In addition, errors must be independent and normally distributed with constant variance. To ensure that these assumptions were met, logarithmic transformation was used [21]: for % of rooting, the analyzed variable was [ln(r + 0.5)], where r is the percentage of rooting (divided by 100). As this transformation requires numerical data above zero, a small number (0.5) was added to this variable before transformation. The other dependent variables were normally distributed.

3. Results

3.1. Rooting Performance

Table 1 shows that IBA pretreatment and substrate composition significantly affected rooting percentage and SRL, whereas season of collection significantly affected all the dependent variables ($p < 0.0001$ in all cases). There were also several significant interactions, but only the pretreatment × season interaction significantly affected the three dependent variables (Table 1). These results confirmed a strong seasonality (harvesting time) of the rooting capacity of stem cuttings.

Table 1. Results of multifactor ANOVA of effect of main factors on rooting performance of cuttings across four seasons. Shown are *p*-values for three principal effects (pretreatment with IBA, substrate, and season, and their two-way interactions) and effects within each season (pretreatment, substrate, and their interaction). Effects were considered significant when $p < 0.05$. A total of 540 stem cuttings were planted each season.

Variables	Effects	Winter	Growing Season Spring	Summer	Autumn	Annual Values
Rooting (log-transformed units)	Pretreatment	0.10	<0.001	0.0005	0.11	<0.0001
	Substrate	0.91	0.03	0.22	0.08	<0.0001
	Season	-	-	-	-	<0.0001
	Pretreatment × Substrate	0.40	0.24	*	*	<0.0001
	Pretreatment × Season	-	-	-	-	<0.0001
	Substrate × Season	-	-	-	-	<0.0001
Root biomass (g)	Pretreatment	<0.0001	0.63	<0.0001	0.04	0.39
	Substrate	0.30	0.01	0.11	0.29	0.08
	Season	-	-	-	-	<0.0001
	Pretreatment × Substrate	0.64	0.04	*	*	0.76
	Pretreatment × Season	-	-	-	-	0.01
	Substrate × Season	-	-	-	-	0.87
SRL (cm g^{-1})	Pretreatment	<0.0001	0.04	<0.0001	0.01	<0.0001
	Substrate	0.08	0.29	0.02	0.95	<0.0001
	Season	-	-	-	-	<0.0001
	Pretreatment × Substrate	0.32	0.86	*	*	0.93
	Pretreatment × Season	-	-	-	-	<0.0001
	Substrate × Season	-	-	-	-	<0.0001

* Summer and autumn: there were not enough living cuttings for analysis because many cuttings dried.

In this sense, the best root-growing season for cuttings of *Juniperus sabina* was spring (Figure 3). In spring, 502 cuttings rooted, and 38 cuttings did not root. In this season, the rooting of cuttings was $51.6 \pm 9.3\%$ (mean $\pm$ LSD interval; Figure 3a) with 4000 ppm IBA, but no significant difference was seen at 1000 and 2000 ppm ($p < 0.05$). Therefore, the best IBA level was 1000 to 4000 ppm. In the other seasons, rooting was very weak. In winter, cuttings in four treatments rooted (99 cuttings), and 441 cuttings did not root; in summer, cuttings in three treatments rooted (89 cuttings), and 451 cuttings did not root; and in autumn, cuttings in three treatments rooted (102 cuttings), and 438 did not root.

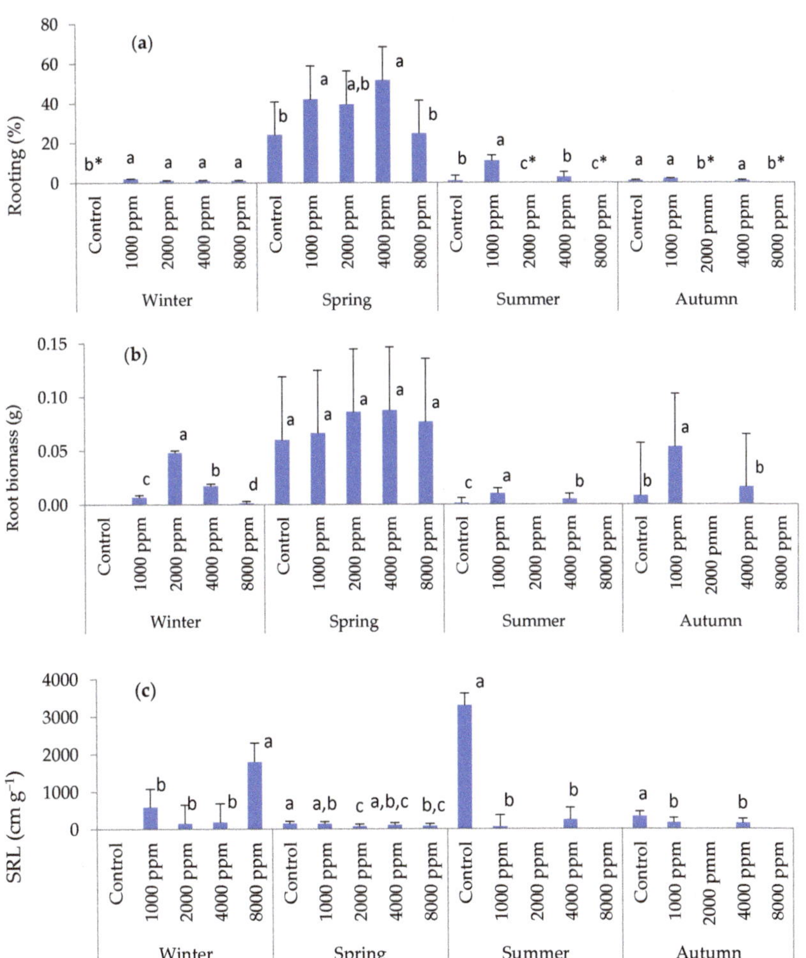

Figure 3. Mean values of rooting performance in rooted cuttings within seasons and with five indole butyric acid (IBA) pretreatments: (**a**) % of rooting, (**b**) root biomass, and (**c**) SRL. Mean values with the same letter do not differ at 0.05 level according to LSD test. Sample data = 540 cuttings in each season. Treatments in which all cuttings dried are not represented in the figure. Error bars: LSD intervals. * Rooting percentage was zero.

The highest biomass of roots also occurred in spring (0.088 ± 0.08 g at 4000 ppm IBA; mean $\pm$ LSD interval; Figure 3b), and there were no significant differences between treatment levels in this season. The lowest root biomass occurred with the control treatment in summer, and with 8000 ppm IBA in winter (Figure 3b). In addition, the highest SRL occurred with the control treatment in summer (3300 ± 314 cm g^{-1}), whereas the lowest occurred with the treatments in spring (67.1 ± 20.3 cm g^{-1} at 2000 ppm IBA (mean $\pm$ LSD interval); Figure 3c). Thus, the combination of no pretreatment of cuttings in the drier season increased SRL as an adaptation to capture water and nutrients under stress.

In several replicates, all of the cuttings dried, thus there were not sufficient cuttings to analyze chemical compounds. In our study, after harvesting the cuttings, a rooting percentage of zero was obtained with pretreatments of control in winter, and 2000 and 8000 ppm in summer and autumn, respectively (Figure 3a). This was due to the cuttings drying during maintenance in the substrates during those seasons.

The best substrate for % of rooting as a function of IBA level was perlite–cocopeat, with a maximum of 62.0 ± 15.6 % (mean $\pm$ LSD interval) at 1000 ppm IBA (Figure 4a). In the pumice substrate, the root biomass was significant with 2000 and 8000 ppm IBA (Figure 4b).

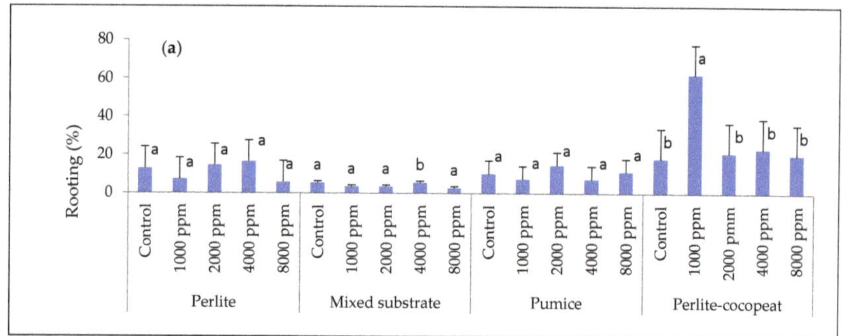

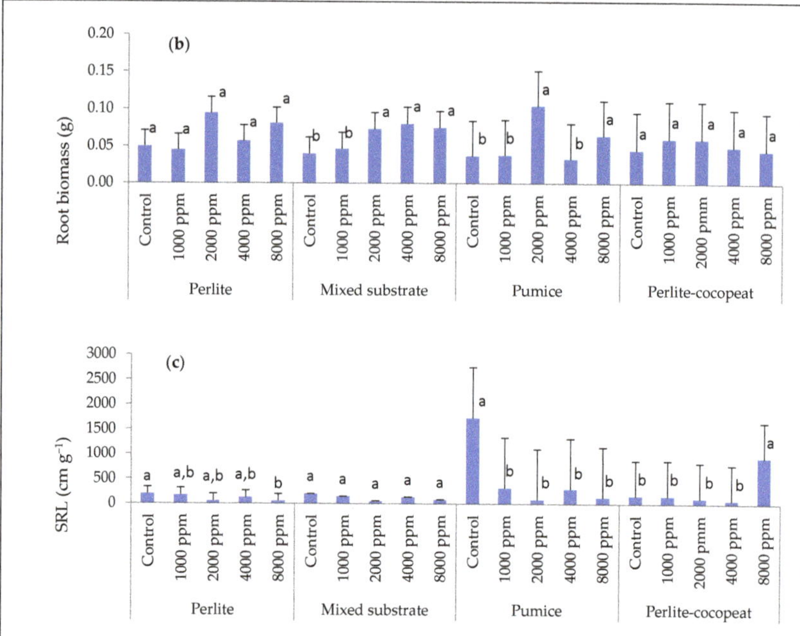

Figure 4. Mean values of rooting performance in rooted cuttings within substrates and with five IBA pretreatments: (**a**) % of rooting, (**b**) root biomass, and (**c**) SRL. Mean values with the same letter do not differ at 0.05 level according to LSD test. Sample data = 540 stem cuttings in each substrate. Error bars: LSD intervals.

There were no significant differences between the levels of root biomass in perlite and perlite–cocopeat substrates. In the mixed substrate, root biomass was significantly higher with 2000 to 8000 ppm IBA than control (Figure 4b). On the contrary, the highest SRL occurred with the control treatment in pumice and with 8000 ppm IBA in perlite–cocopeat (Figure 4c). Thus, it is notable that when rooting reached low levels, these values correlated with high SRL values.

Finally, the best rooting percentage also occurred in spring and in the perlite–cocopeat substrate (65.9 ± 9.3% (mean ± LSD interval); Figure 5a). This result was similar to that obtained in the previous analysis, and the combination of 1000 ppm, perlite–cocopeat, and spring resulted in a higher rooting value (62.3 ± 1.2% (mean ± standard error; third-order interaction value)). With this combination of factors, a rooting percentage of zero was obtained with mixed substrate in winter, perlite and mixed substrate in summer, and mixed substrate and perlite–cocopeat in autumn (Figure 5a).

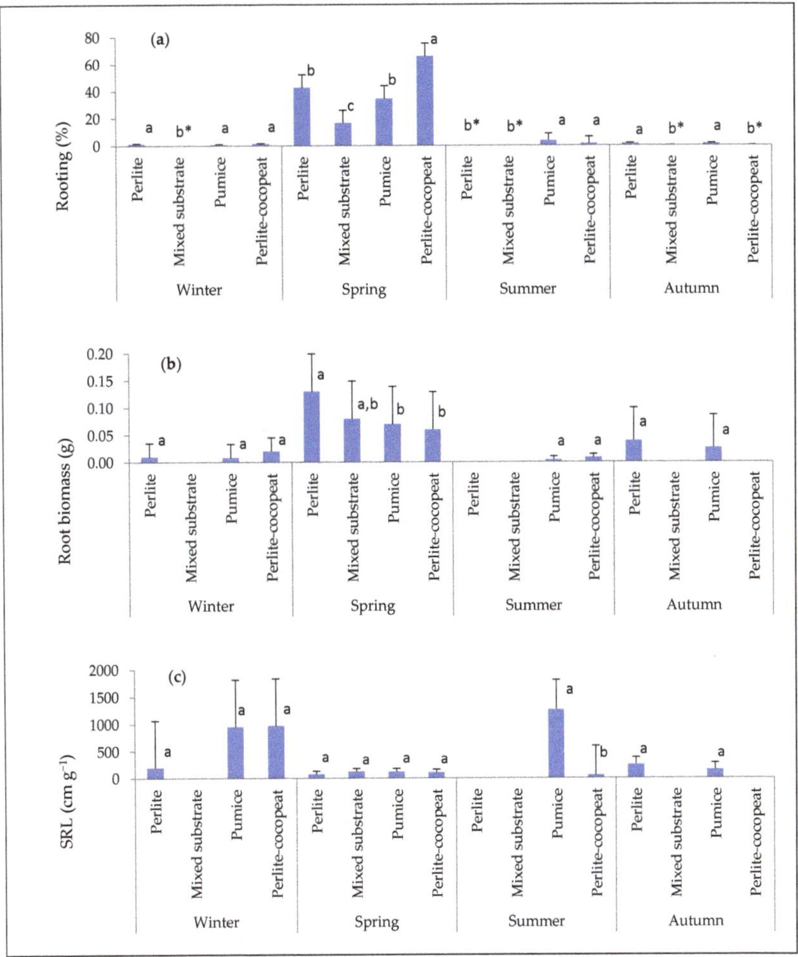

Figure 5. Mean values of rooting performance of rooted cuttings within seasons and with four substrates analyzed: (**a**) % of rooting, (**b**) root biomass, and (**c**) SRL. Mean values with the same letter do not differ at 0.05 level according to LSD test. Sample data = 540 cuttings in each season. Error bars: LSD intervals. *Rooting percentage was zero.

In summer, rooting only occurred in pumice and perlite–cocopeat, with an average of <5%. In autumn, cuttings rooted only in pumice and perlite (<3%), whereas in winter, rooting occurred with substrates of perlite, pumice, and perlite–cocopeat, with an average of <2% (Figure 5a). The greatest root biomass occurred in the spring, and the lowest in summer and winter (Figure 5b). The highest SRL occurred in pumice in summer (again in the warmer season) and perlite–cocopeat and pumice in winter (Figure 5c). On the contrary, in spring, SRL was low in all substrates (Figure 5c). So, in spring, when rooting percentage was the highest, SRL was lowest in all substrates. Additionally, the highest SRL calculated in winter and summer also correlated with a low rooting percentage.

3.2. Auxin and Peroxidase Concentration in Cuttings

Pretreatment with IBA significantly affected the levels of peroxidase enzyme in both rooted ($p < 0.00$) and unrooted ($p = 0.009$) stem cuttings (Table 2).

Table 2. Results of multifactor ANOVA of effects of main factors on auxin and peroxidase composition of stem cuttings across seasons. p-values for principal effects (IBA pretreatment, substrate, and season, and their two-way interactions) and for effects within each season (pretreatment, substrate, and their interaction) are given. Effects were significant at $p < 0.05$.

Chemical Compound	Effects	Winter	Spring	Summer	Autumn	Annual Values
			Growing Season			
Auxin (μg g^{-1}) (Rooted cuttings)	Pretreatment	0.02	0.33	1.00	0.22	0.95
	Substrate	0.69	0.05	0.99	0.94	0.97
	Season	-	-	-	-	0.39
	Pretreatment × Substrate	0.37	0.53	*	0.98	0.99
	Pretreatment × Season	-	-	-	-	0.69
	Substrate × Season	-	-	-	-	0.99
Auxin (μg g^{-1}) (Unrooted cuttings)	Pretreatment	0.01	0.57	0.94	0.11	0.70
	Substrate	0.83	0.07	0.99	0.36	0.66
	Season	-	-	-	-	<0.0001
	Pretreatment × Substrate	<0.0001	0.38	*	0.89	0.92
	Pretreatment × Season	-	-	-	-	0.01
	Substrate × Season	-	-	-	-	0.99
Peroxidase (mg g^{-1}) (Rooted cuttings)	Pretreatment	<0.0001	<0.0001	0.73	<0.0001	<0.0001
	Substrate	0.67	0.43	0.71	0.99	0.91
	Season	-	-	-	-	<0.0001
	Pretreatment × Substrate	0.62	0.86	*	0.80	0.81
	Pretreatment × Season	-	-	-	-	<0.0001
	Substrate × Season	-	-	-	-	0.99
Peroxidase (mg g^{-1}) (Unrooted cuttings)	Pretreatment	0.0001	<0.0001	0.13	0.0003	0.009
	Substrate	0.47	0.14	0.99	0.77	0.86
	Season	-	-	-	-	<0.0001
	Pretreatment × Substrate	0.96	0.0005	*	0.43	0.68
	Pretreatment × Season	-	-	-	-	<0.0001
	Substrate × Season	-	-	-	-	0.77

* There were not enough living cuttings for analysis because many cuttings dried.

It can be seen in Table 2 that season significantly affected the auxin level in unrooted stem cuttings ($p < 0.000$) and peroxidase enzyme in both rooted and unrooted stem cuttings ($p < 0.00$), whereas substrate was not significant ($p > 0.05$). Regarding interactions, only IBA pretreatment × season was significant for the variables auxin (in unrooted cuttings) and peroxidase (rooted and unrooted cuttings) ($p < 0.05$; Table 2). The results within seasons (two-way ANOVA model) also showed several significant effects (Table 2). IBA pretreatment was significant in winter for all dependent variables ($p < 0.05$), and in spring and autumn for peroxidase enzyme in both rooted and unrooted cuttings ($p < 0.05$). For rooted cuttings, there was no significant difference in auxin with IBA pretreatment in spring, summer, and autumn.

Figure 6 shows that the highest amount of auxin (1067 ± 337 µg g^{-1} (mean $\pm$ LSD interval) occurred in unrooted cuttings at 8000 ppm in winter. The amount of internal auxin in this pretreatment was slightly increased compared with fresh samples in unrooted cuttings. However, the amount of internal auxin in the unrooted cuttings did not change significantly during the rest of the seasons (spring, summer and autumn).

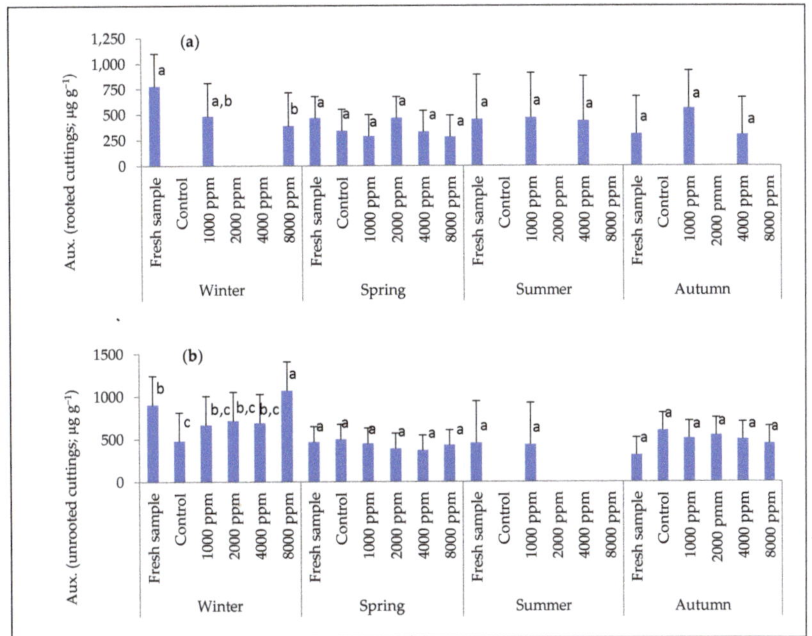

Figure 6. Mean auxin levels within seasons with different IBA treatments for (**a**) rooted and (**b**) unrooted cuttings. Mean values with the same letter were not significantly different at the 0.05 level (LSD test). Error bars: LSD intervals.

Due to the cuttings drying during maintenance in the substrates (mainly in winter and summer), there were not enough samples to measure chemical components in several treatments. Therefore, in those treatments, the response variables were not measured and there were no numerical values (Figure 6).

The highest concentration of peroxidase enzyme occurred in winter for both rooted and unrooted stem cuttings (Figure 7), with a maximum value of 0.108 ± 0.03 mg g^{-1} (mean $\pm$ LSD interval) for 8000 ppm concentration in unrooted cuttings (Figure 7b). In winter, the lowest peroxidase enzyme level occurred in fresh samples (0.04 mg g^{-1}) for both rooted and unrooted cuttings (Figure 7a,b). In spring, the maximum peroxidase enzyme level also occurred in the fresh samples (Figure 7a,b), at the beginning of this season.

In conclusion, we observed interesting relationships between rooting performance, IBA pretreatment, and peroxidase content throughout the seasons. For example, in winter, when rooting % was lower (see Section 3.1), the amount of peroxidase enzyme in rooted cuttings was higher, and in the following season (spring), when rooting was higher, the amount of peroxidase enzyme significantly decreased. In spring, the highest peroxidase enzyme level occurred in fresh cuttings in the early season. On the contrary, in autumn, the lowest concentration occurred in fresh cuttings (early autumn). Therefore, IBA pretreatment seemed to decrease the concentration of peroxidase in spring (coinciding with maximum rooting) and increase the concentration in winter and autumn.

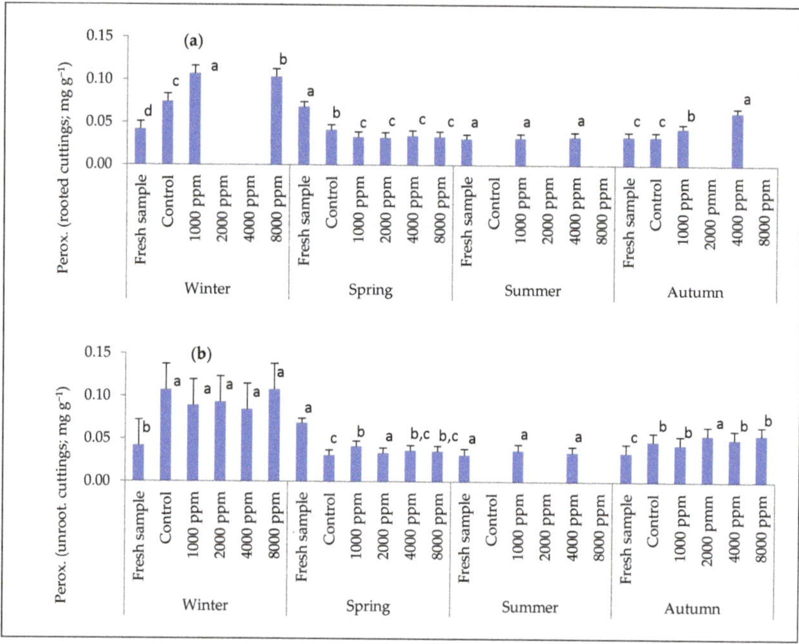

Figure 7. Mean peroxidase within seasons with different IBA treatments for (**a**) rooted and (**b**) unrooted cuttings. Mean values with the same letter were not significantly different at the 0.05 level (LSD test). Error bars: LSD intervals.

4. Discussion

Our results show, as a novelty, an elevated rooting percentage, greater than 60%, in the best protocol (1000 ppm IBA concentration, perlite–cocopeat substrate, and spring as the harvesting time). In previous studies, the success of rooting cuttings of juniper branches was often reported as less than 30% [8]. For example, only 24% of branch cuttings of *Juniperus procera* obtained from 1.5-to-2-year-old stock plants rooted 32 weeks after treatment, and such low rooting success makes the large-scale propagation of this tree impossible [9].

In our research, pretreatment with IBA increased rooting in stem cuttings, in agreement with previous studies [11,34–36]. This is because the compound induces adventitious roots [23,37] and shortens rooting time [38]. Another function of IBA comes from the direct effect of auxin, because it slowly releases a source of IAA [15,39,40]. Thus, IBA has been used for rooting of *Juniperus* species with different types of treatment, but showing diverse results. For example, the results were best with 8000 ppm of IBA in *Juniperus osteosperma* [41], 5000 ppm IBA for *Juniperus virginiana* [13], 1000 to 9000 ppm in *Juniperus scopulorum* [42], and 6000 ppm in *Juniperus excelsa* [43]. In their research on *Juniperus virginiana*, Henry et al. [13] noted that in preliminary studies, IBA concentrations up to 2000 ppm did not stimulate rooting beyond that obtained with 5000 ppm. The capacity of cuttings can vary among species [44], as shown by our results. In our study, the best result was obtained from intermediate levels of IBA (1000–4000 ppm), and we suggest that IBA at 8000 ppm can damage the cuttings and reduce rooting. Therefore, we do not recommend applying this dose as a pretreatment for cuttings of *Juniperus sabina*.

In general, our results are more in agreement with those of Chowdhuri et al. [45], who proposed an IBA concentration of 1000 to 3000 ppm for cuttings of *Juniperus chinensis*, Rifaki et al. [12], who proposed an IBA concentration of 4000 ppm for cuttings of *Juniperus excelsa*, and Esmaeil et al. [7], who proposed 3000 to 6000 ppm for the same. On the contrary, our result was not in agreement with Stuepp et al. [46] and Fragoso et al. [47], because the application of IBA was not efficient at propagating cuttings of *Juniperus chinensis*. Nevertheless, the novelty of our results is that the selected concentration of IBA (1000 ppm) was lower.

The substrate is another relevant factor in the production of seedlings from cuttings in a greenhouse [20]. The bottom heat and mist system used in our experiment was also necessary to favor rooting, in agreement with previous research [21,41,48]. In our study, rooting was more than 60% in the substrate of perlite–cocopeat. In the study by Hong-wei et al. [20] on *Juniperus procumbens*, the best substrate was 1.3 (*v/v*) vermiculite and 2.3 (*v/v*) perlite, with only 36% rooting. The results of our study are also better than those obtained by Cuevas-Cruz et al. [49] in *Pinus*, with 43.5% rooting (in a mixture of peat–perlite–vermiculite substrate); Khoushnevis et al. [35], with 28% rooting in *Juniperus oblonga* (fine and harsh bed); Stuepp et al. [46], with 16% rooting in *Juniperus chinensis* (fine-grained vermiculite and carbonized rice hull, 1:1); and Ayan et al. [50], with 24% rooting in *Juniperus foetidissima*, 31.5% rooting in *Juniperus excelsa*, 38.42% rooting in *Juniperus sabina*, and 31.83% rooting in *Juniperus oxycedrus* (perlite).

In addition to IBA pretreatment, by creating favorable conditions for growth (porosity, cation exchange capacity, salinity, and proper pH), the perlite–cocopeat substrate provides better absorption of water and nutrients for cuttings, resulting in increased maintenance and rooting in comparison with other substrates [51]. This could be due to the differences in substrates in terms of creating suitable conditions for plants, including moisture, soil aeration, and other physical and chemical properties. Due to the high CEC, cocopeat facilitates the storage and distribution of nutrients and improves water management. Perlite also improves aeration in the substrate. In this study, perlite–cocopeat in a 50/50 ratio had moderate water reserve, adequate substrate aeration, and low salinity, and thus could increase the maintenance and rooting of cuttings [52].

Our results reveal that root variables can explain the differences in rooting performance between treatments (and throughout the seasons) because root structure is an adequate indicator of the hydraulic architecture in seedlings [53]. In our research, the worst treatments for rooting percent had the highest values of specific root length (SRL), which occurred in the control treatment in summer, and the lowest occurred in spring. In winter, SLR, especially with 8000 ppm IBA, was high. So, in *Juniperus sabina*, the best time to prepare and plant cuttings for rooting performance is spring. This harvesting time differs from the studies of Guerrero-Campo et al. [54], who found the best rooting of several species of cuttings in different seasons, and Chowdhuri [45], who showed that the best time for rooting *Juniperus chinensis* was summer. On the contrary, our result is in agreement with Fragoso et al. [47] and Tektas et al. [43], who both suggested that the best season for rooting of *Juniperus* is spring.

Roots with a high SRL have a high surface-to-volume ratio for the same C investment, and this can be a strategy to maximize the root–substrate interface, and hence root absorption under water stress, due to the small amount of root biomass available. The persistence and growth of cuttings in seasons with high SRL are compromised, because survival strongly depends on the capacity to take up soil water and nutrients. To counteract their lower root production and maximize seedling survival, cuttings with higher SRL have root traits that confer higher efficiency for the acquisition of soil resources, in agreement with [25]. Thus, the lower SRL values obtained in spring confirm the higher potential of cuttings to take up water and nutrients in this season, and thus the rooting % is higher in spring.

Regarding internal chemical compounds of stem cuttings, Aux/IAA plays various roles in plant growth, such as root development [16,55]. Although Blakesley et al. [56] indicated that auxin content varies by season, in our study, the auxin in rooted stem cuttings was not different between seasons, in agreement with Aliahmad Koruri et al. [18]. Moreover, the results of our study are not opposed to those of Wendling et al. and Blakesley et al. [22,56] linking rooting to IAA concentration. Those studies reported that a decrease in internal auxin in cuttings was accompanied by a decline in the power of rooting.

Several studies on different species have also shown the important role of peroxidase in controlling growth and rooting [57,58]. In our study, the amount of this enzyme varied in different seasons, and when the peroxidase levels were higher (in winter), rooting % was lower. Increasing this enzyme could reduce the rooting of cuttings, which agrees with studies on other forest species, such as *Phoenix dactylifera* and *Populus tremuloides* [58,59]. In fact, major isoperoxidases are considered to be IAA oxidase; therefore, they are able to change the amount of internal auxin. The root appearance phase (the first visible signs of rooting) is accompanied by a gradual decline in peroxidase activity [58].

Increased rooting after the use of hormonal treatments is related to peroxidase activity inhibitors such as polyphenols [58]. In this regard, peroxidase is the most sensitive plant enzyme to environmental stress [60]. On the other hand, many reports have suggested that increasing IAA accumulation by reducing the amount of IAA oxidase/peroxidase can lead to increased rooting [36,61]. Cuttings can show a decrease in the amount of peroxidase after collection, which is associated with an increase in the concentration of IAA. However, the level of IAA can be reduced by increasing peroxidase activity [56], as shown in our results.

5. Conclusions

Based on the results of this experiment, the following protocol is recommended for the vegetative propagation of *Juniperus sabina*: (1) prepare cuttings in spring, (2) treat cutting bases with 1000 ppm IBA, and (3) planting them in a substrate of perlite–cocopeat (1:1). This should yield more than 60% rooting in a few months (approximately 3 months; more efficient than traditional seed production methods). Thus, it is an easy and rapid method for nurseries to propagate *Juniperus sabina*, which is a very interesting shrub for forest restoration and ornamental purposes, worldwide. In the future, our results will be confirmed with new experiments to determine whether spring is the best season to obtain stem cuttings in other juniper species (trees and shrubs).

Author Contributions: Conceptualization, M.A., B.Z., H.Z. and A.R.N.; formal analysis, M.A., F.A.G.-M. and A.R.N.; investigation, M.A. and F.A.G.-M.; methodology, M.A., B.Z. and H.Z.; project administration, B.Z.; resources, B.Z.; supervision, F.A.G.-M., B.Z., H.Z. and A.R.N.; validation, A.R.N.; writing—original draft, M.A.; writing—review and editing, F.A.G.-M. and B.Z. All authors have read and agreed to the published version of the manuscript.

Funding: This research received no external funding.

Data Availability Statement: Data is available upon request to the first author.

Acknowledgments: This study was part of a PhD thesis supported by the University of Lorestan (Iran) and a grant for scientific stay by Maliheh Abshahi at the University of Castilla-La Mancha (Spain) supported by the Ministry of Science, Research and Technology, Iran. We are grateful to three anonymous reviewers for their comments, which allowed us to improve the initial manuscript.

Conflicts of Interest: The authors declare no conflict of interest.

References

1. Farjon, A. *A Monograph of Cupressaceae and Sciadopitys*; Royal Botanic Gardens, Kew: Richmond, Surrey, UK, 2005; p. 648.
2. Adams, R.P. *The Junipers of the World: The Genus Juniperus*, 4th ed.; Trafford Publishing: Victoria, BC, Canada, 2014; p. 422.
3. García Morote, F.A.; Andrés Abellán, M.; Rubio, E.; Pérez Anta, I.; García Saucedo, F.; López Serrano, F.R. Stem CO_2 Efflux as an Indicator of Forests' Productivity in Relict Juniper Woodlands (*Juniperus thurifera* L.) of Southern Spain. *Forests* **2021**, *12*, 1340. [CrossRef]
4. Comino, E.; Marengo, P. Root tensile strength of three shrub species: Rosa canina, Cotoneaster dammeri and *Juniperus horizontalis* Soil reinforcement estimation by laboratory tests. *Catena* **2010**, *82*, 227–235. [CrossRef]
5. Piotto, B.; Di Noi, A. *Seed Propagation of Mediterranean Trees and Shrubs*; Agency for the Protection of the Environment and for Technical Services: Roma, Italy, 2003; p. 108.
6. Lu, D.; Huang, H.; Wang, A.; Zhang, G. Genetic Evaluation of *Juniperus sabina* L. (Cupressaceae) in Arid and Semi-Arid Regions of China Based on SSR Markers. *Forests* **2022**, *13*, 231. [CrossRef]

7. Esmaeilnia, M.; Jalali, S.G.; Tabari, M.; Hosseini, S.M. Influence of plant growth regulator IBA on vegetative propagation of Juniperus excelsa. *Iran. J. For. Poplar Res.* **2006**, *14*, 221–227.

8. Momeni, M.; Ganji-Moghadam, E.; Kazemzadeh-Beneh, H.; Asgharzadeh, A. Direct organogenesis from shoot tip explants of *Juniperus polycarpos* L.: Optimizing basal media and plant growth regulators on proliferation and root formation. *Plant Cell Biotechnol. Mol. Biol.* **2018**, *19*, 40–50.

9. Hazubska-Przybyl, T. Propagation of Juniper Species by Plant Tissue Culture: A Mini-Review. *Forests* **2019**, *10*, 1028. [CrossRef]

10. Amri, E.; Lyaruu, H.; Nyomora, A.; Kanyeka, Z. Vegetative propagation of African Blackwood (*Dalbergia melanoxylon* Guill. & Perr.): Effects of age of donor plant, IBA treatment and cutting position on rooting ability of stem cuttings. *New For.* **2010**, *39*, 183–194. [CrossRef]

11. Nordstrom, A.C.; Jacobs, F.A.; Eliasson, L. Effect of exogenous Indole-3-Acetic-acid and Indole-3-Butyric acid on internal levels of the respective auxins and their conjugation with aspartic-acid during adventitious root-formation in Pea cuttings. *Plant Physiol.* **1991**, *96*, 856–861. [CrossRef]

12. Rifaki, N.; Economou, A.; Hatzilazarou, S. Factors affecting vegetative propagation of Juniperus excelsa bieb. by stem cuttings. *Propag. Ornam. Plants* **2002**, *2*, 9–13.

13. Henry, P.H.; Blazich, F.A.; Hinesley, L.E. Vegetative propagation of Eastern redcedar by stem cuttings. *Hortscience* **1992**, *27*, 1272–1274. [CrossRef]

14. Pacurar, D.; Perrone, I.; Bellini, C. Auxin is a central player in the hormone cross-talks that control adventitious rooting. *Physiol. Plant.* **2014**, *151*, 83–96. [CrossRef] [PubMed]

15. Ludwig-Muller, J. Indole-3-butyric acid in plant growth and development. *Plant Growth Regul.* **2000**, *32*, 219–230. [CrossRef]

16. Daveis, P.J. *Plant Hormones: Biosynthesis, Signal Transduction, Action!* Kluwer Academic Publishers: Dordrecht, The Netherlands, 2004; p. 802.

17. Gao, J.; Xiao, Q.; Ding, L.; Chen, M.; Yin, L.; Li, J.; Zhou, S.; He, G. Differential responses of lipid peroxidation and antioxidants in Alternanthera philoxeroides and Oryza sativa subjected to drought stress. *Plant Growth Regul.* **2008**, *56*, 89–95. [CrossRef]

18. Aliahmad Koruri, S.; Khoushnevis, M.; Matinizadeh, M. *Comprehensive Studies of Juniper Species in Iran*; Publication of Pooneh, Forest & Ranges and Watershed Organization of Iran: Tehran, Iran, 2011; p. 554.

19. Kentelky, E. The Analysis of Rooting and Growth Peculiarities of *Juniperus* Species Propagated by Cuttings. *Horticulture* **2011**, *68*, 380–385.

20. Hong-wei, Y.; Yong-sheng, G.; Hai-jun, S.; Yan-he, R. Experiment on Cutting Propagation of Sabina procumbens. *J. Inn. Mong. For. Sci. Technol.* **2011**, *4*, 6.

21. Sabina, P.; Cornelia, H. Researches concerning the production of planting material using vegetative propagation on *Juniperus horizontalis* mnch. *J. Hortic. For. Biotechnol.* **2009**, *13*, 462–464.

22. Wendling, I.; Warburton, P.; Trueman, S. Maturation in *Corymbia torelliana* x *C. citriodora* Stock Plants: Effects of Pruning Height on Shoot Production, Adventitious Rooting Capacity, Stem Anatomy, and Auxin and Abscisic Acid Concentrations. *Forests* **2015**, *6*, 3763–3778. [CrossRef]

23. Bryant, P.; Trueman, S. Stem Anatomy and Adventitious Root Formation in Cuttings of Angophora, Corymbia and Eucalyptus. *Forests* **2015**, *6*, 1227–1238. [CrossRef]

24. Garcia Morote, F.A.; Lopez Serrano, F.R.; Andres, M.; Rubio, E.; Gonzalez Jimenez, J.L.; de las Heras, J. Allometries, biomass stocks and biomass allocation in the thermophilic Spanish juniper woodlands of Southern Spain. *For. Ecol. Manag.* **2012**, *270*, 85–93. [CrossRef]

25. Paula, S.; Pausas, J. Root traits explain different foraging strategies between resprouting life histories. *Oecologia* **2011**, *165*, 321–331. [CrossRef]

26. Hernandez, E.; Vilagrosa, A.; Pausas, J.; Bellot, J. Morphological traits and water use strategies in seedlings of Mediterranean coexisting species. *Plant Ecol.* **2010**, *207*, 233–244. [CrossRef]

27. Adams, R.P.; Schwarzbach, A.E. Chloroplast capture by a new variety, *Juniperus sabina* var. *balkanensis* R. P. Adams and A. N. Tashev, from the Balkan peninsula: A putative stabilized relictual hybrid between *J. sabina* and ancestral *J. thurifera*. *Phytologia* **2016**, *98*, 100–111.

28. Kottek, M.; Grieser, J.; Beck, C.; Rudolf, B.; Rubel, F. World Map of the Köppen-Geiger climate classification. *Meteorol. Z.* **2006**, *15*, 259–263. [CrossRef]

29. Bohlenius, H.; Fransson, T.; Holmstrom, E.; Salk, C. Influence of Cutting Type and Fertilization in Production of Containerized Poplar Plants. *Forests* **2017**, *8*, 164. [CrossRef]

30. Judd, L.A.; Jackson, B.E.; Fonteno, W.C. Advancements in Root Growth Measurement Technologies and Observation Capabilities for Container-Grown Plants. *Plants* **2015**, *4*, 369–392. [CrossRef]

31. Sridhar, R.; Mohanty, S.; Anjaneyulu, A. Physiology of rice tungro virus-disease-increased cytokinin activity in tungro-infected rice cultivars. *Physiol. Plant.* **1978**, *43*, 363–366. [CrossRef]

32. Hoshmand, H.R. *Design of Experiments for Agriculture and the Natural Sciences*, 2nd ed.; Champan & Hall/CRC: New York, NY, USA, 2006; p. 437.

33. Neter, J.; Kutner, M.; Wasserman, W.; Nachtsheim, C. *Applied Linear Statistical Models*, 4th ed.; McGraw-Hill-Irwin: Chicago, IL, USA, 1996; p. 720.

34. Davies, F.T.; Geneve, R.L.; Kester, D.E.; Hartmann, H.T. Techniques of Propagation by Cuttings. In *Plant-Propagation-Principles and Practices*, 8th ed.; Hartmann, K., Davies, G., Eds.; Pearson Education Limited: Harlow, Essex, UK, 2014; p. 927.

35. Khoushnevis, M.; Teimouri, M.; Matinizadeh, M.; Shirvany, A. Effects of hormones, light and media treatments on rooting of Juniperus oblonga cuttings. *Iran. J. For.* **2012**, *4*, 135–142.

36. Moncousin, C.; Gaspar, T. Peroxidase as a marker for rooting improvement of Cynara scolymus L. cultured in vitro. *Biochem. Physiol. Pflanz.* **1983**, *178*, 263–271. [CrossRef]

37. Kocer, Z.A.; Gozen, A.G.; Onde, S.; Kaya, Z. Indirect organogenesis from bud explants of *Juniperus communis L.*: Effects of genotype, gender, sampling time and growth regulator combinations. *Dendrobiology* **2011**, *66*, 33–40.

38. Negash, L. Successful vegetative propagation techniques for the threatened African pencil cedar (*Juniperus procera* Hoechst. ex Endl.). *For. Ecol. Manag.* **2002**, *161*, 53–64. [CrossRef]

39. Epstein, E.; Ludwig-Muller, J. Indole-3-butyric acid in plants- ocurrence, synthesis, metabolism and transport. *Physiol. Plant.* **1993**, *88*, 382–389. [CrossRef]

40. Poupart, J.; Waddell, C. The rib1 mutant is resistant to indole-3-butyric acid, an endogenous auxin in arabidopsis. *Plant Physiol.* **2000**, *124*, 1739–1751. [CrossRef] [PubMed]

41. Cope, K.; Rupp, L. Vegetative propagation of Juniperus osteosperma (Utah Juniper) by cuttings. *Nativ. Plants J.* **2013**, *14*, 76–84. [CrossRef]

42. Bielenin, M. Rooting and gas exchange of conifer cuttings treated with indole butyric acid. *J. Fruit Ornam. Plant Res.* **2003**, *11*, 99–105.

43. Tektas, I.; Türkoğlu, N.; Cavusoğlu, S. Effects of auxin doses on rooting of *Juniperus L. Prog. Nutr.* **2017**, *19*, 130–136.

44. Ma, W.; Zhang, S.; Wang, J.; Sun, X.; Zhao, H.; Ning, Y. Endogenous hormones, nutritive material and phenolic acid variation in cuttings of japanese larch during rooting. *Acta Bot. Boreali-Occident. Sin.* **2013**, *1*, 19.

45. Chowdhuri, T.K. Performance evaluation of different growth regulators on propagation of Chinese juniper (*Juniperus chinensis* Var. *pyramidalis*) in subtropical zone. *J. Pharmacogn. Phytochem.* **2017**, *6*, 2190–2193.

46. Stuepp, C.A.; Ruffellato-Ribas, K.C.; Macanhão, G.; Fragoso, R.; Rickli, H.C. Rooting of *Juniperus chinensis* var. *kaizuka* for different concentrations of IBA and heights collection. *Rev. Agrar.* **2014**, *7*, 496–503.

47. Fragoso, R.; Zuffellato-Ribas, K.; Macanhão, G.; Stuepp, C.A.; Koehler, H. Vegetative propagation of *Juniperus chinensis. Comun. Sci.* **2015**, *6*, 307–316. [CrossRef]

48. Chong, C. Influence of Bottom Heat and Mulch on Rooting of Evergreen Cuttings. *Comb. Proc. Int. Plant Propagators Soc.* **2003**, *53*, 496–500.

49. Cuevas-Cruz, J.; Jimenez-Casas, M.; Jasso-Mata, J.; Perez-Rodriguez, P.; Lopez-Upton, J.; Villegas-Monter, A. Asexual propagation of Pinus leiophylla Schiede ex Schltdl. et Cham. *Rev. Chapingo Ser. Cienc. For. Ambiente* **2015**, *21*, 81–95. [CrossRef]

50. Ayan, S.; Küçük, M.; Ulu, F.; Gerçek, V.; Sahin, A.; Sivacioğlu, A. Vegetative propagation possibilities of some natural Juniper (Juniperus L.) species. *J. For. Fac. Gazi Uni. Kast.* **2004**, *4*, 1–12.

51. Manios, V.I.; Papadimitriou, M.D.; Kefakis, M.D. Hydroponic culture of tomato and gerbera at different substrates. *Acta Hortic.* **1995**, *408*, 11–15. [CrossRef]

52. Maloupa, E.; Samartzidis, P.; Couloumbis, P.; Komnin, A. Yield quality and photosynthetic activity of greenhouse-grown "Madelom" roses on perlite-zeolite substrate mixtures. *Acta Hortic.* **1999**, *481*, 97–99. [CrossRef]

53. Pratt, R.; North, G.; Jacobsen, A.; Ewers, F.; Davis, S. Xylem root and shoot hydraulics is linked to life history type in chaparral seedlings. *Funct. Ecol.* **2010**, *24*, 70–81. [CrossRef]

54. Guerrero-Campo, J.; Palacio, S.; Perez-Rontome, C.; Montserrat-Marti, G. Effect of root system morphology on root-sprouting and shoot-rooting abilities in 123 plant species from eroded lands in north-east Spain. *Ann. Bot.* **2006**, *98*, 439–447. [CrossRef]

55. Luo, J.; Zhou, J.; Zhang, J. Aux/IAA Gene Family in Plants: Molecular Structure, Regulation, and Function. *Int. J. Mol. Sci.* **2018**, *19*, 259. [CrossRef]

56. Blakesley, D.; Weston, G.; Hall, J. The role of endogenous auxin in root initiation. 1. Evidence from studies on auxin application, and analysis of endogenous levels. *Plant Growth Regul.* **1991**, *10*, 341–353. [CrossRef]

57. Hausman, J. Changes in Peroxidase-activity, auxin and ethylene production during root-formation by Poplar shoots raised in-vitro. *Plant Growth Regul.* **1993**, *13*, 263–268. [CrossRef]

58. Rival, A.; Bernard, F.; Mathieu, Y. Changes in peroxidase activity during in vitro rooting of oil palm (*Elaeis guineensis* Jacq.). *Sci. Hortic.* **1997**, *71*, 103–112. [CrossRef]

59. Yan, S.; Yang, R.; Wang, F.; Sun, L.; Song, X. Effect of Auxins and Associated Metabolic Changes on Cuttings of Hybrid Aspen. *Forests* **2017**, *8*, 117. [CrossRef]

60. Torabian, Y.; Korori, S.; Adeli, A.; Falahchai, M. Quantitative and qualitative study of peroxidase and amylase enzymes of seeds and shoots of Juniper trees (*Juniperus excelsa*) depending on the quality of the bases in the habitat of Razavi Khorasan. *J. Biol. Sci.* **2009**, *3*, 19–27.

61. Moncousin, C. Rooting of in vitro cuttings. In *High-Tech and Micropropagation I*; Springer: Berlin/Heidelberg, Germany, 1991; pp. 231–261.

Article

Castanea crenata Ginkbilobin-2-like Recombinant Protein Reveals Potential as an Antimicrobial against *Phytophthora cinnamomi*, the Causal Agent of Ink Disease in European Chestnut

Maria Belén Colavolpe [1], Fernando Vaz Dias [2], Susana Serrazina [2], Rui Malhó [2] and Rita Lourenço Costa [1,3,*]

[1] Instituto Nacional de Investigação Agrária e Veterinária I.P., Avenida da República, Quinta do Marquês, 2780-157 Oeiras, Portugal

[2] BioISI—Biosystems & Integrative Sciences Institute, Faculdade de Ciências, Universidade de Lisboa, 1749-016 Lisboa, Portugal

[3] Centro de Estudos Florestais, Instituto Superior de Agronomia, Universidade de Lisboa, 1349-017 Lisboa, Portugal

* Correspondence: rita.lcosta@iniav.pt

Abstract: The European chestnut tree (*Castanea sativa* Mill.) is widely cultivated throughout the world's temperate regions. In the Mediterranean region, it has a significant economic role mainly because of the high quality of its edible nuts. The Oomycete *Phytophthora cinnamomi* is one of the most severe pathogens affecting European chestnuts, causing ink disease and significant losses in production. Ginkgobilobin-2 (Gnk2) in *Ginkgo biloba* is a secreted protein with a plant-specific cysteine-rich motif that functions as a lectin, and its carbohydrate-binding properties are closely related to its antifungal activity. The binding of lectins to mannose residues of the cell wall of *Phytophthora* species may disturb and disrup the cell wall structure. This work determined that the amino acid sequence has a signal peptide that directs the final protein peptide to the apoplast. The Cast_Gnk2-like expression was performed and optimized, and different in vitro antagonism tests were done against *P. cinnamomi* using different purified protein concentrations. As a result of one of these assays, Cast_Gnk2-like significantly reduced the mycelia growth of *P. cinnamomi* in liquid medium as shown by the mycelia weight (g) in control treatments was 377% higher than in the treatments. These insights reveal the potential of Cast_Gnk2-like for agricultural uses and biotechnological developments for the pathosystem chestnut/*P. cinnamomi*.

Keywords: Antagonisms; *Castanea sativa*; in vitro biotic interactions; pathosystem chestnut/ *P. cinnamomi*; Phytophthora cinnamomi

Citation: Colavolpe, M.B.; Vaz Dias, F.; Serrazina, S.; Malhó, R.; Lourenço Costa, R. *Castanea crenata* Ginkbilobin-2-like Recombinant Protein Reveals Potential as an Antimicrobial against *Phytophthora cinnamomi*, the Causal Agent of Ink Disease in European Chestnut. *Forests* **2023**, *14*, 785. https://doi.org/10.3390/f14040785

Academic Editors: Ricardo Javier Ordás and José Manuel Álvarez Díaz

Received: 26 February 2023
Revised: 30 March 2023
Accepted: 5 April 2023
Published: 11 April 2023

1. Introduction

The genus *Castanea* belongs to the Fagaceae family, which dominates much of the hardwood forests of the Northern Hemisphere. The European chestnut (*Castanea sativa* Mill.) is considered to be the only native chestnut species in Europe. This species is a multipurpose tree, that is used in the food industry for its high-quality edible fruits (nuts), in the forest industry for timber, and also for ecological and landscaping purposes [1].

Over 100 species have been described within the genus *Phytophthora* [2]. According to many authors [3,4], most survive in the soil for long periods without a host and affect the parts of the plant in contact with the ground, thus causing a destruction of the root system. The impact of this group of organisms is mirrored in its name, as *Phytophthora*, which means, in Greek, plant destroyer [1,4].

The rural economy of the northern mountainous regions of Portugal are mainly based on chestnut culture. Unfortunately, the current area and productivity per hectare are much less than the country's potential, mainly due to root rot caused by *Phytophthora cinnamomi* Rands, the main threat to chestnut orchards in Europe [5]. *Castanea sativa* was severely affected and nut production has declined by 251,549 tons from 1961 to 2015 [1,6]. The economy of mountainous areas in the Mediterranean regions of southwest Europe mainly depends on the significant income that chestnut culture represents to the local farmers, which is sometimes the only income they can get from the land. In recent years, the persistent losses of chestnut trees to *P. cinnamomi* is contributing to a decline in the population of these regions [6].

Phytophthora cinnamomi infects the roots and after reaching the vascular tissues it continues to colonize the roots until it obstructs the xylem vessel, thus restricting root growth and interfering with water and nutrient uptake to the plant shoot. The roots and root collar start to rot, resulting in a progressive decline of the tree. The above-ground symptoms include chlorosis and wilting of leaves, dieback of branches, defoliation, and gradual decline until the tree dies [1]. *Phytophthora cinnamomi*, has an extensive host range, destroying thousands of plant species worldwide and causing devastating impacts on natural ecosystems, agriculture, and forestry [6].

The coexistence of Asian chestnut species from Japan and China (*Castanea crenata* Sieb. & Zucc. and *Castanea mollissima* Blume) with *P. cinnamomi* (originally from the Asian tropics) lead to the development of resistance against the pathogen during the evolution process. The secretion of antifungal proteins and cell wall reinforcement are part of the constitutive defense barriers to pathogen growth in the plants that may explain the difference in resistance to *P. cinnamomi* between *C. crenata* and *C. sativa*. Although the Asian species have proven resistant to ink disease, these species are not adapted to the Atlantic environmental conditions. Therefore, the resistance of Japanese and Chinese chestnut species to this pathogen led to their introduction in breeding programs over the last years as donors of resistance in controlled crosses [7].

Over time, several approaches have been proposed to deal with root diseases (which affect the xylem of trees), such as the intensive application of pesticides and the selection and production of improved plant material with some disease resistance [8]. Phosphite and metalaxyl have been the most frequently used chemicals against *P. cinnamomi*. Nevertheless, the continuous use of these two chemicals has led to the development of tolerance within the pathogen [1]. Effective biological control methods have not been developed to date [1]. More common control approaches involve using resistant rootstocks for propagation or planting of resistant hybrids [1]. Nevertheless, the measures and procedures have not yet been successful, mainly due to the easy development and migration of *P. cinnamomi* zoospores in wet conditions, especially during rainfalls and waterlogging and because of the resistance structures that persist in the soil and are extremely difficult to eradicate [1,9].

A previous study identified a secreted protein with antifungal activity, Ginkbilobin2 (Gnk2), from seeds of the gymnosperm *Ginkgo biloba* [10]. Gnk2 comprises 108 amino acids as a mature protein with a plant-specific cysteine-rich motif. It functions as a lectin [3,11], and its carbohydrate-binding properties are tightly related to its antifungal activity. It binds with high affinity to D-mannose and with less affinity to D-glucose [11], and both exist in the hyphal cell walls of *Phytophthora* species [12].

Our group has been studying the *C. sativa* and *C. crenata* root transcriptome in response to *P. cinnamomi* to elucidate chestnut defense mechanisms. In a previous study [13], we identified candidate genes differentially expressed in the roots of the susceptible species to the pathogen (*C. sativa*), and the resistant one (*C. crenata*) observed after *P. cinnamomi* inoculation. The research demonstrated that both species recognize the pathogen attack, but only the resistant species (*C. crenata*) may involve more genes in the defense response than the susceptible species (*C. sativa*). RNA-seq analysis further enabled the selection of candidate genes for ink-disease resistance in *Castanea* [12]. Following the same rationale to elucidate chestnut defense mechanisms against *P. cinnamomi*, our group selected the Ginkgobilobin-2-like gene (*Cast_Gnk2-like*) from the transcriptomes of *C. crenata*, intending to evaluate the early expression of candidate resistance genes in both species (*C. sativa* and

C. crenata). *Cast_Gnk2-like* was the most expressed gene and the one that best discriminates between susceptible and resistant chestnut genotypes. The highest *Cast_Gnk2-like* expression registered in non-inoculation conditions suggests that *C. crenata* root surroundings may be a hostile environment for fungal and fungal-like pathogens, such as *P. cinnamomi*. On the other hand, *C. sativa* showed a very low *Cast_Gnk2-like* expression level, even after pathogen inoculation [7].

The present work is part of an ongoing research program that aims to find solutions to this important problem regarding the Chestnut/*P. cinnamomi* pathosystem in Europe. It was reported that Gnk2 inhibits the growth of pathogenic fungi such as *Fusarium oxysporum*, *F. culmorum*, and *Candida albicans*; and activates actin-dependent by inducing hypersensitive response (HR-related) plant cell death [10]. The homology of *Cast_Gnk2-like* with *Gnk2* led us to test the encoded protein against *P. cinnamomi*. The present work aims to express and produce the *Cast_Gnk2-like* protein for evaluation of its effectiveness on in vitro assays as a possible antagonist of the *P. cinnamomi* pathogen.

2. Materials and Methods

2.1. Cast_GNK2-like In Silico Analysis

The nucleotide sequence of the C. crenata *Cast_GNK2-like* transcript was obtained from the sequenced root transcriptome after *P. cinnamomi* inoculation [13]. It was annotated as a Cysteine-rich repeat secretory protein 38. After a BLASTn at NCBI and a comparison of the sequences with the highest homology, a prediction of the coding sequence and translation to the amino acid sequence was achieved. A BLAST to *Cast_GNK2-like* was achieved in Uniprot [14] to identify the proteins with higher similarity. Uniprot was also used to predict the protein families and domains of *Cast_GNK2-like* by comparison with the ones of *Ginkgo biloba* (GNK2), *Arabidopsis thaliana*, and *C. mollissima*. Signal peptides for *Cast_GNK2-like* were searched using SignalP—6.0 [15] and TargetP—2.0 [16]. Alignment of the amino acid sequences of *Castanea mollissima* Cysteine-rich repeat secretory protein 38 (A0A8J4V9V8), *Ginkgo biloba* Antifungal protein ginkbilobin-2, C. crenata putative Ginkbilobin-2 protein and *A. thaliana* Putative cysteine-rich receptor-like protein kinase 9, was performed using CLUSTALW [17].

2.2. Bacterial Strains and Growth Condition

Escherichia coli bacteria strains used in this study were TOP10 and BL21 (DE3) pLysS (competent cells from Thermo Fisher Scientific Inc., Waltham, MA, USA). *E. coli* strains containing recombinant plasmids were cultured, at 37 °C, in LB broth medium supplemented with 100 µg/mL ampicillin or 50 µg/mL kanamycin.

2.3. Phytophthora Cinnamomi Strain

Phytophthora cinnamomi (strain PH107) was isolated at the UTAD (the University of Trás-os-Montes and Alto Douro), from Vila Real (Portugal), and it is preserved at the INIAV I.P. (Instituto Nacional de Investigação Agrária e Veterinária I.P.) in Lisbon. It is routinely cultured at 25 °C on Potato-Dextrose Agar (PDA) in Petri plates and maintained in darkness. For all the assays, a fresh culture of *P. cinnamomi* was prepared by transferring a small piece of mycelium from a previous culture to a new plate of PDA nutritive medium and incubated in darkness at 25 °C for 4/5 days (or maximum one week).

2.4. Plasmid Construction

The designed primers for the amplification of the *Cast_Gnk2-like* coding sequence were 5′–CTCCATATGgctgacccattataccatttttgttttag–3′ and 5′– CACGGATCCctaggcatcaacaaag -gggta–3′. The PCR was performed using Phusion™ High-Fidelity DNA Polymerase (2 U/µL) according to the manufacturer protocol. The vectors used for the construction were pET15b and pET28a (Novagen®, Gujarat, India), which carry an N-terminal or an N- and C- terminal oligo-histidine tag (6x His-tags), respectively. The DNA sequence encoding *Cast_Gnk2-like* was inserted into the vectors between *Nde*I and *Bam*HI sites.

Control plasmids were constructed without the *Cast_Gnk2-like* fragment. It is expected to observe a protein fraction (a band in the gels) in the region of 25 kDa. The size of the *Cast_Gnk2-like* coding sequence is represented in Figure 1 (726 bp).

```
1     atg ttg agc tca aaa tat att tct gtc agc ttt cta tta ctc agc   45
1      M   L   S   S   K   Y   I   S   V   S   F   L   L   L   S     15

46    ctc tcc ctc cat gca gtc aat tgt gct gac cca tta tac cat ttt   90
16     L   S   L   H   A   V   N   C   A   D   P   L   Y   H   F     30

91    tgt ttt agc caa gaa agc tac act gcc act agc cgt tat ggt aca  135
31     C   F   S   Q   E   S   Y   T   A   T   S   R   Y   G   T     45

136   aac ttg aat ggc ttg ctc aat ctt ttg tcc acc aaa gtt cct tca  180
46     N   L   N   G   L   L   N   L   L   S   T   K   V   P   S     60

181   aaa ggg ttt ggt ctc agc tcg act ggg caa ggc caa gat cga gca  225
61     K   G   F   G   L   S   S   T   G   Q   G   Q   D   R   A     75

226   aat ggt tta gcc cta tgc cgg ggt gat gtc tca aaa aca aac tgt  270
76     N   G   L   A   L   C   R   G   D   V   S   K   T   N   C     90

271   acg acc tgt gtc att gat gca ggc aaa gag ctt ggt aat cgt tgt  315
91     T   T   C   V   I   D   A   G   K   E   L   G   N   R   C    105

316   cct tat aaa aaa gga gcg ata att tgg tat gat aac tgt ctt ttg  360
106    P   Y   K   K   G   A   I   I   W   Y   D   N   C   L   L    120

361   aag tac tcg aac att gat ttc ttt gga gaa atc gat aac aaa aac  405
121    K   Y   S   N   I   D   F   F   G   E   I   D   N   K   N    135

406   aag ttc tac atg tgg aac gtc caa gat gta gaa aat ccc act tca  450
136    K   F   Y   M   W   N   V   Q   D   V   E   N   P   T   S    150

451   ttc aat cca aaa gtt aag gat ttg tta agc agg tta tct aat aaa  495
151    F   N   P   K   V   K   D   L   L   S   R   L   S   N   K    165

496   gct tat gcc aat cca aaa ttc tat gct acc ggg gac cta aag ctt  540
166    A   Y   A   N   P   K   F   Y   A   T   G   D   L   K   L    180

541   gac tca tca agc aaa cta tat ggt ttg gct caa tgc acc agg gac  585
181    D   S   S   S   K   L   Y   G   L   A   Q   C   T   R   D    195

586   cta tca ggt ctt gat tgt aag aag tgt ctt gat act gcg att agt  630
196    L   S   G   L   D   C   K   K   C   L   D   T   A   I   S    210

631   gaa ctt ccc aac tgt tgc gat gga aaa cga ggt ggg cga gtt gtt  675
211    E   L   P   N   C   C   D   G   K   R   G   G   R   V   V    225

676   ggt ggc agt tgt aac gtt aga tat gaa ctt tac ccc ttt gtt gat  720
226    G   G   S   C   N   V   R   Y   E   L   Y   P   F   V   D    240

721   gcc tag    726
241    A   *
```

Figure 1. Sequences S1. Cast_GNK2-like coding sequence (up) and respective amino acid sequence (below). In red is the prediction of the signal peptide to the apoplast. In blue, the two conserved cysteine motifs attributed to the antimicrobial action. *: stop codon.

2.5. Protein Expression

2.5.1. Preparation of Bacterial Cellular Lysates

The cell culture was centrifuged at 4000 g for 20 min, and the pellet was resuspended in a 50 mM sodium HEPES buffer (4-(2-hydroxyethyl)-1-piperazineethanesulfonic acid), pH 7.5, containing 1 M NaCl and 10 mM imidazole (1,3-diazaciclopenta-2,4-dieno). Cells were disrupted by sonication. The cell debris was removed by centrifugation, and the supernatant was filtrated (0.45 μm) before the protein purification process.

2.5.2. Protein Purification and Electrophoresis (SDS-PAGE)

The supernatant (cell-free extract) obtained after centrifugation was applied to 5 mL Ni^{2+} ion chelating immobilised ion affinity chromatography column (HiTrap, GE Health-care, Chicago, IL, USA). The column was preequilibrated with 50 mM sodium HEPES buffer, pH 7.5, containing 1 M NaCl and 10 mM imidazole. The recombinant protein, Cast_Gnk2-like, was eluted with a linear gradient of imidazole (0–300 mM) in 50 mM sodium HEPES buffer, pH 7.5, containing 1 M NaCl. Protein purity was evaluated by SDS–PAGE (sodium dodecyl sulfate polyacrylamide gel electrophoresis), performed in a Protean II xi cell (Bio-Rad, Hercules, CA, USA) at 480 V and 120 mA. The running buffer contained 25 mM Tris base, 0.2 M glycine, and 0.1% SDS. Gels were stained with Coomassie (Bio-Rad Biosafe, Hercules, CA, USA). The protein concentrations were determined using Bradford reagent (using bright blue Coomassie G-250) from BioRad utilizing bovine serum albumin (BSA) as the standard.

2.6. Western Blot

Identification of antigens recognized by serum antibodies was carried out after the separation of total antigens by SDS-PAGE and Western blotting [18]. Following SDS-PAGE, the separated proteins were transferred to nitrocellulose sheets (Amersham[TM] Protran[®] Premium 0.2 µm pore size NC). The sheets were blocked with phosphate-buffered saline, pH 7.2, containing 0.3% Tween-20 (0.3% PBS-T), and washed twice. Then they were incubated with His-Tag Antibody (AD1.1.10) solution for one hour at room temperature, always in soft shaking. After washing them three times with 0.3% PBS-T, the secondary antibody solution (m-IgGκ BP-HRP) was added for one hour in the same conditions. After that, another three washes with 0.3% PBS-T were done.

Immunoreactive spots were detected with a chemiluminescence-based kit (Amersham Pharmacia Biotech, Amersham, UK). Before immunodetection, the nitrocellulose membranes were stained with 0.2% *w*/*v* Ponceau S in 3% *w*/*v* trichloroacetic acid for 3 min.

2.7. Cast_Gnk2-like In Vitro Antagonist Activity against P. cinnamomi

Two types of antagonist assays were performed. One of them was done in 60 mm diameter Petri dishes to evaluate *P. cinnamomi* mycelium development with different Cast_Gnk2-like pure protein concentrations (0.2 and 0.5 mg/mL). Briefly, in the solid medium assay, mycelial discs of 5-mm diameter were taken from the margins of approx. 1-week-old *P. cinnamomi* cultures and placed on new PDA medium plates. Different protein concentrations were tested and first spread in all the plates homogeneously, forming a thin film. The Petri plates were left for 25 min inside the flow chamber to dry the liquid above the solid medium before inoculation with *P. cinnamomi*. Different volumes of pure protein were also tested (100, 250, and 500 µL/plate). Control PDA medium Petri plates contained buffer solution without Cast_Gnk2-like protein. The plates were kept at 25 °C in darkness, and the growth of *P. cinnamomi* was evaluated during the following days of incubation (mainly 24, 48, and 72 h). Photographs of the mycelial growth were taken and the area determined with Image J software (Version 1.51r, NIH, Rockville, MD, USA). The second assay was performed in a liquid culture using Potato-Dextrose medium (50 mL) in Erlenmeyer flasks. Mycelial discs (10 discs of 5-mm diameter) were added to each flask and growth of *P. cinnamomi* was evaluated by determining mycelial dry weights (g) after two days of shaking (180 rpm) at 25°C. Controls were buffer solution without Cast_Gnk2-like protein.

2.8. Statistics

For fungal inoculation experiments, at least 10 biological replicates were used for each test, and the experiments were performed 3 times with similar results. The results shown in the pictures are examples from representative samples of *n* repetitions. The area of *P. cinnamomi* mycelia growth expressed in mm^2 was determined. Graph Path software was used to prepare the figures, and the results are shown as means ± SEM. Data were analyzed by appropriate *t*-test or ANOVA followed by post hoc comparisons by Tukey or Dunnett's test.

3. Results

3.1. Cast_GNK2-like Predictably Has Two Sites for Microbial Binding

Ginkbilobin-2 protein from G. biloba (GNK2) was the most thoroughly characterized so far. It has one conserved cysteine motif (C-8X-C-2X-C) related to its antimicrobial action [8]. We detected two cysteine motifs in Cast_GNK2-like (Figure 1). The same occurs with A. thaliana CRK9 and C. mollissima A0A8J4V9V8 (Table 1, Pfam column). As expected, the protein with the highest similarity to Cast_GNK2-like is A0A8J4V9V8 (99.6%, E-value 3.3×10^{-180}) from the Chinese chestnut. Additionally, the amino acid sequences of all the proteins start with a signal peptide that guides the final peptide to the apoplast (Table 1). The alignment of the four amino acid sequences also indicates that all the proteins have a signal peptide for the protein to be secreted. Moreover, the alignment shows the existence of one conserved cysteine motif for GNK2 and two motifs for Cast_GNK2-like, as A0A8J4V9V8 and CRK9 (Figure 2).

```
A0A8J4V9V8     MLSSKYISVSFLLLSLS---LHAVNCADPLY--HFCFSQENYTATSRYGTNLNGLLNLLSTKVP--SKGFGLSSTGQGQDRANGLALCRGDVSKTNCTTC
A4ZDL6         MKTMRMNSAFILAFALAAAMLILTEAANTAFVSSAC-NTQKIPSGSPFNRNLRAMLAD----------------------------------------
Cast_GNK2-like MLSSKYISVSFLLLSLS---LHAVNCADPLY--HFCFSQESYTATSRYGTNLNGLLNLLSTKVP--SKGFGLSSTGQGQDRANGLALCRGDVSKTNCTTC
O65469         MSSLISFIFLFLFSFLT---SFKASAQDPFYLNHYCPNTTTYSRFSTYSTNLRTLLSSFASRNASYSTGFQNVTVGQTPDLVTGLFLCRGDLSPEVCSNC

A0A8J4V9V8     VIDAGKELGNRCPYKKGAIIWYDNCLLKYSNIDFFG-EIDNKNKFYMWNVQDVE-NPTSFNPK---VKDLLSRLSNKAYANPKFYATGDLKLDSSSKLYG
A4ZDL6         --------------------------------------------------------------------LRQNTAFSGYDYKTSRAGSGGAPTAYG
Cast_GNK2-like VIDAGKELGNRCPYKKGAIIWYDNCLLKYSNIDFFG-EIDNKNKFYMWNVQDVE-NPTSFNPK---VKDLLSRLSNKAYANPKFYATGDLKLDSSSKLYG
O65469         VAFSVDEALTRCPSQREAVFYYEECILRYSDKNILSTAITNEGEFILSNTNTISPNQSQINQFIVLVQSNMNQAAMEAANSSRKFSTIKTELTELQTLYG

A0A8J4V9V8     LAQCTRDLSGLDCKKCLDTAISELPNCCDGKRGGRVVGGSCNVRYELYPFVDA-----------------
A4ZDL6         RATCKQSISQSDCTACLSNLVNRIFSICNNAIGARVQLVDCFIQYEQRSF--------------------
Cast_GNK2-like LAQCTRDLSGLDCKKCLDTAISELPNCCDGKRGGRVVGGSCNVRYELYPFVDA-----------------
O65469         LVQCTPDLTSQDCLRCLTRSINRMPL---SRIGARQFWPSCNSRYELYSFYNETTIGTPSPPPPPSPSRAN
```

Figure 2. A multiple alignment of *C. crenata* putative Ginkbilobin-2 protein (Cast_GNK2-like), *C. mollissima* Cysteine-rich repeat secretory protein 38 (A0A8J4V9V8), *Ginkgo biloba* Antifungal protein ginkbilobin-2 (A4ZDL6), and *A. thaliana* Putative cysteine-rich receptor-like protein kinase 9 (O65469), with CLUSTALW [17]. Protein accession numbers are indicated (Uniprot Accession). The signal peptide is surrounded by a red box and the conserved cysteine motif (C-8X-C-2X-C) is indicated in red.

Table 1. In silico characteristics of the amino acid sequences homologous to Ginkbilobin-2. N.d.: non-determined. Pos: position of the signal peptide.

Species	Protein Name (Uniprot)	Gene Acronym	Uniprot Accession	Length (AA)	Interpro Family/Domain	Pfam Family, #Hits	Subcellular Localization (Uniprot)	SignalP	TargetP
Ginkgo biloba	Antifungal protein ginkbilobin-2	GNK2	A4ZDL6	134	Gnk2-homologous domain (IPR002902); Gnk2-homologous domain superfamily (IPR038408)	Salt stress response/antifungal (PF01657), 1	Secreted	Secretory	Pos 1–26/27
Arabidopsis thaliana	Putative cysteine-rich receptor-like protein kinase 9	CRK9	O65469	265	Gnk2-homologous domain (IPR002902); Gnk2-homologous domain superfamily (IPR038408)	Salt stress response/antifungal (PF01657), 2	Secreted	Secretory	Pos 1–23/24

Table 1. *Cont.*

Species	Protein Name (Uniprot)	Gene Acronym	Uniprot Accession	Length (AA)	Interpro Family/Domain	Pfam Family, #Hits	Subcellular Localization (Uniprot)	SignalP	TargetP
Castanea mollissima	Cysteine-rich repeat secretory protein 38	N.d.	A0A8J4V9V8	241	Gnk2-homologous domain (IPR002902); Gnk2-homologous domain superfamily (IPR038408)	Salt stress response/antifungal (PF01657), 2	N.d.	Secretory	Pos 1–23/24
Castanea crenata	N.d.	Cast_GNK2-like	N.d.	241	N.d.	N.d.	N.d.	Secretory	Pos 1–23/24

3.2. Protein Isolation and Purification

The results of the recombinant protein expression, following different conditions to optimize the process, showed (according to the electrophoresis gels, SDS-PAGE) that the best expression was obtained using the *E. coli* BL21 (DE3) pLysS, in plasmid pET15, for 16 h at 20 °C, using 1 mM IPTG. We could observe a band in the gels (a protein fraction) in the region of 25 kDa (Figure 3A).

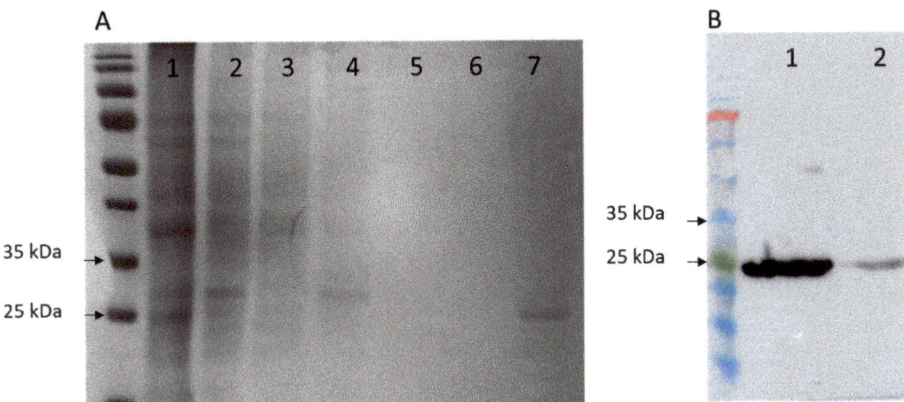

Figure 3. SDS-PAGE gel (14%), panel A, and Western blotting showing the immunodetection of Cast_Gnk2-like protein, panel B. Hyper-expression and purification of Cast_Gnk2-like protein obtained using the E. coli BL21 (DE3) pLysS transformed with pET15b_Cast_Gnk2, lane 1: Cell pellet after sonication; lane 2: cell free extract; lane 3 to 6: fractions of the column washing procedure wash; lane 7: purified recombinant enzyme, 0.5 mg/mL (**A**). Western blotting showing the immunodetection of Cast_Gnk2-like protein using total cell extract after sonication, lane 1—insoluble fraction, lane 2—soluble fraction (**B**). The molecular masses (kDa) of protein standards (NZYTech Ltd., Lisbon, Portugal) are indicated.

The immunoreactive spots in the region of 25 kDa from the Western Blot were detected with a chemiluminescence-based kit (Figure 3B).

3.3. Cast_Gnk2-like In Vitro Activity against P. cinnamomi

Mycelial growth of *P. cinnamomi* in PDA medium containing 0.5 mg/mL Cast_Gnk2-like pure protein was significantly less than the control treatment (without the protein). Figure 4 shows the growth of *P. cinnamomi* mycelia with and without Cast_Gnk2-like pure protein after 24 and 48 h at 25 °C in darkness. Cast_Gnk2-like does not inhibit the growth of *P. cinnamomi*, but it significantly reduces and delays its development.

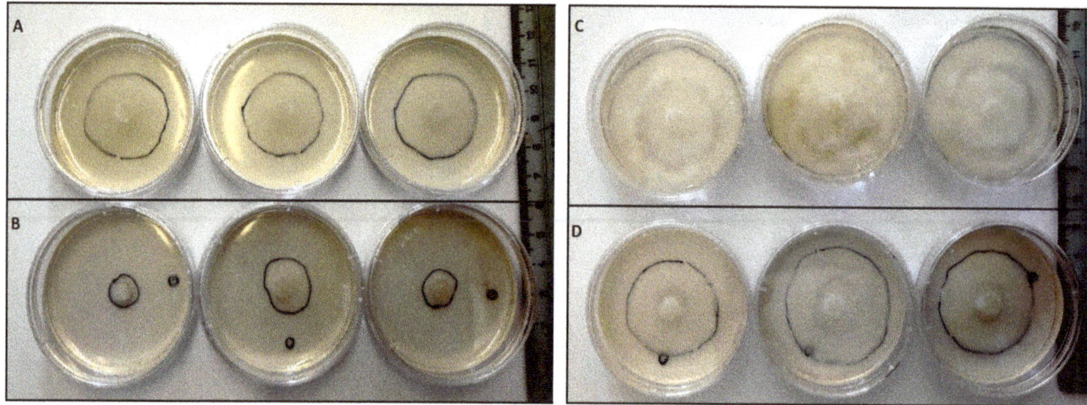

Figure 4. Mycelia of *P. cinnamomi* growth in PDA medium under different conditions (3 Petri plates out of *n* are shown as representative of the assay). (**A**): Control treatment with buffer solution and *P. cinnamomi* mycelial growth after 24 h. (**B**): Cast_Gnk2-like pure protein (250 µL/plate and 0.5 mg/mL) and *P. cinnamomi* mycelial growth after 24 h. (**C**): Control treatment with buffer solution and *P. cinnamomi* mycelial growth after 48 h. (**D**): Cast_Gnk2-like pure protein (250 µL/plate and 0.5 mg/mL) and *P. cinnamomi* mycelial growth after 48 h.

The mycelia area (cm^2) of *P. cinnamomi* growth in PDA medium under control treatment was 7.8 cm^2, while in Cast_Gnk2-like pure protein treatment (0.5 mg/mL), it was 2.3 cm^2, after 24 h. Cast_Gnk2-like pure protein treatment reduced the *P. cinnamomi* growth area by more than 100%. The same pattern was found after 48 h of incubation (Figure 5).

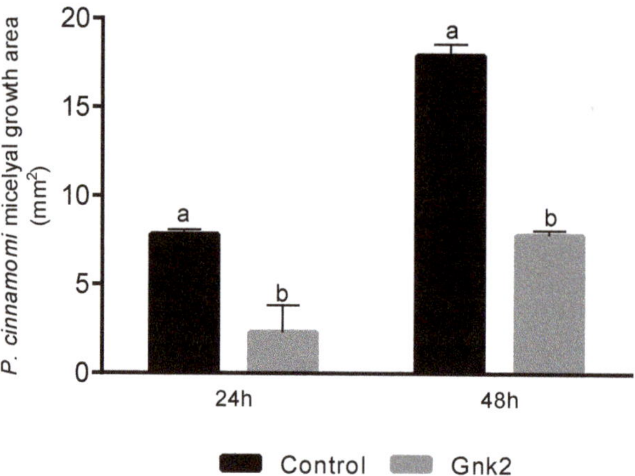

Figure 5. Mycelia area (mm^2) of *P. cinnamomi* growth in PDA under control treatment (only with buffer solution) and Cast_Gnk2-like pure protein treatment (250 µL/plate and 0.5 mg/mL) in Petri plates after 24 and 48 h at 25 °C in darkness. Results are shown as means ± SEM. Significant differences are shown as different letters (*t*-Test, $p < 0.05$).

The dried mycelia of *P. cinnamomi* growing from PDA plugs in potato dextrose liquid medium (after 48 h shaking) under control treatment was visibly higher (Figure 6A) than in Cast_Gnk2-like pure protein treatment, where the mycelia growth was reduced (Figure 6B). The mycelial weight (g) of *P. cinnamomi* under the Control treatment was significantly higher (0.00892 g) than in Cast_Gnk2-like pure protein treatment (0.00187 g) after 48 h in shaking (Figure 7). The Control treatment had an increment in mycelia weight (g) of 377%.

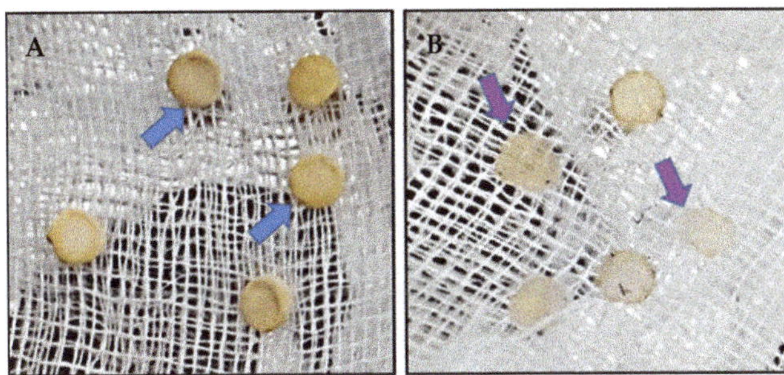

Figure 6. *Phytophthora cinnamomi* mycelia dried (after 24 h at 65 °C) on PDA plugs (5 mm diam.) grown in potato dextrose liquid medium (50 mL) for 48 h in shaking (180 rpm) at 25 °C. Some of *n* plugs are shown as representative of the assay. (**A**): Control treatment (only with buffer solution), and (**B**): Cast_Gnk2-like pure protein treatment (250 µL and 0.5 mg/mL).

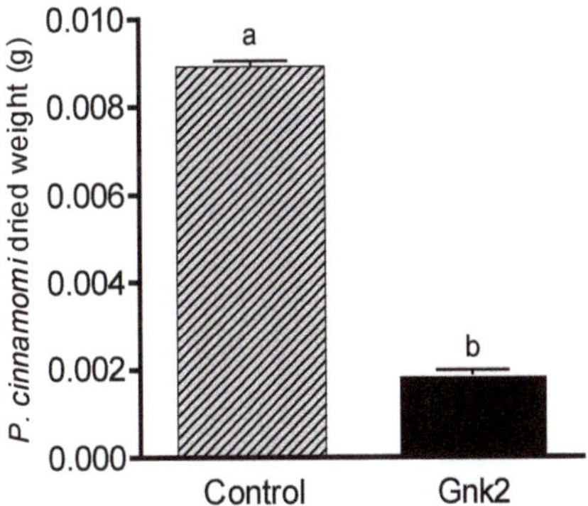

Figure 7. Mycelial dry weight (g) of *P. cinnamomi* growth in potato dextrose liquid medium (50 mL) under control treatment (only with buffer solution) and Cast_Gnk2-like pure protein treatment (250 µL and 0.5 mg/mL) after 48 h in shaking (180 rpm) at 25 °C. Results are shown as means ± SEM. Significant differences are shown as different letters (*t*-test, $p < 0.05$).

4. Discussion

This work is part of the ongoing research program that aims to find solutions to the health problem regarding European chestnuts (*C. sativa*) and the pathogen *P. cinnamomi* [1]. This program involves an integrated approach using genomics, phenomics (precision phenotyping), transcriptomics, association mapping of traits, and histopathology, to reveal the mechanisms of the resistance of Asian species to infection by *P. cinnamomi*, and to transfer the knowledge for the improvement of resistance of European chestnut [7,9,13,18–22]. The program is supported by a breeding program initiated in 2006, based on controlled crosses, from which segregated hybrid populations were obtained with the selection of new genotypes with improved resistance to *P. cinnamomi*, which soon will be disclosed to the market to be used as rootstocks. The research and breeding efforts made in the last decades are having a positive impact since chestnut production in Europe has been increasing since 2015 for the first time in many decades. Nevertheless, more research and techniques are needed to overcome the decline of European chestnut [1].

The secretion of antimicrobial proteins in plants and the reinforcement of the cell wall are part of constitutive defense barriers against pathogens [7,23]. Also, the secretion of compounds toxic to pathogens is an effective chemical defense mechanism in plants. According to Santos et al. (2017) [7], Ginkbilobin-2 (Gnk2) is the most expressed protein, secreted by *Ginkgo biloba* seeds [24], and it was shown to possess antifungal activity. In a previous study where we used a histological approach to observe the responses exhibited by susceptible and resistant chestnuts under *P. cinnamomi* infection, we found that the early accumulation of phenolic-like compounds in cell walls was observed 0.5 h after inoculation in *C. crenata* root tissues which may prevent the spread of hyphae [9]. In the chestnut species resistant to *P. cinnamomi*, *C. crenata*, the Ginkgobilobin-2-like gene (isolated from transcriptomes of *C. crenata* [13] showed relevant expression constitutively and upon *P. cinnamomi* inoculation when compared with the susceptible *C. sativa* [7]. These previous studies of our group focused on the resistance of chestnut and its response to the Oomycete, indicating that *C. crenata* Ginkgobilobin-2-like gene and the encoded protein (Cast_Gnk2-like) might be important in chestnut resistance. In an in vitro tolerance assay with the pathogen *P. cinnamomi*, Serrazina, et al. (2022) [20], observed that transgenic plants of holm oak (*Quercus ilex*) with the Cast_Gnk2-like were able to survive longer than non-transgenic ones [20]. Similar results were obtained in resistance tests against *F. oxysporum* in transgenic cucumber (*Cucumis sativus*) plants, which showed that the expression of Gnk2-1 conferred antifungal activity against the disease [25]. Currently, to further demonstrate Cast_Gnk2-like action, the group is working with somatic embryos of *C. sativa* lines genetically transformed with the overexpression of the gene. These are being maintained for further assays (studies are still in process).

Other authors have also reported the antimicrobial effects of this protein against many plant and human pathogens [10,25]. However, Cast_Gnk2-like was never tested against *P. cinnamomi* pathogen to improve the defense focused in Chestnut/*P. cinnamomi* pathosystem.

The present work focused on expressing the Cast_Gnk2-like protein and testing its efficacy as a control of *P. cinnamomi*. The first step was to optimize the protein expression and test its effectiveness against the pathogen. The recombinant protein expression was performed following different conditions to optimize its process, and the best expression was obtained using the *E. coli* strain BL21 (DE3) pLysS. These results are in correspondence to the previous work of Miyakawa et al. (2007) [26], where the Gnk2 was overexpressed in *E. coli* BL21(DE3) using pET-26b and pDsbABCD1 vectors [25].

Diseases caused by soil-borne pathogens (such as ink disease) are difficult to eradicate, because of their asymptomatic nature during the initial stages of infection and their complex modes of life and dissemination of the pathogens [27]. The possibility of cure or eradication is very low, and it is, therefore, necessary to develop multidisciplinary strategies to reduce and control them. Farmers depend basically on fungicides to control these diseases. For many years synthetic fungicides were used to control plant production decay. However, public concerns about the harmful effects of chemicals on human health and the environment have caused scientists to search for new alternatives. The incidence of pathogens infections might become more severe in the context of climate change with the possible rise of new strains. For that, it is important to start making more efforts toward these problems [1].

Currently, we are developing studies and tests for the *in-silico* protein structure prediction, in vitro 7-day antagonist assays with *P. cinnamomi*, and metabolomics analysis.

This work re-confirmed that the protein studied has significant potential for biotechnological developments that could reduce the destructive impact of these harmful diseases. Moreover, the Cast_Gnk2-like gene is demonstrated to be a valuable candidate for marker-assisted selection of tolerant *C. sativa* genotypes. Finally, methods and strategies for the production of antimicrobial phytopharmaceuticals considering Cast_Gnk2-like (pure protein) against *P. cinnamomi* are also being considered for further research.

5. Conclusions

This work is part of the ongoing research program that aims to find solutions for European chestnut decline mainly due to *P. cinnamomi*, the causal agent of ink disease. The possibility of eradicating this pathogen is extremely low, and it is necessary to develop multidisciplinary strategies to reduce and control it. Our group is studying new approaches and developing novel techniques for this purpose. As a result of this research, we determined that Cast_Gnk2-like has a signal peptide that guides the final protein peptide to the apoplast. Moreover, Cast_Gnk2-like (pure protein) significantly reduced the mycelia growth of *P. cinnamomi* in vitro, indicating the potential antagonistic effect on this pathogen. These insights reveal the potential of Cast_Gnk2-like for agricultural uses and biotechnological developments, which are becoming more necessary and critical in managing the threat this pathogen possess to European chestnut under rapidly changing environmental conditions.

Author Contributions: Conceptualization: R.L.C. and M.B.C.; Methodology: F.V.D., S.S. and M.B.C.; Validation: R.L.C., M.B.C., F.V.D. and S.S.; Formal analysis: R.L.C., M.B.C., F.V.D. and S.S.; Investigation: M.B.C., F.V.D. and S.S.; Resources: R.L.C. and R.M.; Data curation: M.B.C., F.V.D. and S.S.; Writing—original draft preparation: M.B.C.; Writing—review and editing: R.L.C., F.V.D., S.S., R.M. and M.B.C.; Visualization: R.L.C., F.V.D., S.S., R.M. and M.B.C.; Supervision: F.V.D. and R.L.C.; Project administration: R.L.C.; Funding acquisition: R.L.C. and R.M. All authors have read and agreed to the published version of the manuscript.

Funding: This research was funded by Fundação para a Ciência e Tecnologia under the project 414103 FCT-Lx-FEDER-28760, DL57/2016/CP [12345/2018]/CT[2477] to S.S., DL57/2016/CP [12345/2018]/ CT[2470] to F.V.D., UIDB/04046/2020, UIDP/04046/2020, UIDB/04046/2021 and UIDP/04046/2021 to BioISI.

Data Availability Statement: Data from this paper are not included in any other site or repository.

Acknowledgments: We acknowledge Helena Machado (Instituto Nacional de Investigação Agrária e Veterinária, I.P) for providing Phytophotora cinnamomi culture for inoculations and Alan Phillips (BioISI—BioSystems and Integrative Sciences Institute) for English review and valuable suggestions.

Conflicts of Interest: The authors declare that the research was conducted without any commercial or financial relationships that could be construed as a potential conflict of interest.

References

1. Fernandes, P.; Colavolpe, M.B.; Serrazina, S.; Costa, R.L. European and American chestnuts: An overview of the main threats and control efforts. *Front. Plant Sci.* **2022**, *13*, 844–951. [CrossRef]
2. Thines, M. Taxonomy and phylogeny of Phytophthora and related Oomycetes. In *Phytophthora: A Global Perspective*; CABI: Wallingford, UK, 2013; pp. 11–18.
3. Erwin, D.C.; Ribeiro, O.K. *Phytophthora Diseases Worldwide*; American Phytopathological Society (APS Press): St. Paul, MN, USA, 1996.
4. Agrios, G.N. *Plant Pathology*, 5th ed.; Academic Press: Cambridge, MA, USA, 2005; p. 922.
5. Hardham, A.R. Phytophthora cinnamomi. *Mol. Plant Pathol.* **2005**, *6*, 589–604. [CrossRef] [PubMed]
6. Kamoun, S.; Furzer, O.; Jones, J.D.; Judelson, H.S.; Ali, G.S.; Dalio, R.J.; Govers, F. The Top 10 Oomycete pathogens in molecular plant pathology. *Mol. Plant Pathol.* **2015**, *16*, 413–434. [CrossRef]
7. Santos, C.; Duarte, S.; Tedesco, S.; Fevereiro, P.; Costa, R. Expression Profiling of Castanea Genes during Resistant and Susceptible Interactions with the Oomycete Pathogen *Phytophthora cinnamomi* Reveal Possible Mechanisms of Immunity. *Front. Plant Sci.* **2017**, *8*, 515. [CrossRef]
8. Niu, F.; Xing, Y.; Jia, N.; Ding, K.; Xie, D.; Chen, H.; Chi, D. Effectiveness of entomopathogenic fungal strains against poplar/willow weevil (*Cryptorhynchus lapathi* L.) larvae. *J. For. Res.* **2022**, *33*, 1691–1702. [CrossRef]
9. Fernandes, P.; Machado, H.; Silva, M.D.C.; Costa, R.L. A Histopathological Study Reveals New Insights Into Responses of Chestnut (*Castanea* spp.) to Root Infection by *Phytophthora cinnamomi*. *Phytopathology* **2021**, *111*, 345–355. [CrossRef] [PubMed]
10. Mélida, H.; Sandoval-Sierra, J.V.; Diéguez-Uribeondo, J.; Bulone, V. Analyses of Extracellular Carbohydrates in Oomycetes Unveil the Existence of Three Different Cell Wall Types. *Eukaryotic. Cell* **2013**, *12*, 194–203. [CrossRef]
11. Miyakawa, T.; Hatano, K.; Miyauchi, Y.; Suwa, Y.; Sawano, Y.; Tanokura, M. A secreted protein with plant-specific cysteine-rich motif functions as a mannose-binding lectin that exhibits antifungal activity. *Plant Physiol.* **2014**, *166*, 766–778. [CrossRef] [PubMed]
12. Bartnicki-Garcia, S. Chemistry of hyphal walls of Phytophthora. *Microbiology* **1966**, *42*, 57–69. [CrossRef]
13. Serrazina, S.; Santos, C.; Machado, H.; Pesquita, C.; Vicentini, R.; Pais, M.S.; Sebastiana, M.; Costa, R. Castanea root transcriptome in response to *Phytophthora cinnamomi* challenge. *Tree Genet. Genomes* **2015**, *11*, 6. [CrossRef]
14. The UniProt Consortium. UniProt: The Universal Protein Knowledgebase in 2023. *Nucleic Acids Res.* **2023**, *51*, D523–D531. [CrossRef]
15. Petersen, T.; Brunak, S.; von Heijne, G. SignalP 4.0: Discriminating signal peptides from transmembrane regions. *Nat. Methods* **2011**, *8*, 785–786. [CrossRef] [PubMed]

16. Almagro Armenteros, J.J.; Salvatore, M.; Emanuelsson, O.; Winther, O.; von Heijne, G.; Elofsson, A.; Nielsen, H. Detecting sequence signals in targeting peptides using deep learning. *Life Sci. Alliance* **2019**, *2*, e201900429. [CrossRef] [PubMed]
17. Thompson, J.D.; Higgins, D.G.; Gibson, T.J. CLUSTAL W: Improving the sensitivity of progressive multiple sequence alignment through sequence weighting, position-specific gap penalties and weight matrix choice. *Nucleic Acids Res.* **1994**, *22*, 4673–4680. [CrossRef]
18. Fernandes, P.; Tedesco, S.; Vieira da Silva, I.; Santos, C.; Machado, H.; Lourenço Costa, R. A New Clonal Propagation Protocol Develops Quality Root Systems in Chestnut. *Forests* **2020**, *11*, 826. [CrossRef]
19. Santos, C.; Nelson, C.D.; Zhebentyayeva, T.; Machado, H.; Gomes-Laranjo, J.; Costa, R.L. First interspecific genetic linkage map for Castanea sativa x Castanea crenata revealed QTLs for resistance to *Phytophthora cinnamomi*. *PLoS ONE* **2017**, *12*, e0184381. [CrossRef] [PubMed]
20. Serrazina, S.; Martínez, M.T.; Cano, V.; Malhó, R.; Costa, R.L.; Corredoira, E. Genetic Transformation of *Quercus ilex* Somatic Embryos with a Gnk2-like Protein That Reveals a Putative Anti-Oomycete Action. *Plants* **2022**, *11*, 304. [CrossRef] [PubMed]
21. Santos, C.; Zhebentyayeva, T.; Serrazina, S.; Nelson, C.; Cost, R. Development and characterization of EST-SSR markers for mapping reaction to *Phytophthora cinnamomi* in Castanea spp. *Sci. Hortic.* **2015**, *194*, 181–187. [CrossRef]
22. Santos, C.; Machado, H.; Correia, I.; Gomes, F.; Gomes-Laranjo, J.; Costa, R. Phenotyping Castanea hybrids for *Phytophthora cinnamomi* resistance. *Plant Pathol.* **2014**, *64*, 901–910. [CrossRef]
23. Hardham, A.R.; Blackman, L.M. Molecular cytology of Phytophthora-plant interactions. *Australas. Plant Pathol.* **2010**, *39*, 29–35. [CrossRef]
24. Miyakawa, T.; Miyazono, K.; Sawano, Y.; Hatano, K.; Tanokura, M. Crystal structure of ginkbilobin-2 with homology to the extracellular domain of plant cysteine-rich receptor-like kinases. *Proteins* **2009**, *77*, 247–251. [CrossRef] [PubMed]
25. Liu, J.; Huali, T.; Yahong, W.; Aiguang, G. Overexpression of a Novel Antifungal Protein Gene GNK2-1 Results in Elevated Resistance of Transgenic Cucumber to *Fusarium oxysporum*. *Chin. Bull. Bot.* **2010**, *45*, 411–418. [CrossRef]
26. Miyakawa, T.; Sawano, Y.; Miyazono, K.; Hatano, K.; Tanokura, M. Crystallization and preliminary X-ray analysis of ginkbilobin-2 from Ginkgo biloba seeds: A novel antifungal protein with homology to the extracellular domain of plant cysteine-rich receptor-like kinases. *Acta Crystallogr. Sect. F Struct. Biol. Cryst. Commun.* **2007**, *63*, 737–739. [CrossRef]
27. Nelson, B.D.; Mallik, I.; McEwen, D.; Christianson, T. Pathotypes, distribution, and metalaxyl sensitivity of *Phytophthora sojae* from North Dakota. *Plant Dis.* **2008**, *92*, 1062–1066. [CrossRef] [PubMed]

Article

Identification of Candidate Genes Involved in Bud Growth in *Pinus pinaster* through Knowledge Transfer from *Arabidopsis thaliana* Models

José Manuel Alvarez [1,*], Sonia María Rodríguez [1], Francisco Fuente-Maqueda [2], Isabel Feito [2], Ricardo Javier Ordás [1] and Candela Cuesta [1]

[1] Área de Fisiología Vegetal, Departamento de Biología de Organismos y Sistemas, Instituto Universitario de Biotecnología de Asturias (IUBA), Universidad de Oviedo, 33071 Oviedo, Spain; uo270227@uniovi.es (S.M.R.); rordas@uniovi.es (R.J.O.); cuestacandela@uniovi.es (C.C.)

[2] Programa Forestal, Área de Cultivos Hortofrutícolas y Forestales, Servicio Regional de Investigación y Desarrollo Agroalimentario de Asturias (SERIDA), Finca Experimental La Mata, 33820 Grado, Spain; franciscof@serida.org (F.F.-M.); ifeito@serida.org (I.F.)

* Correspondence: alvarezmanuel@uniovi.es

Abstract: *Pinus pinaster* is a plant species of great ecological and economic importance. Understanding the underlying molecular mechanisms that govern the growth and branching of *P. pinaster* is crucial for enhancing wood production and improving product quality. In this study, we describe a simple methodology that enables the discovery of candidate genes in *Pinus pinaster* by transferring existing knowledge from model species like *Arabidopsis thaliana* and focusing on factors involved in plant growth, including hormonal and non-hormonal pathways. Through comparative analysis, we investigated the main genes associated with these growth-related factors in *A. thaliana*. Subsequently, we identified putative homologous sequences in *P. pinaster* and assessed the conservation of their functional domains. In this manner, we can exclude sequences that, despite displaying high homology, lack functional domains. Finally, we took an initial approach to their validation by examining the expression levels of these genes in *P. pinaster* trees exhibiting contrasting growth patterns. This methodology allowed the identification of 26 candidate genes in *P. pinaster*. Our findings revealed differential expression patterns of key genes, such as *NCED3*, *NRT1.2*, *PIN1*, *PP2A*, *ARF7*, *MAX1*, *MAX2*, *GID1*, *AHK4*, *AHP1*, and *STP1*, in relation to the different growth patterns analyzed. This study provides a methodological foundation for further exploration of these genes involved in the growth and branching processes of *P. pinaster*. This will contribute to the understanding of this important tree species and open new avenues for enhancing its utilization in sustainable forestry practices.

Keywords: *Pinus pinaster*; growth regulators; growth; RT-qPCR; branching

Citation: Alvarez, J.M.; Rodríguez, S.M.; Fuente-Maqueda, F.; Feito, I.; Ordás, R.J.; Cuesta, C. Identification of Candidate Genes Involved in Bud Growth in *Pinus pinaster* through Knowledge Transfer from *Arabidopsis thaliana* Models. *Forests* **2023**, *14*, 1765. https://doi.org/10.3390/f14091765

Academic Editor: Keith Woeste

Received: 9 August 2023
Revised: 23 August 2023
Accepted: 29 August 2023
Published: 31 August 2023

1. Introduction

Current climate change and the need to produce wood and other products determine the growing importance of green biology [1] and the urgency of finding roads that allow substantial forest improvement. This goal has to be multidisciplinary, covering applied aspects including plantation management but also other basic ones, i.e., the understanding of the bases of the growth and development of trees.

Plant growth and development are complex processes involving numerous mechanisms, from the formation of the embryo to the attainment of a fully mature individual. All of them combine systems of cell growth and differentiation that give rise to different tissues and organs. During post-embryonic development, growth starts from the meristems, cellular assemblies that remain in an embryogenic state indefinitely and whose division gives rise to the new cells that will constitute the adult plant. Focusing on the aerial part of the plant, it originates in the shoot apical meristem, located at the end of the stem. This, in addition to length growth, plays an organogenic function on the phyllotaxis, forming the leaves and axillary buds from which future branches will arise. The phyllotaxis requires a fine-tuned regulation of the process that involves multiple factors, such as plant hormones, whose spatial control of their distribution leads to different arboreal architectures [2].

In addition to this, other non-hormonal mechanisms that promote growth and development in plants are also known, such as sugars and the red:far red light ratio, among others [3,4]. Sugars not only represent the plant's energy source but also regulate processes including flowering, anthocyanin synthesis, and meristematic proliferation, and light perception represents key information regarding changes in the surrounding environment.

Nowadays, relevant advances have been made to understand individual plant hormones, their mechanisms, and their processes in model species of angiosperms, like *Arabidopsis thaliana* [5]. An example of this is auxins. They were the first plant hormones discovered, synthesized in the stem apex and actively distributed through polar auxin transport, forming a gradient that is related to apical dominance and the inhibition of axillary bud development [3]. Because this phytohormone is involved in virtually all plant development processes, it is considered the most important signaling molecule [6,7]. However, understanding the molecular basis of the different responses offered by molecules related to auxins remains a pending challenge for the future [8].

It has recently been stated that, to be able to understand how plant hormones work together in the regulation of plant growth and development, connections between their pathways (the mechanism named crosstalk) need to be understood [9]. However, despite these advances, there is still much to learn about the molecular mechanisms underlying these processes [10].

Furthermore, it is not yet clear whether the knowledge acquired from studying angiosperm model species, e.g., *A. thaliana*, can be directly applied to gymnosperms such as *P. pinaster*. One of the main reasons is that the identification of candidate genes in *P. pinaster* through forward and reverse genetics poses an extremely challenging task. This is primarily due to the absence of defective mutants, difficulties in applying techniques like T-DNA insertional mutagenesis, and the fact that its genome has not been released yet.

P. pinaster is a species of great ecological and economic importance, with its ability to thrive in a wide range of environments [11] and its potential for use in forestry [12]. In addition, it can withstand cold or temperate climates and all kinds of substrates or environmental factors, including drought, and has high-quality wood [13,14]. Therefore, investigating the role of plant hormones in the growth and development of pine trees, such as *P. pinaster*, could have significant implications for our understanding of plant hormone function and for the development of sustainable forest management practices. Currently, important strides are being made in the field of pine plantations, utilizing a range of techniques from domestication and traditional plant breeding to genetic engineering [15].

Our starting working hypothesis is that stem development is a conserved process in the evolution of seed plants, and therefore the models proposed for *A. thaliana* can be extrapolated to conifers. To contrast this, we need to deepen our knowledge of the physiological and molecular factors involved in the shoot development of gymnosperms by using systems biology as a methodological framework that brings together different organizational levels, thus providing responses to the plasticity and performance of cells and tissues in different environments [16]. Transferring knowledge about growth and development from angiosperms to conifers can have great applicability for the timber industry, where the quality and quantity of wood are economically key. The first step in this knowledge transfer is to obtain the sequence of candidate genes. In this work, we present a simple methodology that enables the discovery of candidate genes in *P. pinaster* by transferring existing knowledge from model species such as *A. thaliana* and focusing on factors involved in plant growth, including hormonal and non-hormonal pathways.

In addition, we assessed the initial validation of the candidate genes obtained by examining the expression levels of these genes in *P. pinaster* trees exhibiting contrasting growth patterns. In Spain, *P. pinaster* has traditionally been divided into two types or subspecies: (i) Atlantic or maritime, mainly located in the northwest; and (ii) Mediterranean or mesogeensis, representing the remaining *P. pinaster* stands. At the time of sampling, late summer (September), the Atlantic origin was in an active growth phase, while the Mediterranean origin was in a resting phase.

The identification of genes involved in plant growth and development in *P. pinaster* will facilitate molecular studies to characterize the function of key genes in gymnosperms.

2. Materials and Methods

2.1. Selection of Candidate Genes in Arabidopsis thaliana and Identification of Homologue Sequences in Pinus pinaster

2.1.1. Selection of Candidate Genes in *Arabidopsis thaliana*

A first selection of genes related to hormonal (abscisic acid, auxins, cytokinins, strigolactones, and gibberellins) and non-hormonal (sugars and red:far red light ratio) key factors was made according to the literature [17,18]. For each group of hormones, members of synthesis, transport, and signaling were represented. Both nucleotide and protein sequences, as the latter are more conserved between species, were compiled in FASTA format from the TAIR database (The Arabidopsis Information Resource, arabidopsis.org).

2.1.2. Identification of Homologue Sequences in *Pinus pinaster*, Sequence Search, and Comparison of Functional Domains

In order to identify homologue sequences in *P. pinaster*, we carried out a screening of the *P. pinaster* transcriptome and proteome data obtained in the frame of the European projects ProCoGen and SustainPine (http://www.scbi.uma.es/sustainpinedb/home_page, accessed on 31 March 2023), and in the PLAZA Gymnosperms (https://bioinformatics.psb.ugent.be/plaza/versions/gymno-plaza/, accessed on 31 March 2023) database.

The TBLASTN and BLASTP algorithms [19], with default settings, were used for the screening using the *A. thaliana* protein sequences as queries. The sequences obtained were compared using the Multiple Sequence Alignment CLUSTALW tool [20], and the protein sequences encoded were analyzed using the InterProScan (http://www.ebi.ac.uk/Tools/pfa/iprscan/, accessed on 31 March 2023), Prosite (http://prosite.expasy.org/, accessed on 31 March 2023), and SMART (Simple modular architecture research tool, /smart.embl-heidelberg.de/) tools. After obtaining the domains, functions, and positions of each sequence, they were plotted with My domains (https://prosite.expasy.org/mydomains/, accessed on 31 March 2023) The similarity degree was represented by the E value. Sequence annotations were performed with the Geneious software v.11 (Biomatters Ltd., Auckland, New Zealand). Interactions between pairs of genes were analyzed using the STRING database (v.11.5).

2.2. Expression Analyses

2.2.1. Plant Material

To validate the candidate genes, samples from basal branches were harvested by the end of the summer (September). We collected the apical and whorled buds of the developing whorl (main apical buds and main whorl buds) and the apical buds from the last fully developed whorl (secondary apical buds) (Figure 1) in eleven year old *P. pinaster* clonal trees, representing the Mediterranean and Atlantic subspecies, characterized by a different model of development. Specifically, the Atlantic origin is characterized by extended and continuous growth, whereas the Mediterranean type presents a shorter growing period with several flushes within the same season. Three trees of each type were grown at the experimental plantation "La Mata" of the "Servicio Regional de Investigación y Desarrollo Agroalimentario de Asturias (SERIDA)" in Grado, Principado de Asturias (SPAIN). Samples were collected, frozen in liquid nitrogen for transport, and stored at $-80\ ^\circ$C until use.

Figure 1. Prototypes of the studied individuals of *P. pinaster* of (**A**) Atlantic and (**B**) Mediterranean origin growing in the experimental plantation "La Mata" of the SERIDA in Grado, Principado de Asturias, Spain; and (**C**) detail of the main apical bud (arrow) and the main whorl buds (star).

2.2.2. RNA Isolation and cDNA Synthesis

The RNA was extracted from the samples using the GeneMATRIX Universal RNA Purification Kit (EURx, Gdańsk, Poland), and its quantity was measured spectrophotometrically at 260 nm. The integrity of the RNA was verified by performing agarose gel electrophoresis. For each sample, 1 µg of total RNA was reverse transcribed with the High-Capacity cDNA Reverse Transcription Kit (Applied Biosystems Inc., Foster City, CA, USA) following the manufacturer's instructions.

2.2.3. Quantitative Real-Time PCR (RT-qPCR)

Gene expression analysis was performed by RT-qPCR in a Bio-Rad CFX96 Real-Time PCR Detection System (Bio-Rad, Hercules, CA, USA), following the standards for this technique, like MIQE [21,22]. *P. pinaster* ubiquitin gene (Acc. AF461687) was used as an endogenous reference gene [23–25]. Specific primers for each gene (list available in Supplementary Table S1) were designed with Primer3 software v.4 [26], following the parameters recommended [21], to amplify an 80–100 bp fragment (amplicon). Three biological replicates and three technical replicates each were analyzed with 5 µL of Fast SYBR Green Master Mix (Applied Biosystems Inc., Foster City, CA, USA), oligonucleotide primers (0.20 µM), and 100 ng of cDNA in a final volume of 10 µL. The protocol used was: 95 °C for 20 s; 45 cycles of 95 °C for 3 s; and 60 °C for 30 s, with a final melting curve to assess for non-specific products. For this purpose, negative controls (no template) and RT-controls (non-retro-transcribed RNA) were also included.

2.2.4. Data Analysis

Analysis of the RT-qPCR data was performed with the qpcR package for R software v4.1.3 (www.dr-spiess.de/qpcR.html, accessed on 31 March 2023), which allows the fitting of the RT-qPCR fluorescence raw data to a five parameter sigmoidal model for obtaining essential PCR parameters such as efficiency, threshold cycle, and transcript abundance [27]. The relative abundance of each transcript was calculated as the mean of the technical duplicates and normalized to the expression value of the reference gene in each sample. Results were expressed as mean normalized expression values ± standard error of three biological replicates. Significant differences in mRNA levels were determined by *t*-test analysis or ANOVA using the Student–Newman–Keuls test for post hoc comparisons (SIGMA-PLOT v11 software, Chicago, IL, USA). In addition, a principal component analysis (PCA) of the gene expression data in both provenances was performed through R software v4.1.3 (https://www.r-project.org/, accessed on 31 March 2023).

3. Results

3.1. Selection of Candidate Genes in Arabidopsis thaliana and Identification of Homologue Sequences in Pinus pinaster

Given the complexity of the networks involved in plant growth processes, a first selection of the main factors involved in plant growth and development was divided into two large groups: hormonal and non-hormonal factors.

Regarding hormonal factors, the main hormonal groups studied were abscisic acid, auxins, cytokinins, gibberellins, and strigolactones. More specifically, within abscisic acid, selected genes were *NCED3* (*NINE-CIS-EPOXYCAROTENOID DIOXYGENASE 3*), *ABCG40* (*ATP-BINDING CASSETTE G*), *NRT1.2* (*NITRATE TRANSPORTER 1.2*), *SNRK2.4* (*SUCROSE NONFERMENTING 1-RELATED PROTEIN KINASE 2*), and *ABI1* (*ABA INSENSITIVE 1*).

In the auxin case, the following genes were analyzed: *PIN1* (*PIN-FORMED 1*), *PP2A* (*SERINE/THREONINE PROTEIN PHOSPHATASE 2A*), *ABP1* (*AUXIN BINDING PROTEIN 1*), *TIR1* (*TRANSPORT INHIBITOR RESPONSE 1*), *ARF7* (*AUXIN RESPONSE FACTOR 7*), and *AXR1* (*AUXIN RESISTANT 1*).

Within the cytokinins group, the genes studied were: *CRE1/AHK4/WOL* (*ARABIDOPSIS HISTIDINE KINASE 4*), *AHP1* (*ARABIDOPSIS HISTIDINE PHOSPHOTRANSFER 1*), *ARR* (*ARABIDOPSIS RESPONSE REGULATOR*), *CRF2* (*CYTOKININ RESPONSE FACTOR 2*), *CYP735A* (*CYTOCHROME 735A*), *IPT* (*ISOPENTENYL TRANSFERASE*), *LOG* (*LONELY GUY*), and *PUPs* (*PURINE PERMEASE*).

In relation to strigolactones, the following genes were studied: *LBO1* (*LATERAL BRANCHING OXIDOREDUCTASE*), *BRC1* (*BRANCHED 1*), *MAX1* (*MORE AXILLARY BRANCHES 1*), *D14* (*DWARF 14*), and *MAX2* (*MORE AXILLARY BRANCHES 2*).

In relation to the gibberellins, the genes studied were: *GID1* (*GIBBERELLIN INSEN-SITIVE DWARF 1*), *GAI* (*GA-INSENSITIVE*), *RGA* (*REPRESSOR OF GA1-3*), and *SLY1* (*SLEEPY 1*).

On the other hand, other important factors involved in plant growth and development are sugar and light. The sugar genes selected were *STP1* (*SUGAR TRANSPORT PROTEIN 1*), *SWEET17*, and *WRKY20*, while concerning light, *PHYB* (*PHYTOCHROME B*) was studied.

3.2. Bioinformatic Analysis of Homologues in Pinus pinaster and Their Functional Domains

For this study, an E value of 10^{-30} was defined as the threshold. Some of the afore-mentioned proteins—IPT, CYP735A, LOG, PUPs, BRC1, and ARR—presented an E value higher than the one set as a threshold and, therefore, were discarded for subsequent experimental study at this analytical stage as they were considered poorly conserved. A set of 26 *P. pinaster* proteins were selected for further analysis (Table 1).

To confirm their possible homology, the functional domains of putative homologue proteins from *A. thaliana* and *P. pinaster* were analyzed using SMART, comparing them in pairs (example given at Supplementary Figure S1). The results showed a high degree of domain conservation in all *P. pinaster* sequences compared to *A. thaliana*, suggesting similar functional capabilities.

To show the relevance of plant hormone crosstalk in the regulation of plant growth and development, interactions between proteins used in this study were evaluated in silico by using the STRING database (Figure 2).

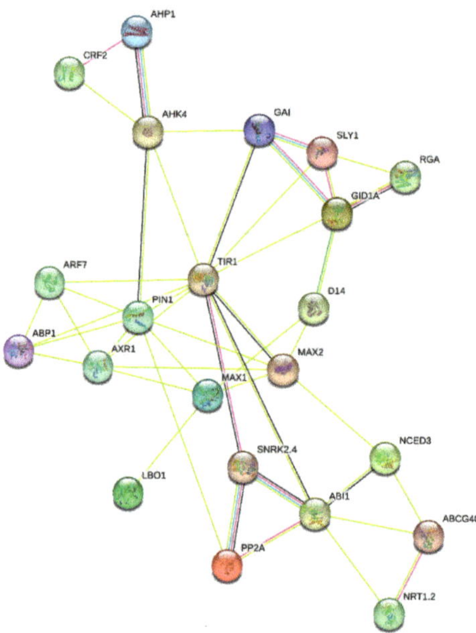

Figure 2. Interactions between proteins were analyzed with STRING. The represented proteins belong to abscisic acid (NCED3, ABCG40, NRT1.2, SNRK2.4, ABI1), auxins (PIN1, PP2A, ABP1, TIR1, ARF7, AXR1), cytokinins (AHK4, AHP1, CRF2), strigolactones (LBO1, MAX1, D14, MAX2), and gibberellins (GID1A, GAI, RGA, SLY1). The edges represent the predicted functional associations. A red line indicates the presence of fusion evidence; a green line—neighborhood evidence; a blue line—co-occurrence evidence; a purple line—experimental evidence; a yellow line—textmining evidence; a light blue line—database evidence; and a black line—co-expression evidence.

Table 1. Classification of the protein sequences object of this study in *Arabidopsis thaliana* and the results of homology in *P. pinaster* (the ID is indicated in TAIR and Gymnoplaza, respectively, and the length of the sequence in amino acids). The E value of the BLAST (from the sequence of *A. thaliana* versus *P. pinaster*) and the identity and positive percentage (BLOSUM62) (from the sequence of *P. pinaster* versus *A. thaliana*) are also indicated. (Only proteins that passed bioinformatics tests with a high degree of homology and/or conservation of their domains are shown.)

| | | | | *Arabidopsis thaliana* | | | *Pinus pinaster* | | | | |
			Protein	ID	Length (aa)	ID	Length (aa)	E value	Identity %	Positive % (BLSM62)
Hormonal	Abscisic acid	Synthesis	NCED3	AT3G14440	599	PPI00015334	412	0.0	68.3	81.1
		Transport	ABCG40	AT1G15210	1442	PPI00033391	1457	0.0	63.1	78.4
			NRT1.2	AT1G69850	585	PPI00013776	598	0.0	48.5	67.3
		Signaling	SNRK2.4	AT1G10940	371	PPI00050983	455	0.0	80.1	90.1
			ABI1	AT4G26080	434	PPI00056219	594	1.0×10^{-111}	49.9	65.9

Table 1. *Cont.*

				Arabidopsis thaliana		*Pinus pinaster*				
			Protein	ID	Length (aa)	ID	Length (aa)	E value	Identity %	Positive % (BLSM62)
Hormonal	Auxins	Transport	PIN1	AT1G73590	622	PPI00011546	695	0.0	57.1	67.8
			PP2A	AT1G69960	307	PPI00008039	306	0.0	90.8	97.4
		Signaling	ABP1	AT4G02980	198	PPI00009705	160	2.0×10^{-53}	49.9	65.9
			TIR1	AT3G62980	594	PPI00012072	574	0.0	65.4	79.3
			ARF7	AT5G20730	1165	PPI00041807	497	1.0×10^{-180}	71.5	83.3
			AXR1	AT1G05180	540	PPI00013097	560	0.0	67.0	82.4
	Cytokinins	Signaling	AHK4	AT2G01830	1080	PPI00064460	1036	0.0	56.7	70.3
			AHP1	AT3G21510	154	PPI00050577	156	4.0×10^{-62}	57.6	80.1
			CRF2	AT4G23750	343	PPI00010365	270	3.0×10^{-31}	43.1	56.9
	Strigolactones	Synthesis	LBO1	AT3G21420	364	PPI00010339	377	8.0×10^{-113}	47.1	67.3
			MAX1	AT2G26170	522	PPI00016810	421	3.0×10^{-134}	53.4	69.3
		Signaling	D14	AT3G03990	267	PPI00017643	267	4.0×10^{-133}	65.8	83.1
			MAX2	AT2G42620	693	PPI00014698	329	6.0×10^{-86}	54.5	67.1
	Gibberellins	Signaling	GID1A	AT3G05120	345	PPI00071627	357	1.0×10^{-162}	66.6	78.2
			GAI	AT1G14920	533	PPI00014310	458	3.0×10^{-174}	62.6	77.8
			RGA	AT2G01570	587	PPI00011857	594	5.0×10^{-174}	49.6	66.0
			SLY1	AT4G24210	151	PPI00016475	219	3.0×10^{-38}	48.9	63.7
Non-hormonal	Sugars	Transport	STP1	AT1G11260	522	PPI00012920	514	0.0	65.0	77.9
			SWEET17	AT4G15920	241	PPI00061954	280	8.0×10^{-75}	52.9	71.7
			WRKY20	AT4G26640	557	PPI00011874	679	4.0×10^{-90}	42.5	58.6
	Light	Signaling	PHYB	AT2G18790	1172	PPI00062444	1139	0.0	67.3	81.5

3.3. Expression Analyses

The expression levels of the genes coding for each of the 26 proteins were then studied in eleven year old individuals with different developmental patterns (Atlantic and Mediterranean shapes). All genes studied showed amplification, and the results indicated significant differences in the expression of 11 genes out of 26 (Table 2; gene expression values are shown in Supplementary Figure S2). All the groups studied, both hormonal and non-hormonal (except for the red:far red light ratio), showed at least one gene with a significant difference in expression (Supplementary Figure S2).

Table 2. Summary of gene expression differences in three types of buds; nd = no significant differences in expression were observed. A = higher expression was observed in that type of bud of Atlantic origin. M = higher expression was observed in people of Mediterranean origin. * = 90% confidence interval; ** = 95% confidence interval.

	Gene	Main Apical Bud	Main Whorl Bud	Secondary Apical Bud
Abscisic acid	NCED3	A **	nd	nd
	NRT1.2	M *	nd	M **
Auxins	PIN1	A *	nd	nd
	PP2A	nd	M *	nd
	ARF7	nd	nd	M *

Table 2. *Cont.*

	Gene	Main Apical Bud	Main Whorl Bud	Secondary Apical Bud
Cytokinins	*AHK4*	nd	M *	nd
	AHP1	nd	A *	nd
Strigolactones	*MAX1*	nd	nd	M *
	MAX2	A **	nd	nd
Gibberellins	*GID1*	M **	nd	nd
Sugars	*STP1*	nd	M *	M **

3.4. Multivariate Analyses of Gene Expression Data

Expression data were analyzed by PCA, which explained 60.1% of the variation observed. The Atlantic and Mediterranean types were separated by principal component two. The differences observed between groups were mainly explained by *AHP1*, *PIN1*, *MAX2*, *CRF2*, *NRT12*, *MAX1*, *ABCG40*, *GID1*, and *LBO* data (Figure 3).

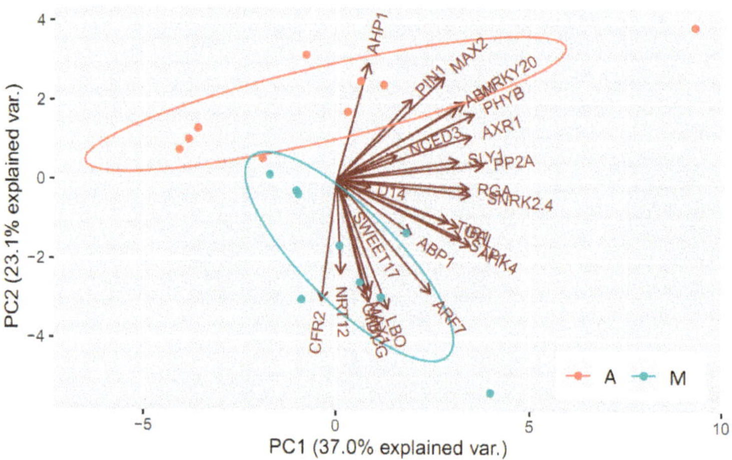

Figure 3. Analysis of gene expression data by principal component analysis (PCA). A: Atlantic type; M: Mediterranean type. Ellipses represent the distribution of projected data points in the space of principal components.

4. Discussion

Molecular studies in *P. pinaster* are still challenging as its genome has not been released yet, which, together with the lack of a collection of defective mutants, contributes to the complexity of conducting studies within this species. Transferring knowledge from angiosperm models can help overcome these difficulties, as has been shown in this study, where we used a straight-forward methodology to obtain candidate genes. To achieve this, we focused our study on factors involved in plant growth, including hormonal and non-hormonal pathways. We used the models described in *A. thaliana* to obtain the sequences of candidate genes. Subsequently, we searched for potential homologous sequences in available *P. pinaster* databases and compared functional domains, which led to the exclusion of sequences that, despite showing high homology, lacked functional domains.

The search for putative homologues revealed a high degree of conservation between angio- and gymnosperms. STRING analysis showed multiple interactions among the genes selected, highlighting the relevance of plant hormone crosstalk in the regulation of plant growth and development.

Once we achieved the main goal of this study, we assessed the initial validation of these candidate genes in *P. pinaster* trees from two subspecies (Atlantic and Mediterranean) exhibiting contrasting growth patterns. With this proof of concept, we could correlate the state of growth with certain hormonal gene dynamics, even determining if the expression of these genes could be critical. At the time of sampling, late summer (September), the Atlantic origin was in an active growth phase, while the Mediterranean origin was in a resting phase. The methodology chosen to carry out this initial validation was RT-qPCR, as it has the capacity to detect and measure minute amounts of nucleic acids in a wide range of samples. Its conceptual and practical simplicity, together with its combination of speed, sensitivity, and specificity in a homogeneous assay, have made it the touchstone for nucleic acid quantification [22]. RT-qPCR results showed that within all groups of hormonal and non-hormonal factors, except light, there are significant differences.

In the case of the study of genes related to abscisic acid, significant differences in expression were observed in *NCED3* and *NRT1.2*. *NCED3* differences were detected in the main apical bud, being superior to their expression in *P. pinaster* of Atlantic origin. This dioxygenase catalyzes key stages in the local biosynthesis of ABA, causing the transformation of violaxanthine to xanthoxin, which is then translocated from the chloroplast to the cytosol [28], which will later lead to ABA. Therefore, a greater expression of *NCED3* would be associated with an increase in ABA.

On the other hand, the expression of *NRT1.2* is differential in both main and secondary apical buds, with greater expression in the Mediterranean origin (resting phase at the time of sample collection). This conveyor is key in transporting signals, and its greater presence is associated with buds in dormancy [28]. This NRT1.2 trend would be consistent with the resting phase associated with Mediterranean provenance at the sampling time.

Three genes related to auxins resulted in having different significant expressions: in the case of *PIN1*, its level at the main apical bud is superior in the origin with active growth (Atlantic origin). *PIN1* encodes a mediator of the active transport of auxins, which promotes the flow of polar transport. Polar auxin transport plays a crucial role in active growth in shoots by regulating various aspects of shoot development and architecture. In addition, it promotes apical dominance and inhibits axillary bud development [3]. Therefore, it can be expected that an increase in expression of this gene would lead to a more elongated phenotype, such as that of Atlantic origin.

A greater expression of *PP2A* has been observed in the main whorl bud of the Mediterranean type. This phosphatase promotes the direction of auxin flow and is key in the control of integrated cell functioning, in cell development in the face of stress, and in the membrane interactions of plant cells [29]. The results indicate that the main whorl bud has higher auxin transport and might therefore compete with the main bud for growth, suggesting that this strategy could lead to reduced growth in Mediterranean samples. Similarly, the positive regulator of auxin-mediated transcription, ARF7, presents higher expression in the secondary apical buds of Mediterranean origin.

Apart from that, two of the studied genes related to cytokinins showed significant differences in expression. *AHK4* encodes a signaling receptor protein [30,31], and the results showed significantly higher expression in the main whorl bud of the Mediterranean origin. Considering the antagonist effect that cytokinin exerts on auxin dynamics, the results here presented endorsed the higher expression of certain cytokinin-related genes in the non-active apical growth samples (Mediterranean). From a crosstalk perspective, AHK4 has been proposed as responsible for mediating greater stability of 1-aminocyclopropane-1-carboxylic acid, which is the ethylene precursor. As it is known, ethylene promotes inhibition of elongation [32], so the experimental results obtained are consistent, as individuals with higher expression of *AHK4* are in a resting phase.

On the contrary, the higher expression of *AHP1* in the Atlantic region could be associated with its involvement in cytokinin signaling and its influence on processes such as cell division, elongation, and differentiation.

Strigolactone-related genes also resulted in different significant expressions. A significantly higher expression of the biosynthesis-related gene *MAX1* has been observed in the secondary apical bud of *P. pinaster* of Mediterranean origin (with lower growth). This cytochrome transforms the precursor of strigolactones, carlactone, together with LBO [33], into a mobile, bioactive strigolactone.

A greater expression of *MAX2* was found in the main apical bud of the phenotype with greater growth (Atlantic origin). The protein encoded by this gene is involved in the polyubiquitination complex that, in the presence of strigolactones, activates the response to them [33,34]. In this study, a greater expression in the main apical bud was observed in individuals from the Atlantic model, fitting with the higher auxin content proposed and supporting the second messenger model, with a connection between auxin, strigolactone signaling, and promoted apical growth.

In the case of giberellins, a significantly higher expression of the gene coding its receptor, *GID1*, was observed in the main apical bud of the Mediterranean origin, pointing out the discrepancy between the expected active growth assumed with high GID1 levels and the lower growth phenotype assumed on this Mediterranean origin [2].

Significant differences in *STP1* (sugar-related gene) expression were observed in both secondary and main whorl buds, being significantly higher in individuals of Mediterranean origin. The function of this sugar carrier is key in the regulation of the absorption of monosaccharides from the environment, promoting growth [35,36]. In light of these data, it is not clear if the expression of *STP1* reflects where the resources are allocated. Indeed, even if its higher levels indicate a higher sugar content, STP1 cannot be considered an active growth marker, as the phenotype that presents less growth (the Mediterranean one) has significantly higher values.

Multivariant analysis by PCA showed that both origins were clearly separated by principal component two, suggesting that the different expression patterns of the genes studied could be in part responsible for the different growth phenotypes observed between Atlantic and Mediterranean provenances.

It can therefore be claimed that the methodology used in this study allowed the acquisition of 26 candidate genes in *P. pinaster* from *A. thaliana* models. By taking this approach, genes identified in this study will facilitate molecular studies to characterize the function and correlation of key genes in gymnosperms. In addition, the methodology described in this study may be applied to identify candidate genes involved in other processes in this species.

5. Conclusions

This study analyzed the hormonal and non-hormonal factors involved in plant growth and branching, using *A. thaliana* as a reference model. The methodology used in this study allowed the identification of 26 candidate genes in *P. pinaster*. In addition, we designed an experimental system for the initial validation of the candidate genes by studying their expression levels in three types of buds in individuals with contrasting growth. Our results revealed numerous significant differences in gene expression related to differential growth phenotypes. This methodology facilitates the transfer of knowledge from model plants to *P. pinaster*.

Supplementary Materials: The following supporting information can be downloaded at: https://www.mdpi.com/article/10.3390/f14091765/s1, Figure S1: Representation of the functional domains of some of the studied proteins; Figure S2: Relative gene expression analyzed by RT-qPCR; Table S1: Specific primers for each gene.

Author Contributions: Conceptualization, J.M.A., I.F., R.J.O. and C.C.; methodology, J.M.A., S.M.R., F.F.-M., I.F., R.J.O. and C.C.; writing—original draft preparation, J.M.A., S.M.R., R.J.O. and C.C.; writing—review and editing, J.M.A., S.M.R., F.F.-M., I.F., R.J.O. and C.C. All authors have read and agreed to the published version of the manuscript.

Funding: This research was funded by the "Instituto Nacional de Investigación y Tecnología Agraria y Alimentaria" (INIA) and the "Fondo Europeo de Desarrollo Regional" (FEDER) cofounding (RTA2017-00063-C04-04).

Data Availability Statement: Sequences used in this study are openly available in TAIR (https://www.arabidopsis.org/) and Gymnoplaza (https://bioinformatics.psb.ugent.be/plaza/versions/gymno-plaza/) databases.

Acknowledgments: We would like to thank Ismael Aranda and Maite Cervera (Instituto de Ciencias Forestales ICIFOR (INIA-CSIC)) for letting us sample the trees used in this work.

Conflicts of Interest: The authors declare no conflict of interest.

References

1. Vennetier, M.; Girard, F.; Taugourdeau, O.; Cailleret, M.; Caraglio, Y.; Sabatier, S.-A.; Ouarmim, S.; Didier, C.; Thabeet, A.; Vennetier, M.; et al. Climate Change Impact on Tree Architectural Development and Leaf Area. In *Climate Change—Realities, Impacts Over Ice Cap, Sea Level and Risks*; InTech: London, UK, 2013. [CrossRef]
2. Anfang, M.; Shani, E. Transport Mechanisms of Plant Hormones. *Curr. Opin. Plant Biol.* **2021**, *63*, 102055. [CrossRef] [PubMed]
3. Mason, M.G.; Ross, J.J.; Babst, B.A.; Wienclaw, B.N.; Beveridge, C.A. Sugar Demand, Not Auxin, Is the Initial Regulator of Apical Dominance. *Proc. Natl. Acad. Sci. USA* **2014**, *111*, 6092–6097. [CrossRef]
4. Rameau, C.; Bertheloot, J.; Leduc, N.; Andrieu, B.; Foucher, F.; Sakr, S. Multiple Pathways Regulate Shoot Branching. *Front. Plant Sci.* **2015**, *5*, 741. [CrossRef]
5. Emenecker, R.J.; Strader, L.C. Auxin-Abscisic Acid Interactions in Plant Growth and Development. *Biomolecules* **2020**, *10*, 281. [CrossRef] [PubMed]
6. Weijers, D.; Wagner, D. Transcriptional Responses to the Auxin Hormone. *Annu. Rev. Plant Biol.* **2016**, *67*, 539–574. [CrossRef]
7. Cherenkov, P.; Novikova, D.; Omelyanchuk, N.; Levitsky, V.; Grosse, I.; Weijers, D.; Mironova, V. Diversity of Cis-Regulatory Elements Associated with Auxin Response in *Arabidopsis thaliana*. *J. Exp. Bot.* **2018**, *69*, 329–339. [CrossRef]
8. Enders, T.A.; Strader, L.C. Auxin Activity: Past, Present, and Future. *Am. J. Bot.* **2015**, *102*, 180–196. [CrossRef]
9. Skalický, V.; Kubeš, M.; Napier, R.; Novák, O. Auxins and Cytokinins—The Role of Subcellular Organization on Homeostasis. *Int. J. Mol. Sci.* **2018**, *19*, 3115. [CrossRef]
10. Alvarez, J.M.; Bueno, N.; Cuesta, C.; Feito, I.; Ordás, R.J. Hormonal and Gene Dynamics in de Novo Shoot Meristem Formation during Adventitious Caulogenesis in Cotyledons of *Pinus pinea*. *Plant Cell Rep.* **2020**, *39*, 527–541. [CrossRef]
11. López-Hinojosa, M.; de María, N.; Guevara, M.A.; Vélez, M.D.; Cabezas, J.A.; Díaz, L.M.; Mancha, J.A.; Pizarro, A.; Manjarrez, L.F.; Collada, C.; et al. Rootstock Effects on Scion Gene Expression in Maritime Pine. *Sci. Rep.* **2021**, *11*, 11582. [CrossRef]
12. Valledor, L.; Guerrero, S.; García-Campa, L.; Meijón, M. Proteometabolomic Characterization of Apical Bud Maturation in *Pinus pinaster*. *Tree Physiol.* **2021**, *41*, 508–521. [CrossRef]
13. De Miguel, M.; Sánchez-Gómez, D.; Cervera, M.T.; Aranda, I. Functional and Genetic Characterization of Gas Exchange and Intrinsic Water Use Efficiency in a Full-Sib Family of *Pinus pinaster* Ait. in Response to Drought. *Tree Physiol.* **2012**, *32*, 94–103. [CrossRef] [PubMed]
14. Alía, R.; Martín, S. *Technical Guidelines for Genetic Conservation and Use for Maritime Pine (Pinus pinaster Ait.)*; EUFORGEN: Rome, Italy, 2003.
15. Bueno, N.; Cuesta, C.; Centeno, M.L.; Ordás, R.J.; Alvarez, J.M. In Vitro Plant Regeneration in Conifers: The Role of *WOX* and *KNOX* Gene Families. *Genes* **2021**, *12*, 438. [CrossRef] [PubMed]
16. Valledor, L.; Carbó, M.; Lamelas, L.; Escandón, M.; Colina, F.J.; Cañal, M.J.; Meijón, M. When the Tree Let Us See the Forest: Systems Biology and Natural Variation Studies in Forest Species. *Prog. Bot.* **2018**, *81*, 353–375. [CrossRef]
17. Lee, Z.H.; Hirakawa, T.; Yamaguchi, N.; Ito, T. The Roles of Plant Hormones and Their Interactions with Regulatory Genes in Determining Meristem Activity. *Int. J. Mol. Sci.* **2019**, *20*, 4065. [CrossRef] [PubMed]
18. Luo, Z.; Janssen, B.J.; Snowden, K.C. The Molecular and Genetic Regulation of Shoot Branching. *Plant Physiol.* **2021**, *187*, 1033–1044. [CrossRef]
19. Altschul, S.F.; Madden, T.L.; Schäffer, A.A.; Zhang, J.; Zhang, Z.; Miller, W.; Lipman, D.J. Gapped BLAST and PSI-BLAST: A New Generation of Protein Database Search Programs. *Nucleic Acids Res.* **1997**, *25*, 3389–3402. [CrossRef]
20. Thompson, J.D.; Higgins, D.G.; Gibson, T.J. CLUSTAL W: Improving the Sensitivity of Progressive Multiple Sequence Alignment through Sequence Weighting, Position-Specific Gap Penalties and Weight Matrix Choice. *Nucleic Acids Res.* **1994**, *22*, 4673–4680. [CrossRef]
21. Udvardi, M.K.; Czechowski, T.; Scheible, W.R. Eleven Golden Rules of Quantitative RT-PCR. *Plant Cell* **2008**, *20*, 1736–1737. [CrossRef]
22. Bustin, S.A.; Benes, V.; Garson, J.A.; Hellemans, J.; Huggett, J.; Kubista, M.; Mueller, R.; Nolan, T.; Pfaffl, M.W.; Shipley, G.L.; et al. The MIQE Guidelines: Minimum Information for Publication of Quantitative Real-Time PCR Experiments. *Clin. Chem.* **2009**, *55*, 611–622. [CrossRef]

23. Álvarez, J.M.; Cortizo, M.; Ordás, R.J. Characterization of a Type-A Response Regulator Differentially Expressed during Adventitious Caulogenesis in *Pinus pinaster*. *J. Plant Physiol.* **2012**, *169*, 1807–1814. [CrossRef] [PubMed]
24. Alonso, P.; Cortizo, M.; Canton, F.R.; Fernández, B.; Rodríguez, A.; Centeno, M.L.; Cánovas, F.M.; Ordás, R.J. Identification of Genes Differentially Expressed during Adventitious Shoot Induction in *Pinus pinea* Cotyledons by Subtractive Hybridization and Quantitative PCR. *Tree Physiol.* **2007**, *27*, 1721–1730. [CrossRef] [PubMed]
25. Alvarez, J.M.; Cortizo, M.; Bueno, N.; Rodríguez, A.; Ordás, R.J. CLAVATA1-LIKE, a Leucine-Rich-Repeat Protein Receptor Kinase Gene Differentially Expressed during Adventitious Caulogenesis in *Pinus pinaster* and *Pinus pinea*. *Plant Cell Tissue Organ. Cult.* **2013**, *112*, 331–342. [CrossRef]
26. Rozen, S.; Skaletsky, H. Primer3 on the WWW for General Users and for Biologist Programmers. *Methods Mol. Biol.* **2000**, *132*, 365–386. [CrossRef]
27. Ritz, C.; Spiess, A.N. QpcR: An R Package for Sigmoidal Model Selection in Quantitative Real-Time Polymerase Chain Reaction Analysis. *Bioinformatics* **2008**, *24*, 1549–1551. [CrossRef]
28. Gonzalez-Grandio, E.; Pajoro, A.; Franco-Zorrilla, J.M.; Tarancon, C.; Immink, R.G.H.; Cubas, P. Abscisic Acid Signaling Is Controlled by a BRANCHED1/HD-ZIP I Cascade in Arabidopsis Axillary Buds. *Proc. Natl. Acad. Sci. USA* **2017**, *114*, E245–E254. [CrossRef]
29. Máthé, C.; M-Hamvas, M.; Freytag, C.; Garda, T. The Protein Phosphatase PP2A Plays Multiple Roles in Plant Development by Regulation of Vesicle Traffic-Facts and Questions. *Int. J. Mol. Sci.* **2021**, *22*, 975. [CrossRef]
30. Qiu, Y.; Guan, S.C.; Wen, C.; Li, P.; Gao, Z.; Chen, X. Auxin and Cytokinin Coordinate the Dormancy and Outgrowth of Axillary Bud in Strawberry Runner. *BMC Plant Biol.* **2019**, *19*, 528. [CrossRef]
31. Hallmark, H.T.; Rashotte, A.M. Review—Cytokinin Response Factors: Responding to More than Cytokinin. *Plant Sci.* **2019**, *289*, 110251. [CrossRef]
32. Li, S.M.; Zheng, H.X.; Zhang, X.S.; Sui, N. Cytokinins as Central Regulators during Plant Growth and Stress Response. *Plant Cell Rep.* **2021**, *40*, 271–282. [CrossRef]
33. Walker, C.H.; Siu-Ting, K.; Taylor, A.; O'Connell, M.J.; Bennett, T. Strigolactone Synthesis Is Ancestral in Land Plants, but Canonical Strigolactone Signalling Is a Flowering Plant Innovation. *BMC Biol.* **2019**, *17*, 70. [CrossRef] [PubMed]
34. Waters, M.T.; Gutjahr, C.; Bennett, T.; Nelson, D.C. Strigolactone Signaling and Evolution. *Annu. Rev. Plant Biol.* **2017**, *68*, 291–322. [CrossRef] [PubMed]
35. Yamada, K.; Kanai, M.; Osakabe, Y.; Ohiraki, H.; Shinozaki, K.; Yamaguchi-Shinozaki, K. Monosaccharide Absorption Activity of Arabidopsis Roots Depends on Expression Profiles of Transporter Genes under High Salinity Conditions. *J. Biol. Chem.* **2011**, *286*, 43577–43586. [CrossRef] [PubMed]
36. Lemonnier, P.; Gaillard, C.; Veillet, F.; Verbeke, J.; Lemoine, R.; Coutos-Thévenot, P.; La Camera, S. Expression of Arabidopsis Sugar Transport Protein STP13 Differentially Affects Glucose Transport Activity and Basal Resistance to *Botrytis cinerea*. *Plant Mol. Biol.* **2014**, *85*, 473–484. [CrossRef] [PubMed]

MDPI
St. Alban-Anlage 66
4052 Basel
Switzerland
www.mdpi.com

Forests Editorial Office
E-mail: forests@mdpi.com
www.mdpi.com/journal/forests

* 9 7 8 3 7 2 5 8 0 1 2 1 3 *